I0868697

The Astronomical System of Aristotle: An Interpretation

History of Science and Medicine Library

VOLUME 58

The titles published in this series are listed at *brill.com/hsml*

The Astronomical System of Aristotle

An Interpretation

By

Gerardo Botteri
Roberto Casazza

Translation by

Agustina Casero
María Sara Loose

Revision and Edition by

Jillian Tomm

BRILL

LEIDEN | BOSTON

The research work of Gerardo Botteri and Roberto Casazza, as well as the edition and publication of the original version of the book in Spanish (2015) and its translation into English, were financed by three Argentine institutions: Universidad Tecnológica Nacional, Universidad Nacional de Rosario and Biblioteca Nacional Mariano Moreno.

This book is a revised English translation from the original Spanish publication *El sistema astronómico de Aristóteles: Una interpretación*, published by Ediciones Biblioteca Nacional, Buenos Aires, 2015.

Cover illustration: Aristotle's astronomical system (see also Figure 30, p. 92).

The Library of Congress Cataloging-in-Publication Data is available online at https://catalog.loc.gov
LC record available at https://lccn.loc.gov/2022052331

Typeface for the Latin, Greek, and Cyrillic scripts: "Brill". See and download: brill.com/brill-typeface.

ISSN 1872-0684
ISBN 978-90-04-52552-8 (hardback)
ISBN 978-90-04-52553-5 (e-book)

Copyright 2023 by Gerardo Botteri and Roberto Casazza. Published by Koninklijke Brill NV, Leiden, The Netherlands.
Koninklijke Brill NV incorporates the imprints Brill, Brill Nijhoff, Brill Hotei, Brill Schöningh, Brill Fink, Brill mentis, Vandenhoeck & Ruprecht, Böhlau, V&R unipress and Wageningen Academic.

Prologue (*Horacio González*); Copyright 2023 by Florencia González and Ana Lucía González.

English translation: Copyright 2023 by the Universidad Nacional de Rosario.

All rights reserved. No part of this publication may be reproduced, translated, stored in a retrieval system, or transmitted in any form or by any means, electronic, mechanical, photocopying, recording or otherwise, without prior written permission from the publisher. Requests for re-use and/or translations must be addressed to Koninklijke Brill NV via brill.com or copyright.com.

This book is printed on acid-free paper and produced in a sustainable manner.

τὸ γὰρ αὐτὸ νοεῖν ἐστίν τε καὶ εἶναι
For thinking and being are the same
PARMENIDES, Fragment B3

• • •

εἰ οὖν οὕτως εὖ ἔχει, ὡς ἡμεῖς ποτέ, ὁ θεὸς ἀεί, θαυμαστόν: εἰ δὲ μᾶλλον, ἔτι θαυμασιώτερον. ἔχει δὲ ὧδε
If, then, God is always in that good state in which we sometimes are, this compels our wonder; and if in a better this compels it yet more. And God is in a better state.
ARISTOTLE, *Metaphysics*, Λ, 7, 1072b24–26

To our ten unmoved movers

Contents

Appendices

Acknowledgements to the Original Spanish Edition

We owe special thanks to the Escuela de Filosofía of the Facultad de Humanidades y Artes of the Universidad Nacional de Rosario and its director, Prof. Beatriz Porcel, who actively promoted the original 2008 seminar and the creation of the study group; to Silvana Filippi, Professor of History of Medieval and Renaissance Philosophy and director of the Centro de Estudios Patrísticos y Medievales, who favoured the inclusion of cosmological topics in the syllabus of her chair and who provided valuable feedback on the manuscript; to Alejandro Gangui and his group *Didáctica de la Astronomía* (CONICET-IAFE-UBA), with which some of the questions treated in this work were developed and broadened; to Marcelo Boeri, who generously made keen observations of our early work; to Analía Sapere, who made a thorough revision of the Greek texts included in this volume; to Sebastián Scolnik for his valuable advice as editor, to Alejandro Truant for his careful graphic design, which enlivened our work, and also to the prolific publications team of the Biblioteca Nacional Mariano Moreno of Argentina; to María Elena Díaz, María Angélica Fierro, Ana María Sardisco, Antonio Natolo, Julio Castello Dubra, Daniel Di Liscia, Silvia Magnavacca, Antonio Tursi, Gabriel Bengochea, Verónica Casazza, Marcos Ruvituso, Jazmín Ferreiro, Lucas Oro Hershtein, Ezequiel Ludueña, María José Coscolla, Claudia D'Amico, Alejandro Ranovsky, Eduardo Glavich, Sergio Manterola, Ricardo Sassone, Ricardo Avenburg, Rodolfo Gómez, Aníbal Szapiro, José Emilio Burucúa, Silvina Vidal, Miguel de Asúa and Francisco Bertelloni for their comments and bibliographical suggestions; to Lindsay Judson for making available some of his works upon request; to Cecilia Colombani, Raúl Lavalle and Guido Fernández Parmo for their friendly invitations to fruitful academic gatherings; to Alicia Chiesa, Lucía Casasbellas Alconada and Gustavo Míguez for typing some of the texts; to Marcelo Ferrari for his help with the difficult task of processing the digital Greek texts; to Miguel Saralegui and Nicolás Kwiatkowski for their editing suggestions; to Haroldo Tomás Avetta, Ezequiel Grimson and Horacio González for supporting our work at the Universidad Tecnológica Nacional and the Biblioteca Nacional; and to Constantino Baikouzis, who literally opened up the gates of heaven for us with his indispensable *Carta celeste para la latitud de Buenos Aires*.[1] We would also like to acknowledge all members of the *Grupo de*

1 The figures in this book were developed by the authors but draw from a figure in the work of Norwood Russell Hanson: *Constelaciones y conjeturas* (Madrid: Alianza, 1978), p. 83. The original English edition: *Constellations and Conjectures*, ed. W. C. Humphreys,

Estudio del Cielo (Alejandra Buzaglo, Guillermo Barbieri, Pablo Polito, Marco Bortolotti, Khamil Nazer, Sabina Piazza, Ángeles Rivarola, Facundo Fagioli, Carolina Sosa, Paola Rocha, Gisela Cadirola, Germán Tessmer), with whom we have shared this enriching and enjoyable endeavour of recognising constellations and studying texts from the European tradition of cosmology.

Finally, we would like to remember our dear fathers, both departed, who through their inquisitiveness and eagerness to learn laid the ground for our enthusiasm for such unusual matters; and our mothers, for their constant love and support. A special thanks to our spouses, Sonia Sauer (GB) and Carolina Carman (RC), who through love and understanding became the iron pillars supporting this work; and also to our children, Julián, Victoria, Gerónimo, Joaquín, Pilar (GB), Manuel, Jerónimo and Hilario (RC), who by giving us joy—as well as a lot of work—have intriguingly fed the content of these pages. We also wish to express our gratitude to all our colleagues, relatives, babysitters and friends for their help and encouragement.

Gerardo Botteri & Roberto Casazza
Rosario, Argentina
August 2014

Jr, (Dordretch/Boston: Reidel, 1973). Hanson's figure inspired our fundamental scheme of Aristotle's systems, here broadened and significantly reformulated. The late Renaissance engravings belong to astronomical works of the Biblioteca Nacional of Argentina, and are reproduced here with permission. We are grateful to the staff of the Sala del Tesoro and Mapoteca of the Biblioteca Nacional, and would like to thank especially photographer Marcelo Huici for his images of our regular solids, and Viviana Azar for reproducing several Renaissance engravings.

Acknowledgements for this Translated Edition

The long path towards an English edition of our book involved many stages. The most important milestone was the confidence invested in our book by Brill Science Collections editors Stephen Soehnlen, Rosanna Woensdregt, Stefan Einarson, Rianne van Zanen and Gera van Bedaf. We thank them for their generous reception of the idea, and we are deeply grateful to Els van Egmond, who we had the good fortune to meet in 2017 at the 52nd International Congress on Medieval Studies in Kalamazoo, USA, for introducing our book to her colleagues. Alberto Manguel, Elsa Barber and Juan Sasturain, as directors of the Biblioteca Nacional of Argentina, provided steadfast support for the project and granted the editorial copyright to Brill. Once the translation became possible, it became a reality only through the great intellectual effort of Agustina Casero and María Sara Loose, English Translators at the Cuerpo de Traductores of the Universidad Nacional de Rosario, under the guidance of their director, María Gabriela Piemonti. Thanks to their tremendous patience and work we arrived at a translation of the text that pleases us very much. No praise is enough for Agustina and Sara, who confronted challenges in the text that turned up line by line, as well as many non-linguistic difficulties. And finally, Jillian Tomm's review, editing and suggested improvements increased the quality of the whole. Our gratitude to all of them is huge, but surely still unequal to what they gave to us.

Gerardo Botteri & Roberto Casazza
Rosario, Argentina
May 2022

Notes on the Authors

Gerardo Bartolomé Botteri
(1961–) / (BPhys, Universidad Nacional de Rosario; BPsy, Universidad Nacional de Rosario; MBA, Universidad de Belgrano) is currently professor of physics at the Universidad Tecnológica Nacional (Facultad Regional San Nicolás) and assistant professor of philosophy in the Faculty of Psychology of the Universidad Nacional de Rosario.
E-mail: g_barto@hotmail.com

Roberto Fabián Casazza
(1968–) / BA (Universidad de Buenos Aires); MA (The Warburg Institute, University of London); PhD (Universidad Nacional de Rosario) is currently professor of Medieval and Renaissance Philosophy at the Universidad Nacional de Rosario, and Research Fellow at the Biblioteca Nacional of Argentina, where he leads the project "Incunabula in Latin American Collections not yet registered in the ISTC."
E-mail: casazza.roberto@gmail.com.

Both are founding members of the Grupo de Estudio del Cielo (Facultad de Humanidades y Artes, Universidad Nacional de Rosario), devoted to naked-eyed observation and understanding of the apparent motions of stars and planets, and to the study of astronomical-cosmological texts of the classical tradition.

Prologue

Reflection on the heavens is perhaps the most eloquent point at which the acts of observing and theorising meet, revealing their common origin. In the remote origins of the word *theory* we find the practice of looking. It is impossible to differentiate some form of immediate perception from the configuration of theories that exhibit a high degree of abstraction and are strengthened by well-defined and specific language. Perhaps the self-confidence of theories has produced a comprehensible abandonment of interest in the perceptive, optic, and intuitive genealogy upon which they are actually built; a basis that, in the case of the observation of atmospheric and cosmic realms, is directly linked to the kind of pleasure that derives from a direct and naked observation, devoid of instruments apart from the simple, natural human imagination. This book by Gerardo Botteri and Roberto Casazza shows that a rigorous study of Aristotle's *Metaphysics* is not simply an exercise in the history of astronomy but constitutes a broad inquiry into our germinal ideas about speed, motion, and the spherical nature of real entities, as well as the relation between theology and gnoseology.

The theses presented here advance with erudition rooted in several branches of universal knowledge, some sheltered in specialised sectors of the university and in a few little-known bibliographies—the authors rightly draw attention to Werner Jaeger's exceptional *Aristotle*, but it would be unjust to overlook the remarkable research efforts of these two authors that reveal a profound intellectual curiosity that is rare in our current context. It is, after all, the heavens that are the source of the first questions about human existence and its astounding (and necessary) projections in all forms of mythical-poetical and scientific narratives; it is the same sky that inspires the questions of the child, the believer, and the researcher. Packed with its specific elaborations, boldly substantiated arguments, and sound research into the summit of epistemology (whatever its tenor)—each page of this book brings us back to the graceful youthful impulse with which any person, even someone with an untrained eye, yet armed with secrets and elusive thoughts, may turn their gaze from time to time to the modest changes in the sky in an exercise that could be qualified as leisure. And in a way it is; the kind of surprising insightful leisure from which arises the most elaborated science and also the possibility of recreating those ancient times in which the astral conceptions of the Babylonians, the Chaldeans, and the Greeks imagined a rhetoric in the heavens aimed at understanding the human world, the animal world, and the divine world, expressed through myth, poetry dedicated to the gods, and the birth of empirical science.

The enigma of the Aristotelian Prime Mover—addressed in imaginative ways within this book—remains to this day. Philosophical questions of this nature are never solved. There are theoretical constructions that belong to their own particular language, subsequently transmitted, together with that language, to other ages and times to be adapted, contested, or forgotten. The works of Aristotle, which were, as we know, carefully scrutinised by Thomas Aquinas, were subject to an unparalleled transference that can, itself, very well be studied alongside his works on poetic art and rhetoric, which still today lack a consensual understanding and bear the stamp of unceasing interest in the origin of time and space, subject and thought. This formidable book is a testimony of these permanences, fidelities, and efforts.

Horacio González
Director of Biblioteca Nacional
Buenos Aires, 2014

Abbreviations

DK Diels, Hermann and Walther Kranz, eds. *Die Fragmente der Vorsokratiker.* 3 vols. 6th ed., revised and expanded. Berlin: Weidmann, 1951–1952 (first published 1903).

GCS *Die Griechischen Christlichen Schriftsteller der ersten Jahrhunderte*. Leipzig: Hinrichs, 1897–1969.

KRS Kirk, G. S., J. E. Raven, and M. Schofield. *The Presocratic Philosophers*. 2nd ed. Cambridge: Cambridge University Press, 1983.

LFP *Los filósofos presocráticos*. 3 vols. Edited by Conrado Eggers Lan et al. Madrid: Gredos, 1978–1980. References here are to reissued editions: Vol. 1 (2000); Vol. 2 (2003); Vol. 3 (1997).

Illustrations

Introduction

The textual and graphical reconstructions of ancient astronomical systems offered in this work seek to synthesise Aristotle's ideas about the whole of planetary motions. As those acquainted with Aristotelian cosmology will recognise, the difficulties of such a project are mainly three:

1. Aristotle did not leave a clear record of his thought on celestial motions (his most significant work on this topic, *On the Heavens*, specifically addresses the structure and organisation of the cosmos, and presents some general statements on circular motion and the distinctive character of ether, but proves insufficient to a full elucidation of the motion of celestial bodies).
2. The Stagirite's main passage on planetary motions (*Metaphysics*, Λ [Book XII], 8, 1073a17–1074b14), in which he increases the number of Unmoved Movers from one to fifty-five, is—and we subscribe to this hypothesis—a late addition to other texts, written over twenty years earlier, that address the Prime Mover (Λ, 7 and Λ, 9–10). In those earlier texts, the Aristotelian-monotheist-theology unfolds quite differently from its treatment in Λ, 8, in which astronomical references pave the way for a sort of meta-astral oligotheism. The problem brought about by this addition is truly significant, as it forces a reassessment of the signification of the entire theory of the Prime Mover and leads to conclusions not easily reconciled with some central points of Aristotelian thought.
3. Acknowledging the abovementioned limitations and contradictions, the brief and completely synthetic text of Λ, 8 (only twenty lines long) forces us into the realm of *interpretation*; hence the subtitle of this book.
4. Aristotle left an outline rather than an explanation of his astronomical system, so if the aim is a full understanding of his thoughts, one must also consider concepts and schemata beyond what is strictly available in his textual sources, granted that they be taken as hypothetical.
5. That is precisely what we have done here: build a model of the behaviour of the heavens based on Aristotelian principles that respects, on the one hand, astronomical phenomena and, on the other, Aristotle's philosophical texts. We aim not to diverge overmuch from what we consider to be the strictly logical consequences of his philosophical matrix.

In short, aiming at understanding the implications of chapter Λ, 8, we have sought to broaden the Aristotelian exposition by offering a tentative doctrine, compatible with the general lines of Aristotelian thought, that accounts for the way the heavens function. The resulting system is—we believe—coherent

© GERARDO BOTTERI AND ROBERTO CASAZZA, 2023 | DOI:10.1163/9789004525535_002

and plausible, and its fundamental aspects coincide with the rare fragments of Aristotle's texts that address astronomy (the most relevant of which are provided here in Appendix 1: "Text Fragments"). A study of this selection reveals to what extent Aristotle's doctrine of planetary motions was left incomplete in the *corpus* that has come down to us. Notwithstanding the possibility that ideas embodied in fragment Λ, 8 may have been further orally formulated at the Lyceum, we unfortunately lack any such records.

The following pages tread a somewhat precarious path, and we endeavour to follow the balance necessary to prevent oversimplified or too-ambitious conclusions. The two main problems addressed are 1) to account for the need of multiple Unmoved Movers—as postulated by Aristotle in *Metaphysics*, Λ, 8—to explain celestial motions, using apparent heavenly motions as a guide; and 2) to offer a systematic view of what we have postulated to be the available fragmentary textual base for Aristotle's celestial system. Our aim is to shed some light on these two themes, which hold, as ideas of Aristotle, a privileged place in the history of Western thought. We offer our hypotheses, regarding this *crux interpretum*, hoping that fellow inquirers will be able to draw from, build upon, and reformulate them so that they may be improved upon.

Whatever it brings, this study of Aristotelian thought—along the boundaries between physics and metaphysics—has underlined what we believe to be a somewhat disregarded philosophical truth of our time: that attempting to climb the ladder of metaphysics without previously following the path of the particular sciences has its dangers—may even involve an element of complacency—and has consequences regarding philosophical understandings and misunderstandings about the world.

In summary, our work seeks to make the case that within Aristotelian uranology, one must admit the existence of numerous Unmoved Movers in order to account for both the planetary motions and the generation/corruption processes in the sublunary domain. To this end, we offer, first, a review of Aristotle's cosmology as part of a larger tradition that considers the form of the Whole as limited and spherical. We then describe the apparent heavenly motions, which were the same for ancient thinkers as they are for us today. Once described, these motions become the leading thread in an analysis of the doctrines of Eudoxus, Callippus, and Aristotle—the doctrines so briefly outlined in fragment Λ, 8, 1073a17–1074b14 (§ 15) of *Metaphysics*, the mystery at the centre of our enquiry.

The perspective we propose is characterised by at least the following three points:

1) It takes into account the problem of the chronological evolution of Aristotelian thought, especially when it comes to the consequences of introducing the idea of numerous Unmoved Movers, from a late text, into his cosmology and ontotheology (as suggested by Werner Jaeger and later supported by different interpreters).[1]
2) It is focused on elucidating a problematic text and offering a systematic synthesis of what might have been the Aristotelian system of the heavens. Our interpretation therefore minimises some apparent doubts that Aristotle himself had on this matter and does not focus on or attempt to resolve possible contradictions between the doctrine of multiple Unmoved Movers and some other points of his metaphysics.
3) It offers a series of considerations that will permit us to accept: a) The *continuity* of the heavens proposed by Aristotle; b) the *integration* of planetary subsystems, which Eudoxus and Callippus thought of as independent;[2] and c) the *relationship* between the Prime Mover and the Unmoved Movers of Planetary Spheres (a relationship we believe to be generally misunderstood in the Aristotelian tradition, and particularly in interpretations of Λ, that underestimate the importance of the astronomical issue in the internal structure of the text).

On a different note, we would like to mention that this book originates from an experience shared by the authors. It can be summed up as follows: we had both read the *Metaphysics*, Λ, where Aristotle presents his well-known doctrine of the Prime Mover, before we had each personally come to understand the observable motions of the sky in a scientific sense. We then found, upon re-reading the same text, a new interpretation. Taking into account what can be seen in the sky at night, Aristotle's text suddenly became another. Consequently, his words about the Prime Mover, and especially about the multiple Unmoved Movers of Planetary Spheres, took on a new meaning. Our surprise became curiosity, which became a quest for systematisation (never to be fully successful, and less so in such a text of Aristotle), which led eventually

1 W. Jaeger, *Aristoteles: Grundlegung einer Geschichte seiner Entwicklung* (Berlin: Weidmann, 1923); published in English as *Aristotle: Fundamentals of the History of His Development*, trans. R. Robinson (Oxford: Clarendon Press, 1934). See particularly chapter 14, "Revision of the Theory of the Prime Mover."

2 When we employ the notion of "subsystem" here, we refer to the systems of independent spheres of each planet, conceived by Eudoxus and Callippus. This distinction aims at reserving the notion of "system" for any complete model that includes all planetary subsystems. Otherwise, it would be difficult to differentiate between part and whole, especially when addressing the models of Eudoxus and Callippus, but also in Aristotle's reformulation. In some cases, when referring to a full account of celestial motions, we have assigned to the more generic notion of "model" a meaning akin to that of "system."

to the desire to verify whether the interpretation that now occurred to us, if diaphanous, would find agreement in the philosophical community. The following pages are the product of that journey.

This book is to a certain extent also the corollary of a seminar given at the Facultad de Humanidades y Artes of the Universidad Nacional de Rosario (UNR) entitled "Estudio filosófico del cielo: Hacia una reuranización de la experiencia presente," ("Philosophical studies of the sky: Towards a re-uranization of present cultural experience"), out of which emerged the Grupo de Estudio del Cielo, created in 2009, on the initiative of the Facultad de Humanidades y Artes of the Escuela de Filosofía.[3]

3 This research was conducted between 2008 and 2013. During that period, partial versions of Aristotle's system of the heavens were presented in various forums. Our first presentation, "El sistema astronómico de Aristóteles: *Metafísica*, Λ, 8, 1073a14–1074a34," was published as part of *Actas del XXI Simposio Nacional de Estudios Clásicos* (Santa Fe: Universidad Nacional del Litoral, 2010). The papers "Necesidad onto-cosmológica de la 'pluralidad' de motores inmóviles aristotélicos," and "'Y el número de todas las esferas, cincuenta y cinco': la teoría aristotélica de los movimientos planetarios," were presented at *II Jornadas de Pensamiento Antiguo*, organised by the Escuela de Humanidades of the Universidad Nacional de San Martín (13–14 May 2011), and published in *Expresar la* phýsis: *Conceptualizaciones antiguas sobre la naturaleza*, ed. Esteban Bieda and Claudia Mársico (Buenos Aires, UNSAM Edita, 2013). Similarly, sections of the following works were incorporated into this book: "Cómo mueve y qué mueve el Primer Motor Inmóvil," and "El sistema astronómico de Platón: República 616a–617e," delivered at *II Jornadas de Filosofía Antigua*, organised by the Chair of Historia de la filosofía Antigua of the Facultad de Humanidades of the Universidad Nacional de Mar del Plata (3 December 2011). Other papers, namely, "La esfericidad del Todo en el pensamiento presocrático" (presented by Roberto Casazza), and "La animación de las esferas celestes aristotélicas: una (inadecuada y recurrente) lectura platónica" (presented by Gerardo Botteri) were discussed at the *VI Jornadas sobre el Mundo Clásico*, organised by Universidad de Morón (12–13 October 2012). In addition, we should mention that the material from chapters 2 and 9, as well as some from chapter 1 originated in the PhD Dissertation of Roberto Casazza, at the Universidad Nacional de Rosario, entitled, *Sphaericus ordo—La fundamentación del marco cosmológico esférico en la Tradición Clásica* (2016). Otherwise, the equations for planetary positions of chapter 4 where entirely developed by Gerardo Botteri.

CHAPTER 1

The Spherical, Limited, and Hierarchical Cosmology of Aristotle

> What do you see, Walt Whitman?
> Who are they you salute, and that one after another salute you?
> I see a great round wonder rolling through space,
> I see diminute farms, hamlets, ruins, graveyards, jails, factories, palaces,
> hovels, huts of barbarians, tents of nomads upon the surface,
> I see the shaded part on one side where the sleepers
> are sleeping, and the sunlit part on the other side,
> I see the curious rapid change of the light and shade,
> I see distant lands, as real and near to the inhabitants
> of them as my land is to me.
>
> WALT WHITMAN, "Salut au monde!", *Leaves of Grass*

Aristotle (384–322 BC) was undoubtedly the greatest systematic thinker of the ancient world, his thought presenting an articulated whole based primarily on the concept of nature (φύσις). Interpretations of Aristotelian astronomy involve, however, a great deal of uncertainty on account of the fragmentary nature of the source base available for its reconstruction. No specific astronomical work by the Stagirite has survived, nor seems to have been written; consequently, a complete picture of Aristotle's star system can be grasped only by piecing together disparate passages found in *On the Heavens, Meteorology, Movement of Animals, On Generation and Corruption, Physics* and, especially, from a fragment in his *Metaphysics*.[1] This last will be examined in detail in what follows.

A disciple of Plato and a member of the Academy, Aristotle was nonetheless critical of the idealising character of Platonic thought. He left treatises on a

1 Full bibliographic information for references in footnotes will be provided for the first citation of a work only, with abbreviated entries provided for subsequent citations. Complete bibliographic information is also provided in the Sources and Bibliography sections.

© GERARDO BOTTERI AND ROBERTO CASAZZA, 2023 | DOI:10.1163/9789004525535_003

vast variety of subjects—logic, rhetoric, botany, zoology, mineralogy, meteorology, physics, metaphysics, ethics, politics—composed for the most part in the succinct prose of lecture notes, called λόγοι. His works as we know them today consist largely of such texts and other notes, gathered together by Aristotle and his disciples, but also by numerous scholars of the Lyceum through the next centuries. This is significant for our study, as the basis of his astronomical system, as we have reconstructed it, is a chapter from the *Metaphysics*, namely Λ, 8, which, according to most critics—followers of Bonitz, and especially of Jaeger's influential *Aristoteles* (1923)—would have been a late writing awkwardly inserted by this later Aristotelian *traditio* into the treatises on the first philosophy, or metaphysics.

Of the many different aspects of Aristotle's complex cosmology, we will focus primarily on the three most essential to an understanding of his astronomical system: 1) his proposal of a limited spherical universe, 2) his hierarchical characterisation of different cosmic strata and, especially, 3) his account of celestial motions.

The influence of Aristotle's cosmological thought has been enormous: it dominated thinking during the Middle Ages—in both the Arab and Latin worlds—but even during the European Renaissance the main lines of his celestial system (consisting of numerous homocentric spheres) and, particularly, the dual nature of his physics (both sublunary and superlunary), formed the core matrix of associated scientific discussion and development.

Aristotelian cosmology, in its mature formulation, can be summarised as follows: the universe, or totality of all that is—called without distinction τὸ ὅλον or τὸ πᾶν—is conceived as One and Unique, being a Whole that is both limited and spherical. This tremendous sphere contains two principal regions: the first, called the *sublunary domain*, extends from the centre of the universe (the centre of the Earth) outward to the lower limit of the lunar sphere; the second, called the *superlunary domain*, expands from the sphere of the Moon to the outer limit, also spherical, of the universe. Hence, the heavens of Aristotelian cosmology constitute a region limited by two concentric spheres. The sublunary and superlunary domains are furthermore distinguished by a number of essential differences: 1) the elements of the one (earth–water–air–fire) do not produce compounds with the other (ether); 2) they are characterised by different motions; 3) the composite elements of each correspond to—to put it in modern terms—different physics; 4) sublunary entities are more likely to be perceived, as they can be experienced through all senses, while celestial bodies can only be experienced through sight; and 5) each has a different relationship to the eternal and divine.

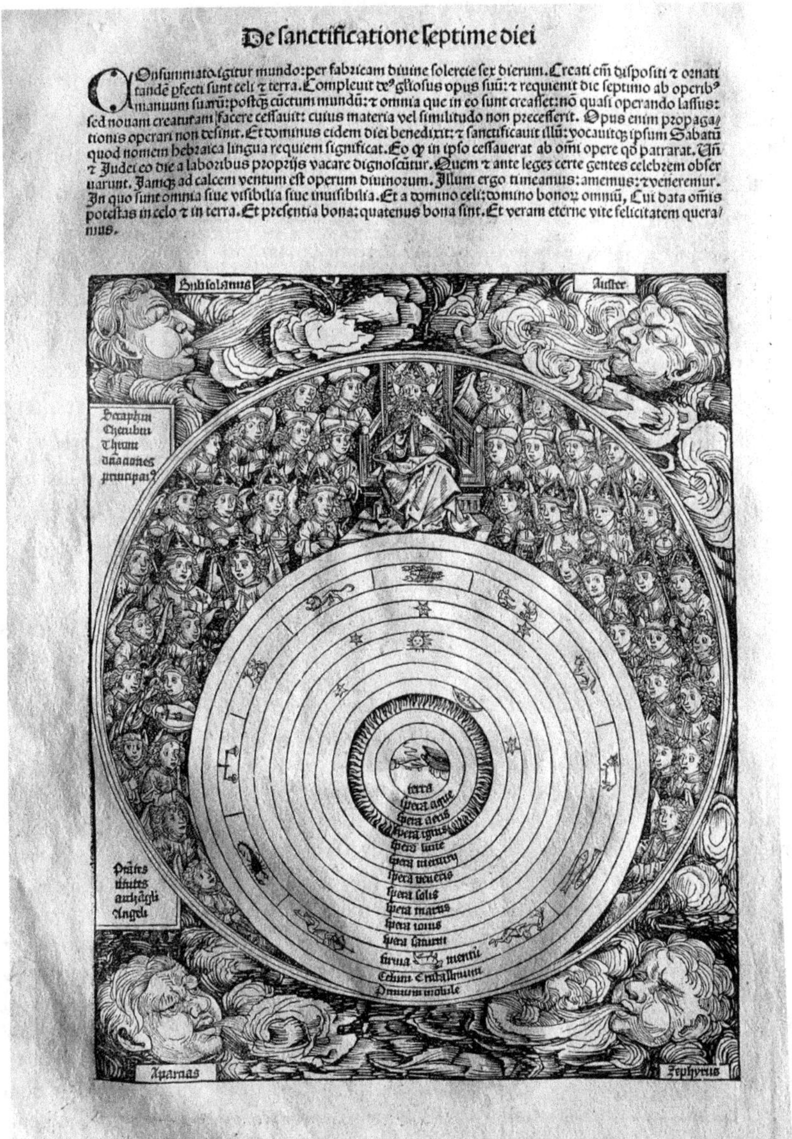

FIGURE 1 A late medieval representation of the cosmos through a scheme of homocentric spheres. The Earth, at the centre, is surrounded by spheres of water, air, and fire, which are themselves enveloped by planetary spheres. Beyond the sphere of the fixed stars (or the firmament)—marked by the symbols of the zodiac—is the angelic domain (the Christian translation of Aristotle's "intelligible movers" from *Metaphysics*, Λ, 8). Pictured are the nine legions of angels: Seraphim, Cherubim, Thrones, Dominions, Principalities, Powers, Virtues, Archangels, and Angels. Engraving from Hartmann Schedel's *Liber chronicarum, cum figuris et ymaginibus ab initio mundi*, published in Nuremberg by Anton Koberger in 1493.

The sublunary domain is characterised by a concentric distribution of its four elements arranged singly in sphaerical layers such as those we find in an onion. Hence, in the upper layer is placed "what we are accustomed to call fire, though it is not really fire" (ὃ διὰ συνήθειαν καλοῦμεν πῦρ, οὐκ ἔστι δὲ πῦρ),[2] and then, moving inwards, we find layers of air, of water, and, finally at the centre, of earth. Each of these elements possess, furthermore, an equivalent total mass: according to Aristotle, given equal masses of water and air, for example, the volume of the air would be ten times greater than that of the water[3] ("if, e.g., one pint of Water yields ten of Air, both are measured by the same unit; and therefore both were from the first an identical something"[4]). Extending this kind of relationship to all elements of the sublunary domain would imply that the volume of earth would therefore be less than that of water, just as the volume of water is less than that of air, and again the volume of air would constitute a smaller portion of the total volume than would the empyrean layer.

Moreover, the characteristic motion of the sublunary domain is rectilinear, moving to and from the centre of the spheres, each cycle of motion having a clear beginning and ending; while the characteristic motion of the superlunary domain is circular, without end or beginning, and therefore eternal and lacking

2 *Meteorology*, I, 3, 340b23 (§ 11); English trans. of *Meteorologica* by H. D. P. Lee, Loeb Classical Library (London: W. Heinemann; Cambridge, MA: Harvard University Press, 1952) p. 19.

3 In order to clarify Aristotle's idea of the proportion between elements, we use the modern notion of mass in its chemical sense, that is, "state of matter" (solid, liquid, gas, etc.). However, the notion of mass as a quantity that can be measured is closely linked to the principle of inertia, a principle unknown to ancient thinkers. The other notion of mass—i.e., that which determines the force of gravitational attraction (gravitational mass), which is demonstrably equivalent to the former—has a higher degree of agreement with Aristotelian considerations of the heaviness or lightness of terrestrial bodies, but is equally inappropriate for understanding the Aristotelian idea of material quantity and, finally, even his ideas of heaviness and lightness. Aristotle does not reason strictly in terms of equal *masses* but of equal *amounts*; that is, if air is generated from a given amount of water, the air will occupy a volume ten times greater than that occupied by the water from which it was generated, *caeteris paribus*. In short, his comparison of volumes assumes a principle of conservation of matter, from which it logically follows that the equivalence ratio between two types of elemental matter is determined by the comparison of the quantity (volume) of two elements, one quantity being generated by the other.

4 *On Generation and Corruption*, II, 6, 333a23–26; English trans. H. H. Joachim in *The Complete Works of Aristotle, Revised Oxford Translation*, 2 vols., ed. J. Barnes (Princeton: Princeton University Press, 1984; Past Masters Electronic Edition, InteLex Corp, 1992), vol. 1, p. 545. Aristotle's 1-to-10 proportion of water to air probably comes from the Pythagorean assumptions rather than from empirical knowledge, as the proportion between water and air densities is closer to 1 to 1000. In view of this theoretical premise, it is possible that he expected the same relationship between earth and water, and air and fire (*On the Heavens*, III, 5, 304a6), although these estimations are also empirically wrong.

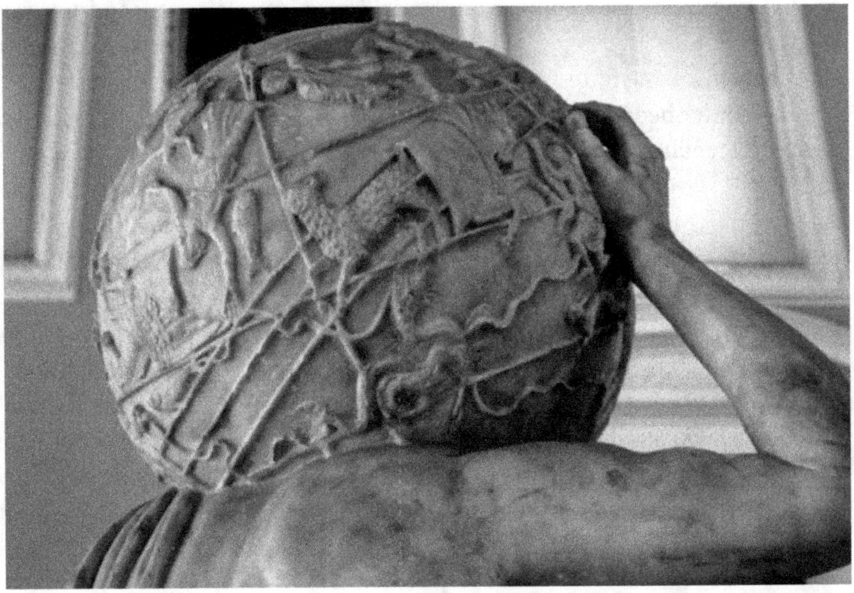

FIGURE 2 The Farnese Atlas, 2nd century sculpture, located in the Museo Archeologico Nazionale, Naples. Visible on the surface of the sphere is the celestial equator, the ecliptic, the First Point of Aries and the zodiac band. Among the constellations that can be seen are Pisces, Aries, Taurus, Orion, Andromeda, Hydra and Perseus.

an opposite motion.[5] The particular nature of these motions is due to the material composition of each domain: the entire superlunary realm is composed of ether (ἡ αἰθήρ), the element nearest to the divine; while the abovementioned composite of four elements—earth (ἡ γῆ), water (τὸ ὕδωρ), air (ἡ ἀήρ), fire (τὸ

5 The rectilinear motion of sublunary bodies has an opposite or contrary motion whose end points are spatially and qualitatively different on account of the body's natural resting place; for example, if the natural movement of a flame is upwards, then any downwards motion of the same would be forced and is, therefore, its opposite (the natural place of compound bodies corresponds to their dominant element). On the other hand, in circular motion the end and the beginning are one and the same, and this absence of a spatial and qualitative difference between endpoints determines its continuous and eternal nature since 1) the lack of an opposite inevitably leads to this consequence; and 2) any given point of the circular trajectory is, without distinction and simultaneously, the end and the beginning of the motion. Therefore, celestial bodies in motion are already and always in their natural place—a place that might be considered where their souls (at least during the Platonic stage of Aristotelian exploration) "wish" to remain. Aristotle clearly bases this conception on the assumption that celestial bodies are in eternal movement, as apparent in the regularity of the heavens, and the idea of circular motion in the superlunary domain offers no contradiction to this, as would the rectilinear motion of the lower spheres (*Physics*, VIII, 8, 264b9–19 –§ 31–).

πῦρ)—prevails in the sublunary domain. These last elements tend towards separate, homogeneous layers, though there remain always mixed residuals, especially involving terrestrial substances and, above all, living organisms.

Ether, described by Aristotle as an extremely subtle homogenous matter of noble and regular behaviour, is the element that fills the entirety of the spherical celestial layers, but is also, in a denser form, the matter of celestial bodies themselves.[6] Regarding celestial bodies—the generic term ἄστρον refers both to the myriad of stars and to the seven wandering bodies—he claims that they are always spherical in shape (*On the Heavens*, II, 8, 290a7–24 –§ 10–) and that they are not capable of self-motion, but are fixed to ethereal spheres (*On the Heavens*, II, 8, 289b2–290a7 –§ 27–), spheres moved by intelligible immaterial movers, at least according to *Metaphysics*, Λ, 8.[7] Such spherical layers of ether have regular circular motion, which explains the recurrence of celestial motions and supports their *intelligibility*. This aspect—that is, understanding through concepts—was precisely the principal and shared goal of the most distinguished members of Plato's Academy, in which Aristotle spent over

6 Aristotelian cosmology, built upon a geometrising conceptual basis drawn from Pythagoras, Plato, and Eudoxus, assigns to celestial motions a perfection typical of the ideal domain: geometry operates in the intelligible plane, fed by representations of perfect solids. Aristotle assigns *perfect* circular orbits to celestial bodies, which—as we should remember—are material just like the spheres. Moreover, he conceives the celestial sphere as both sensible and perfect. This idea was strongly criticised by Nicholas of Cusa in *De docta ignorantia*, book II, §§ 158–159, where, deeply horrified by the notion of "material perfection," he presents a devastating critique of Aristotle's proposal on the possibility of a synthesis between the sensible and the perfect. Cusa's critique is born of a conviction of the utter impossibility of any form of perfection in the corporeal domain, and for this same reason he rejects the stability of celestial poles—and consequently, the perfection of orbits—as well as the notion of an absolute centre of the Earth. This appeal for the acceptance of imperfection was embraced by Giordano Bruno, who depicted an *infinite* and *homogenous* cosmos, deeply anti-Aristotelian in its conceptual opposition to the Stagirite's *finite, heterogeneous* and *hierarchical* cosmos.

7 Among the main empirical sources of evidence that led Aristotle to the idea that celestial bodies do not rotate on their own axes was direct observation of the Moon, as recorded in *On the Heavens*, II, 8, 290a7–290b12, where he states: "but the Moon always shows us its face (as men call it)"; English trans. W. K. C. Guthrie, Loeb Classical Library (London: W. Heinemann; Cambridge, MA: Harvard University Press, 1939), p. 189. Indeed, because of the tidal friction in regions where the oceans meet the Earth's crust, the moon has accommodated its axial rotation to its synodic orbital period and, as a result, we always see from Earth the same side. As Aristotle considers the Moon to be fixed to the upper sphere of the sublunary domain, he can account for this phenomenon without including axial rotation as part of the motions of the celestial body.

twenty years at the side of his famous teacher, and where he met Eudoxus and possibly also Callippus, learning their astronomical doctrines.

There are experiential (optical), epistemic, and historical reasons (§§ 5–6, 9–10) behind the prevalence of the circle and the sphere in all orders of Aristotelian cosmology. First—the optical reason: a naked-eye examination of the night sky reveals the circumpolarity of star motion. Thus, the notion of a spherical universe, foundation of ancient astronomy, can be explained by the fact that the sky is perceived by the observer as a whole that rotates on the axis of a sphere, a line known still today as the *celestial axis* in positional astronomy. As Aristotle expressed it: "but we ourselves see the heaven revolving in a circle" (τὸν δ' οὐρανὸν ὁρῶμεν κύκλῳ στρεφόμενον).[8] Second—the epistemic reason: the geometric regularity of the celestial sphere facilitates prediction and calculation using the spherical proto-trigonometry of the schools of Antiquity, that was such a powerful tool for the imposition of intelligibility on reality. Third—the historical reason: the paradigm that conceives the cosmos as a sphere dates back to the Babylonian astronomical tradition, consolidated by the late eighth century BCE, later embraced by Chinese and Egyptian astronomies, and then much enriched by Greek astronomers, as we will illustrate. Even at the dawn of Greek cosmology, in the works of some pre-Socratic philosophers such as Anaximander, Xenophanes, Parmenides, and Empedocles, one notices a certain interest in the sphere as a geometric figure and finds attempts to interpret fundamental cosmological coordinates as spherical. However, the sphericity of the cosmos was definitively consolidated by Aristotle's work, and this aspect of his doctrines remained influential up to the days of Giordano Bruno and Kepler.

Finally, it is worth noting that the analysis of the properties of celestial spheres by some pre-Socratics coincided with 1) the appearance of the first systematic enquiries (including experimentation) into the natural world (φύσις); 2) the development of geometry and its possibilities; and 3) the emergence of enquiry into Being and its multiple manifestations; all of which are deeply entwined in Aristotle's work. Thus, in so far as it may enrich the historical meaning of his astronomical system, we will analyse the process that led to the consolidation of the paradigm conceiving the cosmos as a sphere in pre-Aristotelian thought, and then review the original ideas introduced by Aristotle into that tradition.

8 *On the Heavens*, I, 5, 272a5; English trans. Guthrie, p. 37.

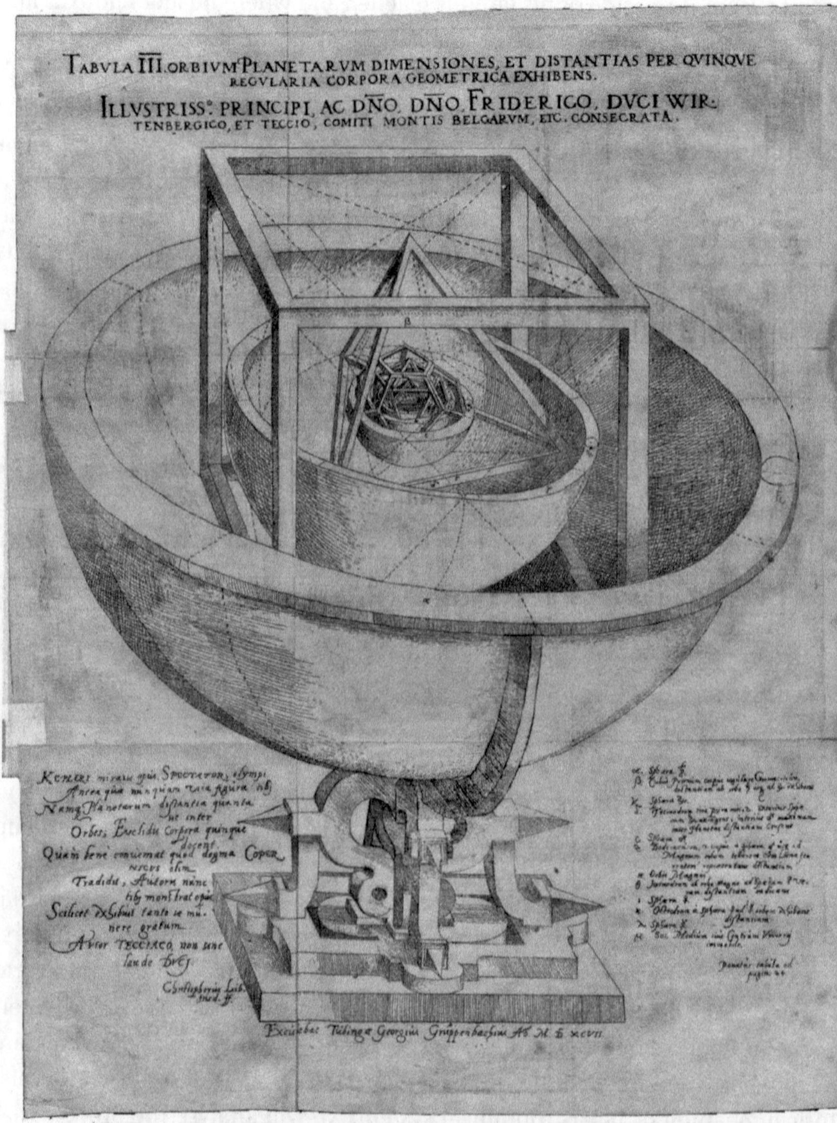

FIGURE 3 Diagram from the *Mysterium cosmographicum* by Johannes Kepler (1596), illustrating his polyhedral hypothesis. Kepler intended to build a scale model from the drawing but never succeeded in doing so. This image summarises the last attempt of a tradition involving homocentric spheres that had Pythagorean origins. In his youth, before discovering the elliptical nature of Mars's orbit, which led him to articulate his First Law, Kepler imagined a homocentric system of spheres in which interplanetary distances followed a plan of God, the exalted cosmic geometer, who designed the solar system by replicating the distances generated by the five Platonic solids inscribed in successive concentric spheres and ordered so as to preserve harmony.

CHAPTER 2

The Spherical Whole in Pre-Socratic Philosophy

> Εἰ τοίνυν ὅλον ἐστίν, ὥσπερ καὶ Παρμενίδης λέγει, Πάντοθεν εὐκύκλου σφαίρης ἐναλίγκιον ὄγκῳ, μεσσόθεν ἰσοπαλὲς πάντῃ· τὸ γὰρ οὔτε τι μεῖζον οὔτε τι βαιότερον πελέναι χρεόν ἐστι τῇ ἢ τῇ, τοιοῦτόν γε ὂν τὸ ὂν μέσον τε καὶ ἔσχατα ἔχει, ταῦτα δὲ ἔχον πᾶσα ἀνάγκη μέρη ἔχειν· ἢ πῶς

> If then the Whole is, as Parmenides says, *on all sides like the mass of a well-rounded sphere, equally weighted in every direction from the middle; for neither greater nor less must needs be on this or that,* then being, being such as he describes it, has a centre and extremes, and, having these, must certainly have parts, must it not?
>
> PLATO, *Sophist*, 244e

∵

Several pre-Socratic philosophers chose the sphere to represent certain core ideas, and Aristotle, influenced by this tradition, ultimately consecrated the sphere—on solid geometric grounds—as the unique shape of the Whole (τὸ πᾶν). In this section, we will summarise the main doctrines that link the form of the sphere with concepts of Being, God, the One, and the Whole, in testimonies and fragments from Xenophanes, Parmenides, Empedocles, Leucippus, and Democritus. A comparative assessment of these philosophers' central doctrines reveals a fascination with the spherical by pre-Socratic thinkers and permits us to outline the principal reasons behind the association of the sphere with ideas of totality, stability, and fullness in the first stage of Western philosophy. Among the reasons for these associations, two, especially, emerge as significant: 1) the characteristic attributes of the sphere as perfection, simplicity, and homogeneity; and 2) the aptitude of the sphere to represent, across cultures, the Absolute in metaphorical and spatial terms.[1]

1 The creation of *mandalas* (literally "circles" in Sanskrit) is practised in various cultures, especially in Buddhist and Hindu contexts, as part of a striving toward spiritual elevation. The concrete symmetries of a *mandala* and the possibilities of new and infinite forms are

© GERARDO BOTTERI AND ROBERTO CASAZZA, 2023 | DOI:10.1163/9789004525535_004

FIGURE 4 The Mainz Globe, a Roman celestial globe of the 2nd or 3rd century CE. Among the constellations visible here are Leo, Perseus, Cancer, Hydra and Draco. Römisch-Germanisches Zentralmuseum, Mainz

Inquiry into the sphere (both in geometrical and cosmological terms) emerged in ancient Greece.[2] According to Diogenes Laertius, it was Anaximander of Miletus (*c.*610–546 BCE), disciple of Thales, who "was the first to draw on a map the outline of land and sea, and he constructed a [celestial] globe as well" (καὶ σφαῖραν κατεσκεύασε).[3] This information corresponds to the testimony of the Suida, according to which Anaximander "wrote *On Nature, Circuit of the Earth* and *On the Fixed Stars* and a *Celestial Globe* and some

---

thought to facilitate subtle metaphysical experiences; as a bridge between the self and the cosmos, the mandala favours introspection and reveals the intimate harmonies between microcosm and macrocosm. A similar idea can be found in the Taoist Cosmic Wheel: the centre-periphery relationship highlights the tension between what is unchanged (the centre or law), and what is ephemeral and contingent (peripheral or daily activity). The wise man knows how to remain at the centre, without being overwhelmed by constant motion, while understanding the cyclic course of the real. Medieval *rotae* (circular graphics used for pedagogical and synoptic purposes) are further examples of how the circular (or spherical) has been considered well suited to expressions of a whole, and suggestive of connections between the superficialities of daily reality and deeper layers of meaning or existence. The strength and longevity of the tradition can be seen today in the use made in self-help literature of the Wheel of Life, a device that illustrates, again, the balance required for health or different forms of success. See R. Casazza, *Iconology of the Medieval and Renaissance Iconography of voluntas* (MA Thesis, The Warburg Institute, University of London, 1995).

2 A great many words in classical Greek are related to the notion of the sphere, as a brief compilation of terms will show (we follow here the 2007 electronic version of Liddell & Scott, *A Greek-English Lexicon*, published by the Perseus Project). The term σφαῖρα, for instance, signifies a sphere, balloon, ball, or hole. In Empedocles's cosmology, σφαῖρος denotes exclusively the state of absolute universal harmony. The word σφαιρίον, a neutral form of σφαῖρα, is used for a molecule, atom, small ball, the tip of the nose, etc. A σφαίρωσις is a process through which a sphere is created, and σφαιρωτός refers to what is round. The adverb σφαιρηδόν refers to things that behave, physically, like a sphere. A σφαῖρα is also a type of boxing glove; also related to boxing are the words σφαιρομαχία (fight) and σφαιράρχης (referee). A σφαιρομαχία is ball game, and the words σφαιρομάχος or σφαιριστικός or σφαιροπαίκτης refer to the player. The verb meaning "to play with a ball" is expressed by σφαιρομαχέω or σφαιρίζω or σφαιροπαικτέω. A court or field is designated by σφαιριστήριον. A ballgame fan is a σφαιροπαικτικός, and while the verb σφαιροποιέω means to manufacture balls or balloons, the similar term σφαιροποιία refers to divine creation of celestial spheres. The adjective σφαιρικός assigns a spherical quality to an object, while σφαιροειδής connotes that which looks like a sphere. In astronomy, σφαιρογραφία is the "celestial globe" (a study model), and σφαιροθεσία is the position of a star in the sky. Any rounded or globe-shaped object is a σφαίρωμα, and when obliquity is attributed to this term, it refers to the zodiac. The verb σφαιρόω means "to make balls or rolls," or, metaphorically, "to be focused on something." A type of rounded fishing net is called a σφαιρών.3.

3 DK 12 A 1; KRS 94; LFP I 67: Diogenes Laertius (3rd c. CE); this English trans. from *Lives of Eminent Philosophers* [*Vitae Philosophorum*]2 vols., trans. R. D. Hicks, Loeb Classical Library (London: W. Heinemann; Cambridge, MA: Harvard University Press, 1975 [1925]), II, 1.

other works."[4] According to Eusebius of Caesarea, Anaximander was "the first designer of gnomons for distinguishing the solar tropics, and times and seasons, and equinox."[5] According to Pliny, Anaximander was the first Greek to understand, by the time of the fifty-eighth Olympic Games, the obliquity of the zodiac. This means, as Pliny added, that he opened the way to a correct understanding of astronomical phenomena.[6] Anaximander did not, however, assign to the Earth a spherical shape but, rather, a cylindrical one. And although he seems to have conceived the idea of a celestial globe (related to the zodiac if we are to believe Pliny's testimony), his interest in the concept of the sphere can most accurately be said to represent a suggestive intellectual pointing or *direction*—something like Plato's pointing finger in Raphael's famous *School of Athens*. In the following pages, we will attempt to show the development of the concept, which, we argue, is manifest (if only in embryo) already in the sixth century.

Variously complex and elaborated developments involving the spherical shape can be found amongst a broad range of early philosophical doctrines,

4 KRS 95; DK 12 A 2; LFP I 66: *Suidae Lexicon* [*Suda* or *Souda*] (a tenth-century encyclopaedia of the ancient Mediterranean world, prepared by Byzantine scholars), ed. Ada Adler (Leipzig: Teubner, 1928–1938), s.v. Anaximander son of Praxiades.

5 DK 12 A 4; LFP I 73: Eusebius of Caesarea (3rd–4th c. CE), *Praeparatio Evangelica*, ed. K. Mras, GCS 21, vol. 8 (Berlin: Akademie Verlag, 1954), X, 14, 11; English trans. from *Eusebii praeparatio evangelica*, 5 vols., trans. E. H. Gifford (Oxford: Oxford University Press, 1903), X, XIV, 504a. It should be noted, however, that in *The Republic*, 600a, Plato argues the same for Thales of Miletus (DK 11 A 3; LFP I 46).

6 DK 12 A 5; LFP I 152: Pliny (23–79 CE), *Naturalis historiae libri II*, VI, 31. This text is rich in technical post-Anaximandrean astronomical terms. In addition, Theon of Smyrna—quoting Eudemus—attributes the discovery of the zodiac's obliquity to Oenopides (DK 11 A 17; LFP I 47), while Aetius (2nd–1st c. BCE) states that the concepts of the celestial sphere and the zodiac's obliquity were already known by "Thales [Pythagoras and their followers, who] hold that the sphere of the entire heaven is divided into five circles which they call 'zones'; and of these the first is called the 'arctic zone,' and is always visible, the next is the 'summer solstice,' the next is the 'equinoctial,' the next the 'winter solstice,' and the next the 'Antarctic,' which is invisible. And the ecliptic in the three middle ones is called the 'zodiac' and is projected to touch the three middle ones. All these are cut by the meridian at a right angle from the north to the opposite quarter"; DK 11 A 13= Aetius II, 12, 1 (D. 340); LFP I 48; English trans. from *The First Philosophers of Greece*, ed. and trans. *A. Fairbanks* (*London:* K. Paul, Trench, Trübner & co., 1898), p. 6. The passage is clearly influenced by Hellenistic astronomical terms. Therefore, the fact that it attributes this knowledge to Thales and the Pythagoreans makes it unreliable. Strabo also attributes this division to Parmenides, although without any justification (DK 28 A 44a; LFP I 985). This group of testimonies illustrates the opaque complexity of this doxography.

including those of Xenophanes (with a theological turn);[7] of Parmenides (metaphysical in nature); of Empedocles (with its cosmogonic bias); and finally of Leucippus and Democritus, where we find clear astrophysical characteristics. These approaches to and uses of the sphere, different as they may be, share some common features, as we shall see.

The philosophical doctrines of Xenophanes of Colophon (*c.*580–570 CE; *c.*475–466 CE) are, as we know, part of a theological discourse that systematically mocked the anthropomorphism of Homer's gods (DK 21 B 14 and DK 21 B 15; KRS 166; LFP I 500 and LFP I 502). Xenophanes's single and unique god (εἷς θεός) is perhaps the first philosophical god in the Greek tradition (DK 21 B 23; KRS 174; LFP I 504); embracing the entire cosmos, it coincides in extension with the Whole, and, being imperishable, it is the only legitimate divinity. Xenophanes called this god One and, according to Simplicius, "always he remains in the same place, moving not all; nor is it fitting for him to go to different places at different times, but without toil he shakes all things by the thought of his mind":[8] the deity's single activity seems to be—anticipating Aristotle—*thought*.

Although Xenophanes's reasoning remains within the characteristic pre-Socratic inability to conceive the spiritual as such, he postulates that the subtle materiality of divine thought embraces the sensible Whole, which remains homogeneous and immobile. This passivity contrasts with and complements the all-embracing activity of the god, as "all of him sees, all thinks, and all hears" (KRS 172; DK 21 B 24; LFP I 511). In this way, Xenophanes makes the three central intuitions of ancient Greek thought explicit: "the whole is divine"; "the whole is One"; "the whole is thought," thus consolidating the image that became a key philosophical starting point (Whole = One = θεός).

Briefly, the most significant testimonies that attribute sphericity to Xenophanes's god are the following ones. Theodoret, Bishop of Cyrrhus, affirms that:

> [Xenophanes of Colophon] said that the whole is one, spherical, and limited, not generated but eternally and totally motionless (ἓν εἶναι τὸ

---

7 This is the version consecrated by doxographical tradition, rather than Xenophanes's own development. In any case, the *process of spherification* of Xenophanes's spheric god shows that the sphere is, according to the main principles of Greek philosophy, the sensible intuition that best suits the unity of what is Divine and Absolute.

8 KRS 171 (unless otherwise specified, English translations of quotations having KRS references are from this source); DK 21 B 25 and DK 21 B 26; LFP I 512 and LFP I 513: Simplicius of Cilicia (6th c. CE), *In Aristotelis Physica commentaria*, ed. H. Diels, Commentaria in Aristotelem graeca 9–10 (Berlin: Reimer, 1882 & 1895), 23, 11 and 23, 20.

πᾶν ἔφησε σφαιροειδὲς καὶ πεπερασμένον, οὐ γενητὸν ἀλλ' ἀίδιον καὶ πάμπαν ἀκίνητον);[9]

while a text attributed to Hippolytus of Rome also expresses that, according to Xenophanes:

> the god is eternal and one and similar to all, finite, spherical, and perceptive in every part (καὶ τὸν θεὸν εἶναι ἀίδιον καὶ ἕνα καὶ ὅμοιον πάντηι καὶ πεπερασμένον καὶ σφαιροειδῆ καὶ πᾶσι τοῖς μορίοις αἰσθητικόν).[10]

In the treatise *On Melissus, Xenophanes and Gorgias*, the author, usually identified as Pseudo-Aristotle, attributes to Xenophanes the idea that god is spherical.[11] Aristotle, for his part, notes simply that Xenophanes—focusing on the cohesion of the firmament—was the first to state that the One is god, although he does not mention anything about its spherical shape.[12]

Significantly, mentions of Xenophanes's spherical god are found only in late testimonies. The main historical grounds for crediting this idea to Xenophanes appear to be the relationship established by Parmenides between Being and the notion of the spherical (which we will describe in a moment), and the idea—neither definitively proved nor refuted—that Xenophanes was Parmenides's teacher. From the doxographical interpretations that relate characteristics of Parmenides's Being to Xenophanes's god, everything indicates that, as Kirk has suggested, the notion of this god animating the Whole gradually

---

9 Theodoretus, *Treatment of Greek Afflictions* 4.5; from Aetius (Diels, 284n); English trans. from J. H. Lesher's edition of *Xenophanes of Colophon: Fragments: A Text and Translation with Commentary* (Toronto: University of Toronto Press, 1992), p. 216; see also, DK 21 A 36; LFP I 463: Theodoret, Bishop of Cyrrhus (*c.*393–458/466 CE), *Therapeutiques des maladies helleniques* (*Graecarum affectuum curatio*), French trans. and ed. P. Canivet, Sources Chrétiennes, 57, (Paris: Cerf, 1958), IV, 5. Furthermore, Diogenes Laertius, in his *Vitae philosophorum*, IX, 19 (DK 21 A 1; LFP I 509), also affirms that, according to Xenophanes, God's substance is spherical (οὐσίαν θεοῦ σφαιροειδῆ).

10 DK 21 A 33; LFP I 464: Hippolytus of Rome (3rd c. CE) [attribution questioned], *Refutatio omnium haeresium*, ed. P. Wendland, GCS 26 (Leipzig: Hinrichs, 1916), I, 14, 2; English trans. from M. D. Litwa's *Refutation of All Heresies, Translated with an Introduction and Notes* (Atlanta: SBL Press, 2016), p. 49.

11 DK 21 A 28; LFP I 465: Pseudo-Aristotle (1st c. CE), *Aristotelis qui fertur de Melisso, Xenophane, Gorgia libellus*, ed. H. Diels (Berlin: Abhandlungen der Königlichen Akademie der Wissenschaften, 1900), 977b20–22.

12 Aristotle, *Metaphysics*, I, 5, 986b23–25. This passage suggests that Parmenides could have been Xenophanes's disciple. This doctrine is certainly debated, but it is considered possible by many critics.

adopted over the centuries a spherical shape.[13] By late Antiquity it had become difficult to divorce the idea of god from that of the sphere, a pairing that is, furthermore, consistent with Xenophanes's *thinking god* that is the Whole. In short, we are dealing with a Parmenidean development that emerged within a context in which the sphericity of the cosmos was already a widely accepted notion, at least since Aristotle. And so, in the broad eclecticism of post-Aristotelian thought, a spherical shape was attributed to Xenophanes's god: this best and simplest geometrical figure was the most appropriate for a stable, homogenous, divine and perfect being.

The doctrine of Parmenides of Elea (*c*.510–*c*.450 CE) clearly legitimates the pseudo-Xenophanean equation of sphere and whole. It is found in his famous ontological meditation, whose proem announces that truth is "well-rounded" (ἀληθείης εὐκυκλέος).[14] This statement, not a mere occasional characterisation, strongly links Parmenides's theory of knowledge to his ontology. After all, "Thinking and Being are the same" (τὸ γὰρ αὐτὸ νοεῖν ἐστίν τε καὶ εἶναι).[15] Similarly, if the only possible sensible representation of Absolute Being is a perfectly spherical figure, then that of the single truth—namely the unity of the invariable Whole—will be closed, complete, and finished as well (i.e., εὐκυκλέος).

Parmenides's most fundamental passage is gathered by Simplicius, who appears to have been the last thinker to have access to the entire Περὶ φύσεων. There, Parmenides states:

> it [namely, what-is] is complete from all directions like the bulk of a ball well-rounded [εὐκύκλου σφαίρης ἐναλίγκιον ὄγκωι] from all sides equally matched in every way from the middle [μεσσόθεν ἰσοπαλὲς πάντηι]; for it is right for it to be not in any way greater or lesser than in another. For neither is there what-is-not—which would stop it from reaching the same—nor is there any way in which what-is would be more than what-is

13 On this point, Kirk affirms (KRS 174, note 1]), that "It was probably because of its motionless unity that Xenophanes's god was identified with Parmenides's Being, and later absorbed some of its properties. As early as Timon of Phlius it is called 'equal in every way' (ἴσον απάντηι, cf. μεσσόθεν ἰσοπαλὲς πάντηι in Parmenides, 299), and so becomes credited with spherical shape. Xenophanes may have described it as 'all alike' (ὁμοίην in Timon, Fr. 59, DK 21 A 35), since this is implicit in the whole of it functioning in a particular way as in KRS 172 [also DK 21 B 24; LFP I 511]; its sphericity goes beyond the fragments and is debatable."

14 KRS 288; DK 28 B 1; LFP I 938: Simplicius, *In Aristotelis de Caelo commentaria*, 557, 25–29.

15 DK 28 B 3; LFP I 935; English trans. from Plotinus, *Enneads*, vol. 5: *Enneads V. 1–9*, trans. A. H. Armstrong Loeb Classical Library (London: W. Heinemann; Cambridge, MA: Harvard University Press, 1984, p. 42.

> in one way and in another way less, since it is all inviolable; for equal to itself from all directions, it meets uniformly with its limits.[16]

The image of the sphere in this passage simply provides, we believe, a sensible representation, through metaphor, of the primary quality of Parmenides's Being: its homogeneity, consistent with its isotropy and indestructibility.

The relationship between Being and the sphere has led to widely debated philosophical reflection. As Virginia Guazzoni-Foà summarises in her 1966 article, already in Antiquity two antithetic stances arose regarding the meaning of Parmenides's sphere: on the one hand we have the stance of Plato (*Sophist*, 244e), for whom Being has, literally, a spherical shape; on the other hand, we find the later position that culminates in Simplicius (*In Aristotelis Physica commentaria*, 146, 29–33), which understands the sphere as metaphor.[17]

It is worth repeating that there is a close relationship between the realms of being and knowledge in Parmenides, and to interpret him we need to take a closer look at this. With a somewhat Christian existentialist bias, Guazzoni-Foà takes as the starting point Parmenides's position on the limits of human knowledge, whose main characteristic is, *more husserliano*, intentionality; that is, its noetic-noematic composition. It must be reminded that, according to Husserl, νόημα is always a partial and limited conception of the known, constantly accessed by the activity of consciousness or, in a more technical sense, the νόησις or "noetic act." From this perspective, the individual's contingent and circumscribed thinking necessarily knows *some* of the truth but never the *whole Truth*. From verse 34 of B 8 (LFP I 936; KRS 299) Guazzoni-Foà underlines

---

16 DK 28 B 8, vv. 42–49; LFP I 927: Simplicius, *In Aristotelis de Caelo commentaria*, 146, 15–22; English trans. from *A Presocratics Reader: Selected Fragments and Testimonia*, 2nd ed., ed. P. Curd, trans. R. D. McKirahan and P. Curd (Indianapolis: Hackett Publishing Company, 2011), pp. 60–61.

17 According to Virginia Guazzoni-Foà, in her "Un ripensamento sulla ΣΦΑΙΡΗ di Parmenide," *Giornale di Metafisica* 21 (1966): 344–354, as regards the meaning of Parmenides's sphere, there are realist (Zeller, Diels, Patin, Burnet, Levi, Stefanini) and symbolist (Natorp, Kinkel, Coxon, Herbertz, Joël, Zeppi) interpreters as well as geometers (Diano, Enriques), neo-Thomists (Sciacca, Mazzantini), etc. To complete the picture, the author adds other isolated stances corresponding to realism-symbolism (Gomperz), eidetism (Stenzel), and logical-ontological dynamism (Calogero). Diano and Enriques's geometrising interpretations emphasise that it is not feasible to conceive the sphere as actual; its three-dimensional nature would be ideal, as is the case of every geometric entity. Thus, Deano suggests understanding the sphere as static, without dimensions; therefore, as the sphere does not occupy space, nothing would be left outside of it. Enriques, for his part, understands the sphere as strictly metaphysical. Guazzoni-Foà's own noematic interpretation has the distinctive neo-Thomist phenomenological stamp of her time. Therefore, it presents Parmenides's sphere as a privileged representation of the Absolute.

a key note on the relationship between *truth* and *Truth*: What can be thought is only the thought that *it is* (ταὐτὸν δ' ἐστὶ νοεῖν τε καὶ οὕνεκεν ἔστι νόημα).

This "what can be thought" is Being (ἐὸν), and *by virtue of Being*, limited human thinking exists (νόημα): the only thought possible (that of Being) exists by virtue of Being. But at the same time this thought, while it rests on the Whole (Being), does not belong *to the Whole*, but *to the part*. The image of the sphere shows, precisely, that our thinking is limited because, as Guazzoni-Foà has pointed out, "it is the limit of the sphere that indicates intentionality, that is the nóema" (*e il limite della sfera che indica l'intenzionalità, cioé il noema*).[18]

Thinking is, therefore, an infinite searching that presupposes Being; at the same time, Being is infinite for human intentionality, although individuals can only represent it finitely. The sphere is then a *representation* of the Absolute, and as the thought of the individual does not coincide with absolute Truth, which is Being, its representation can only be "hemispheric," just as when a sphere (such as the moon) is viewed, only a section of the whole can be seen. In Paul's words, "for now we see only a reflection as in a mirror" (1Co 13:12). For Parmenides, then, the sphere is simply the chosen representation of the primal intuition of Being.

This metaphor complements others that appear in the poem: Being "was caught by *Dike*," and "is held by *Ananke* and the *Moira*." All these images evoke, in mythical terms, a metaphor for holding as regards what cannot be held, so as to understand what holding implies.[19] Thus, for Guazzoni-Foà, Parmenides's sphere represents, in a *static sense*: 1) absolute Being, in fact eternal and infinite, which can only be represented by individuals as finite and 2) the eternal and infinite truth which is identical with absolute Being (Being = Truth); while, in a *dynamic sense*, it shows: 3) the aspiration of finite, relative beings to perceive the Absolute, and in so doing find its closest representation in the image of the sphere, and 4) the circular and recurring flow of limited human thought which, in aspiring to know the Absolute, *becomes* in the very heart of Being, which is itself free of any becoming.

---

18 Guazzoni-Foà, p. 351.

19 The Thomist tradition suggests that Parmenides distinguished the dualism of absolute being and relative being, but did not resolve it. From this point of view, it was not until the appearance of a Thomist solution (the difference between *essence* and *existence* in created entities, but not in God, in whom both principles can be identified) that philosophy rested on solid ground. According to this school, Aquinas corrects the univocal predication of Being in the Parmenidean scheme by means of a radical clarification of the metaphysical structure of the contingent. See Raúl Echauri, "Parménides y el ser," *Anuario filosófico*, Vol. 6, Nº 1, 1973, pp. 98–115.

FIGURE 5 The famous 'cosmic-edge riddle' signalled by Simplicius, among others, proposes the following problem: if a man reaches the limits of the universe, could he stretch his arm or staff beyond the edge? Here, the scene is illustrated in an engraving from Camille Flammarion's *L'atmosphère: météorologie populaire*, Paris, 1888.

The image presented in the quoted verses (42–49) of B 8 inevitably coincides with several testimonies derived from that same source. Among them, to quote only one, is the testimony attributed to Hippolytus of Rome that reports Parmenides's view that:

> … the totality is eternal, not generated, spherical, all alike (but having no place in itself), motionless, and limited.
>
> … ἀίδιον εἶναι τὸ πᾶν καὶ οὐ γενόμενον καὶ σφαιροειδὲς καὶ ὅμοιον, οὐκ ἔχον δὲ τόπον ἐν ἑαυτῶι, καὶ ἀκίνητον καὶ πεπερασμένον.[20]

20 DK 28 A 23; LFP I 905: Hippolytus of Rome, *Refutatio omnium haeresium*, I, 11, 2; English trans. from *The Texts of Early Greek Philosophy: The Complete Fragments and Selected*

At this point in the development of philosophy, concepts of Being were in a process of development amidst the burgeoning foundations of categories concerning both ontology and the extension of the geometric universe—already present in ancient Greek thought—which today we call "Euclidean." It is within this context that Parmenides's intuition of the unity and homogeneity of Being leads him to visualise the "Whole Being" or, more simply, "Being," as spherical. Assessed from its historical basis, one can see that, in effect, no other material form offers such an apposite amalgam of attributes with which to depict Being. The sphere is, for instance, the most economical of objects, since for a given volume of matter it represents the smallest possible surface of all shapes; it is, furthermore, the symmetrical figure *par excellence*, for all points of its surface are equidistant from the centre. And its face is isotropic, such that one can trace a straight line from any point to complete a circle.[21] Moreover, the simplicity, beauty, and purity of the sphere credibly prompted an association with the divine, as noted in the testimony of Aetius, in which, for Parmenides, "[the god] is immobile, limited and spherical" (τὸ [θεὸν εἶναι] ἀκίνητον καὶ πεπερασμένον σφαιροειδές).[22]

Parmenides's conception of Being as "limited" (πεπερασμένον) also calls attention, especially as, *a priori*, the remaining characteristics of Being (as one, motionless, and homogeneous) appear to be compatible with the idea—developed by other thinkers such as Melissus—that the Whole is infinite. That he conceives it as limited supports the hypothesis of a Pythagorean influence behind Parmenides's spherical representation of Being. If we pay special

---

Testimonies of the Major Presocratics, 2 vols, ed. and trans. D. W. Graham (Cambridge: Cambridge University Press, 2010), vol. 1, p. 221.

21 This condition of identity maintained by every point in a perfect sphere is the basis of Aristotle's affirmation that, in its infinite symmetry, motion and rest coincide in the sphere (*Physics*, VIII, 9, 265b1–2): "so that a sphere is in a way both in motion and at rest; for it continues to occupy the same place"(διὸ κινεῖταί τε καὶ ἠρεμεῖ πως ἡ σφαῖρα· τὸν αὐτὸν γὰρ κατέχει τόπον); English trans. R. P. Hardie and R. K. Gaye in Barnes, *Complete Works of Aristotle*, vol. 1, p. 443.

22 DK 28 A 31; LFP I 914: Aetius, *Placita*, I, 7, 26; English trans. A. H. Coxon, ed., *The Fragments of Parmenides. A Critical Text with Introduction and Translation, the Ancient Testimonia and a Commentary*, rev. and expanded ed. with new translations by R. McKirahan (Las Vegas: Parmenides Pub., 2009). This testimony's authenticity has been questioned; we accept it, however, and take this text as the starting point for Xenophanes's spherical god. See also DK 28 A 23; LFP I 891: Hippolytus of Rome, *Refutatio omnium haeresium*, I, 11, 1 "Moreover, Parmenides supposed that everything is one, eternal, unborn, and spherical"; English trans. from Litwa, *Refutation of All Heresies*, p. 43. And see KRS 297; DK 28 B 8, 22–25; LFP I 903: Simplicius, *In Aristotelis Physica commentaria*, 145, 23–26: "Nor is it divided, since it all exists alike; nor is it more here and less there, which would prevent it from holding together, but it is all full of being. So it is all continuous: for what is draws near to what is."

attention to the Pythagoreans' Table of Opposites (called ἀρχαί by Aristotle, as if to show that it is not appropriate to conceive them as principles), we find the concepts of Limit (πέρας), Unity (ἕν), and Rest (ἠρεμοῦν) in the left-hand column—the "good" column. This would explain, if we are to accept the idea of Pythagoras's influence, how Parmenides could see limitation as an acceptable attribute of Being.[23] And if we seek a three-dimensional expression of this limitation, one that agrees with the remaining characteristics of Being, the sphere naturally and immediately comes to mind. We do not affirm absolutely that Parmenides's Being is spherical; however, given that it is both limited and whole (i.e., lacking fissures), it could be spherical. We believe that this is precisely how Aristotle later reasoned when he postulated that only a spherical figure could be the container of the Whole: he is convinced that the universe must be limited, and therefore concludes that it is spherical. This does not seem to be the case with Parmenides. For him, the sphere is not a *container*, but rather a *representation*, that is, a kind of poetic depiction of his discovery: the inviolability of the *being* of what is.

Finally, regardless of the precise relationship between Xenophanes and Parmenides, it seems clear that from these two early thinkers emerged a matrix that eventually brought together the monistic and divine conceptions of Being with the paradigm that conceives the cosmos as a sphere. These philosophical frames, which are in themselves closely linked, will find their apogee in Aristotle's onto-cosmological development.

With Empedocles (*c.*484–*c.*425 BCE) we acquire the notion of a sphere that, although based on Parmenides's ontology, stands out on account of its *dynamism*. One could say that Empedocles subjects Parmenides's static Being to the powerful force of Time. In fact, Empedocles's "Sphairos" (Σφαῖρος) preserves almost every characteristic of Parmenides's Being, as one, perfect, motionless, homogeneous, and divine; it is indivisible and embraces all. However, it is not eternal, so it comprises only a supracosmic phase in the eternal becoming of the Whole.[24]

---

23 It is worth bearing in mind the initiatic aspects found in the Proem of Parmenides's *Peri physeos*, an aspect that links it with Pythagoreanism.

24 The legend of Empedocles's death, which has him throw himself into the volcano Etna (DK31 A1; LFP II 268: Diogenes Laertius, *Vitae*, VIII, 67–69; DK 31 A 16; LFP II 269: Strabo, *Geography*, VI, 274), agrees with the central note of his philosophy: the presence of an anti-cosmic will in the world ("cosmic" must be understood here as the instances of balance and order between alterities, for example when the scales are even between Love and Strife). The triumph of Love (an anti-cosmic moment) implies the fusion of the limited with the Whole, such as we find in the Sphairos.

The special character of the Sphairos will be better understood if we introduce here a brief schematic outline of Empedocles's cyclic cosmogony. Empedocles considers that there are two universal principles of motion, and he calls them Love (Φιλία)—also called Aphrodite or Harmony—and Strife (Νεῖκος). These two metaphysical principles act on the four "roots" (ῥιζώματα), or on four kinds of living beings (ὄντα) according to Isocrates (DK 24 A 3; LFP II 323),[25] whose names correspond to the four traditional elements of earth, water, air, and fire (but which must not be mistaken for the Aristotelian elements). The dispersive effect of Strife acting upon these roots produces intermediary configurations in which earth, water, air, and fire interpenetrate into new states to create the concrete entities of the world. The end of this active motion of Strife is achieved when, overcoming Love's resistance, it separates all to a maximum degree. At this point, and reflexively, Love's work begins, drawing together disaggregated parts until the universe recovers its cosmic state, a fleeting moment of order and harmony, which is constantly being "updated," in a manner of speaking. The impulse of Love does not end there, however, but continues to act until a state of absolute peace is reached, in the form of a giant and homogeneous sphere, which is Empedocles's Sphairos. And there the universe rests for a time under Love's dominance.

This situation is of course unstable. Once the perfect and homogeneous sphere has been achieved, Strife resumes its task until, through great labour, it achieves anew the triumph of dispersion. The cycle is reproduced indefinitely, as expressed in these verses:

> A twofold tale I shall tell: at one time they [i.e. the roots] grew to be one alone,
> out of many, at another again they grew apart to be many out of one.
>
> δίπλ' ἐρέω· τοτὲ μὲν γὰρ ἓν ηὐξήθη μόνον εἶναι,
> ἐκ πλεόνων, τοτὲ δ' αὖ διέφυ πλέον' ἐξ ἑνὸς εἶναι.[26]

In Empedocles, the fight between unity and multiplicity reaches a stirring drama between these two principles, Φιλία and Νεῖκος, religious names—we believe—for two metaphysical entities that could be termed "Force towards the One" and "Force towards the Multiple." These counter-principles interweave in a kind of cosmic breathing that, with each exhalation and inhalation, gives life to all that emerges from and returns to the divine, spherical Absolute:

25 Isocrates, *Antidosis*, 268.
26 KRS 348; DK 31 B 17; LFP II 293: Simplicius, *In Aristotelis Physica commentaria*, 158, 1.

> For they are as they were previously and will be, and never,
> I think, will endless time be empty of both of these.
>
> ἧι γὰρ καὶ πάρος ἔσκε(?), καὶ ἔσσεται, οὐδέ ποτ', οἴω,
> τούτων ἀμφοτέρων κενεώσεται ἄσπετος αἰών.[27]

That said, while the peak of Strife's triumph is believed to be an instantaneous flash, like the snap of a spring, critics do not agree on the duration of the peaceful Sphairos. La Croce suggests that it could be an acosmic phase (the "cosmic" being a state of balance between Φιλία and Νεῖκος), and therefore timeless, a kind of absolute metaphysical limit at which the hemicycles come together, the end of the dominion of Love and the beginning of Strife's ascendancy.[28] Guthrie, for his part, proposes that the duration of the Sphairos accounts for an entire third of the total cycle (equal to the cosmic periods that shift from the Sphairos to Strife, and then again from Strife to Unity). However, he does not say exactly how long it would last. Other authors such as O'Brien[29] suggest comparing the Sphairos with the period of soul purification described in Empedocles's ethico-anthropological poem *Purifications*. This, says Empedocles, lasts thirty thousand seasons, which equals ten thousand years, taking an ancient value of ὥρα (DK 31 B115; LFP II 445).[30] Such an equa-

27 DK 31 B 16; LFP 301: Hippolytus of Rome, *Refutatio omnium haeresium*, VII, 29.9; English trans. from Curd, *A Presocratics Reader*, p. 83.

28 See LFP II 285, p. 86, to LFP II 339, p. 120 and LFP II 445, p. 167 for Ernesto La Croce's comments on this matter.

29 See chapter 4 of D. O'Brien, *Empedocles' Cosmic Cycle: A Reconstruction from the Fragments and Secondary Sources* (Cambridge: Cambridge University Press), 1969.

30 There are early testimonies, especially in Homer and Hesiod, which split the year into three ὧραι or seasons (spring, summer, and autumn). See *The Online Liddell-Scott-Jones Greek-English Lexicon*, s.v. ὥρα, at The Perseus Project (source on-line) for an account of those ancient passages. However, other authors, preceding and succeeding Empedocles (such as Alcaeus of Mytilene, Hippocrates of Kos, and Aristophanes), claim that there are four seasons, referring—we deduce—to the astronomical periods that travel from the equinox to the solstice and vice versa. Homer's denial of autumn seems rather doubtful, since the season is mentioned in the *Iliad*, V, 5 ("δαῖέ οἱ ἐκ κόρυθός τε καὶ ἀσπίδος ἀκάματον πῦρ / ἀστέρ' ὀπωρινῷ ἐναλίγκιον, ὅς τε μάλιστα / λαμπρὸν παμφαίνῃσι λελουμένος ὠκεανοῖο" ["She kindled from his helm and shield flame unwearying, like to the star of harvest-time that shineth bright above all others when he hath bathed him in the stream of Ocean" (trans. A. T. Murray, 1924)]), and also at XVI, 385 ("ἤματ' ὀπωρινῷ, ὅτε λαβρότατον χέει ὕδωρ" ["as beneath a tempest the whole black earth is oppressed, on a day in harvest-time" (trans. Murray); or "As under a great storm black earth is drenched on an autumn day" (trans. Fitzgerald, 1974)]), at XXI, 346 (ὀπωρινός) and at XXII, 27 (ὀπώρης). We see it also in the *Odyssey*, V, 328 ("ὡς δ' ὅτ' ὀπωρινὸς Βορέης φορέῃσιν ἀκάνθας" ["As when in autumn the North Wind bears the thistle-tufts over the plain" (trans. Murray,

tion would suit the Pythagoreans nicely, as one can then relate the passage to the journey of souls described by Plato in *Phaedrus*, 248e, which lasts precisely ten thousand years, and assign an ethical significance to Empedocles's cosmological doctrines. But as Empedocles did not define the duration of the Sphairos, we believe that it would be unwise to venture one.

How is it possible to conceptualise the situation of these forces—Love and Strife—during the polar stages of the cycle? One could reasonably assume, in this search for a systematic understanding (that is, again, repeatedly frustrated by the fragmented text base), that during the period of the Sphairos, Strife would be constrained far from the sphere's limit, while during Strife's triumph, Love would be penned within the geometrical centre of the disaggregation process. Yet following such a scheme that assigns space-time limits to these cosmic forces, Strife can never be trapped at the centre, because there would no longer be a reason for a new cycle to begin. Although Empedocles does not tell us so, one can consider the possibility that tiny asymmetries enable Strife to permeate the Sphairos—analogous to the balance between matter and antimatter before the Big Bang—leading always to new cosmic revolutions.

Happily for us, the description of the Sphairos has survived to our day in the form of charming verse:

> thus it was set in place with the tight covering of Harmony,
> a rounded Sphere rejoicing in circular solitude.
>
> οὕτως Ἁρμονίης πυκινῶι κρύφωι ἐστήρικται
> Σφαῖρος κυκλοτερὴς μονίηι περιηγέι γαίων.[31]

1919)]); XI, 192 ("αὐτὰρ ἐπὴν ἔλθῃσι θέρος τεθαλυῖά τ' ὀπώρη" ["But when summer comes and rich autumn" (trans. Murray]), XII, 76 and XIV, 384. The four seasons appear clearly enumerated by Hippocrates in his treatise *Regimen*, III, 68, 12–18: "I divide the year into the four parts most generally recognised—winter (χειμών), spring (ἦρ), summer (θέρος) and autumn (φθινόπωρον). Winter lasts from the setting of the Pleiads to the spring equinox; spring from the equinox to the rising of the Pleiads; summer from the Pleiads to the rising of Arcturus; autumn from Arcturus to the setting of the Pleiads" (trans. W. H. S. Jones, 1931). Here the temporal value attributed to ὥρα is crucial, as the Sphairos period would last either 10,000 or 7,500 years, dismissing, in the second case, a Phythagorean assimilation. Empedocles's *Purifications* clearly shows a period of spiritual purge ("τρίς μιν μυρίας ὥρας ἀπὸ μακάρων ἀλάλησθαι" ["he wanders for thrice ten thousand seasons away from the blessed ones" (trans. Inwood, 2001)] DK 31 B 115, 4–5; LFP 445: Hippolytus of Rome, *Refutatio omnium haeresium*, VII, 29), although its relation to his cosmogonic doctrines seems rather forced. We greatly appreciate Constantino Baikouzis's observations on this issue.

31 DK 31 B 27; LFP II 285: Simplicius, *In Aristotelis Physica commentaria*, 1183, 28. English trans. from Graham, *The Texts of Early Greek Philosophy*, vol. 1, p. 363 [text 55].

And then,

> two branches do not spring from its back,
> nor feet, nor nimble knees, nor productive genitals,
> but it was a Sphere both unique and equal to itself.
>
> οὐ γὰρ ἀπὸ νώτοιο δύο κλάδοι ἀίσσονται,
> οὐ πόδες, οὐ θοὰ γοῦν(α), οὐ μήδεα γεννήεντα,
> ἀλλὰ Σφαῖρος ἔην καὶ <πάντοθεν> ἶσος ἑαυτῶι.[32]

Some clarifications should be made here: the indivisibility of the Sphairos should not be confused with a state of mixture (μῖξις) in the Aristotelian sense of the term. John Philoponus (490–566 BCE) clearly pointed this out when he described the Sphairos as ἀποιόν (i.e., lacking quality).[33] This consideration forces us to take a qualitative leap from reality—suspending the laws of space and time—when we try to grasp this notion of sphericity. One wonders, then: has Empedocles conceived this instance as actually spherical, that is, subjected to the same space-time laws that are then denied when postulating the Sphairos's interior indeterminacy? Or should we think that Empedocles is simply using Parmenides's sphere—the primary representation at the time of the absolute nature of the Whole—to represent the fullest instance of his recurrent cosmic cycle?[34] In this case, the Sphairos's harmonic balance would be metaphorical rather than effective.

The four roots are, furthermore, never in a stable situation except when there is an absolute dominance of Strife. According to Plutarch's interpretation (*c.*50–*c.*120 BCE),[35] which is tinged with Aristotelianism, Strife's dominance would culminate in a sphere of fours layers, much like the four Aristotelian elements in the sublunary realm, except for the component of living beings on the Earth's surface that we find in Aristotle.[36] Each layer, for Plutarch, comprises

32 DK 31 B 29: Hippolytus of Rome, *Refutatio omnium haeresium*, VII, 29; English trans. from Litwa, *Refutation of All Heresies*, p. 547.

33 DK 31 A 41; LFP II 288: John Philoponus, *In Aristotelis de generatione et corruptione commentaria*, 19, 3. For Aristotle's interpretation of the Sphairos as μῖξις, see H. Cherniss, *Aristotle's Criticism of Presocratic Philosophy* (Baltimore: Johns Hopkins University Press, 1935).

34 Moreover, following in the steps of Xenophanes and Parmenides, Empedocles believed that the Sphairos was the most elevated form of the divine.

35 DK 31 B 27; LFP II 340: Plutarch, *De facie in orbe lunae*, 962 D.

36 *On Generation and Corruption*, II, X, 333a23; in the following passage, Aristotle says that it is not Strife but Love what separates the elements, which are, by nature, prior to god (the Sphairos) and, according to Empedocles, they are also gods themselves (DK 31 A 40;

one of the four elements: fire in the outermost layer, followed by air, then water and, at the centre, earth. In the intermediate stages of the cycle, the four roots are constantly jostling each other—because of the dynamics between the two cosmic forces—arriving at contingent states of dominance and equilibrium (sensible entities). These periods of dominance occur at specific moments of the cycle's rotation, which must be understood in terms of both the annual microcycle, which engenders the flourishing and decay of living beings, and also the universal macrocycle, which forges ahead blindly and repeatedly, ruled by Necessity, on a path (for Parmenides unattainable) between the one and the multiple. True births and deaths are, therefore, only apparent, being part of a permanent flux of rebalancing forces (DK 31 B 8, DK 31 B 13, DK 31 B 14; LFP II 324, LFP II 326, LFP II 327), a doctrine that can be interpreted as a sort of anticipated formulation of the law of the conservation of energy.[37]

Almost contemporary with Empedocles, the atomists Leucippus (*c.*500 BCE) and Democritus (*c.*470/460 BCE) showed some surprisingly profound cosmological insights based on a radical rejection of the Whole as limited. They did maintain, however, the limited shape of the sphere to explain the macro- and, to some extent, micro-physical constitution of reality. Leucippus and Democritus's guiding cosmological idea was, in simple terms, that we inhabit one of the infinite spherical worlds that drift in a sea of endless emptiness. These worlds, for which generation and corruption is not cyclically repeated, are necessarily infinite because the original quantity of atoms in the universe is equally infinite. At the same time, for materialistic atomism, the relationships between atoms are inevitably governed by Necessity (Ἀνάγκη) (DK 67 A 10 and DK 68 A 83; LFP II 430 and LFP II 435). This cosmological scheme anticipates the one elaborated by Thomas Wright in the eighteenth century (also accepted by Immanuel Kant and Johann Lambert), which proposes the coexistence of infinite "island universes," each of them also spherical.[38]

In atomist terminology, the expression τὸ πᾶν is preserved for what we now call the "universe," while the notion of κόσμος refers to the infinite contingent "worlds" that are randomly distributed in emptiness. Each cosmos or world evolves—according to the text attributed to Hippolytus of Rome, summarising

LPF II 320). Here, the contradictory nature of Aristotle's conception of the Sphairos as mixture or μῖξις is quite clear: if the Sphairos is god, and he is a mixture of the roots, then one must affirm that the roots too are gods, and even more original than the cosmic forces. But this is completely impossible from Empedocles's view.

37 The first law of thermodynamics by Clausius (1865) states that the energy of the universe is constant (*Die Energie der Welt ist konstant*).

38 A. Gangui, *El Big-Bang: La génesis de nuestra cosmología actual* (Buenos Aires: Eudeba, 2005), pp. 155–159.

Democritus's ideas—"until it is no longer able to receive anything in from the outside" (ἀκμάζειν δὲ κόσμον, ἕως ἂν μηκέτι δύνηται ἔξωθέν τι προσλαμβάνειν).[39] This idea is complemented by Aetius's testimony: "Leucippus and Democritus say that the world is spherical" (Λεύκιππος καὶ Δημόκριτος σφαιροειδῆ τὸν κόσμον).[40]

The outstanding feature of the atomist thinkers is, as has often been signalled, their inquiry into questions of natural philosophy from a *mechanistic* perspective. They hold a privileged standing in the history of physics for their endeavours to account for the observable using strictly natural explicative principles. The following passage, transmitted by Diogenes Laertius, IX, 31, illustrates this orientation and describes the process of the emergence and "spherification" of worlds:

> He says that the universe [κόσμους] is infinite [ἀπείρους], as has been said. Part of it is a plenum, and part void, which he says are the elements. There are infinitely many worlds composed of this, and they are resolved into those elements. The worlds come into being in this way. A large number of bodies of every shape become separated from the infinite [ἐκ τῆς ἀπείρου] into a great void [εἰς <μέγα κενόν>], congregate together and form a single swirl [δίνη], in which, as they collide and circle in all sorts of ways, they are separated out, like to like. Because of their number they can no longer rotate in equilibrium, but the small ones are as it were sifted out into the external void; the rest remain and, becoming entangled with one another, move round together, making up a primary spherical structure [πρῶτόν τι σύστημα σφαιροειδές]. (32) There separates off from this a sort of membrane [ὑμήν], containing bodies of every kind; as these swirl round the surrounding membrane becomes thin through the resistance of the central mass [ἐπὶ τὸ μέσον], as the bodies on its inner surface are continually flowing off into the centre because of the contact within the swirl. In this way the earth comes into being, as the bodies which have been carried into the middle remain there; and on the other hand the surrounding membrane [τὸν περιέχοντα οἷον <ὑμένα>] grows by separating off bodies from the outside, adding to itself any which it touches as it whirls round. Some of these fasten together into a structure which is at

---

39 DK 68 A 40; LFP II 460; English trans. from Litwa, *Refutation of All Heresies*, p. 47.

40 DK 67 A 22; LFP II 464; English trans. from C. C. W. Taylor, *The Atomists: Leucippus and Democritus. Fragments: A Text and Translation with a Commentary* (Toronto: University of Toronto Press, 1999), p. 97 [Fr. 81].

> first moist and muddy, but which dries as it rotates in the universal swirl, finally catching fire and constituting the nature of the stars.[41]

This description of the process of cosmic generation fails to account, however, for one issue about which Leucippus and Democritus hypothesised: the physical quality of the very *limit* that contains each world. As Aetius tells us, "Leucippus and Democritus stretch round the world a coat or membrane [ὑμήν] made of hook-shaped atoms,"[42] and this "coat" acts as a spherical sac inside of which entities are distributed according to atomic similitude.[43] Aetius then describes the process by which the external membrane is formed:

> The present world [κόσμος] was formed in an curved shape [περικεκλασμένωι σχήματι] in the following way. The atomic bodies were in continuous, chance, undesigned motion of extreme rapidity, and many, of a variety of shapes and sizes (because of their number), collected together in the same place. Once collected, all the larger and heavier began to sink down, and the small, round [περιφερῆ], smooth, slippery ones began to be squeezed out by the pressure of the atoms and to travel upwards. And when the force which was driving them upwards began to be exhausted, but they were prevented from travelling downwards, they were pushed into places where there was room for them. These were at the circumference, and in these the mass of bodies became curved [οὗτοι δὲ ἦσαν οἱ πέριξ, καὶ πρὸς τούτοις τὸ πλῆθος τῶν σωμάτων περιεκλᾶτο]; and by tangling together in that curved shape they formed the heavens [τὸν οὐρανόν].[44]

It is worth bearing in mind that the atomists also chose the spherical shape in their propositions at the atomic level, just as they conceived minute fire corpuscles as spherical. Aristotle mentions this briefly when he regrets their

41 DK 67 A 1; LFP II 453: Diogenes Laertius, *Vitae*, IX, 30; English trans. from Taylor, *The Atomists: Leucippus and Democritus*, pp. 94–95 [Fr. 77].

42 DK 67 A 23; LFP II 467: Aetius, *Placita*, II, 7, 2; English trans. from Taylor, *The Atomists: Leucippus and Democritus*, p. 95 [Fr. 77b].

43 This fact is very significant if we consider that, for example, in spite of his systematic approach, Aristotle never accounted for what happens at the limit of the celestial sphere, between the ethereal interior and the exterior "emptiness."

44 DK 67 A 24; LFP 454: Aetius, *Placita*, I, 4; English trans. from Taylor, *The Atomists: Leucippus and Democritus*, p. 96 [Fr. 79]. One of the guiding ideas of the prolific and lucid book *Spheres*, 3 vols., by P. Sloterdijk (Cambridge: MIT Press, 2011, 2014, 2016), namely, that men are essentially sphere-makers, is based on the fact that the very first human habitation is the maternal gestational sac, which is more or less spherical, and that this is the only cavity in which humans, throughout their entire lives, are neither afraid nor hungry.

silence regarding shapes with respect to the other three elements (Aristotle, as we know, rejects Platonic micro-physics, in which the elements are made up of regular minute solids: earth of cubes or hexahedrons; water of icosahedrons; air of octahedrons; and fire of tetrahedrons). In *On the Heavens*, III, 4, 303a, he points out that: "Yet they [Leucippus and Democritus] never went so far as to define or characterise the shape of each element, except to assign the sphere to fire [τῶι πυρὶ τὴν σφαῖραν ἀπέδωκαν]. Air and water and the rest [of the substances] they distinguished by [the] greatness and smallness [of their atoms]…."[45]

When we consider the subsequent development of the Greek science, we see that the empirical scientific legacy of the atomists was then pushed aside by the idealist tradition that dominated inquiry into nature in both the classical and Hellenic periods. This tradition proceeded *a priori*, and used Euclidean geometry as a model, as we shall see in the review of the astronomical doctrines of Plato, Eudoxus, Callippus, and Aristotle that follows.

---

45 DK 67 A 15; LFP II 472; English trans. Guthrie, p. 291.

CHAPTER 3

The Platonic Mandate: Reducing Celestial Phenomena to Circular Motions

> *Quod motus corporum coelestium sit aequalis ac circularis, perpetuus, vel ex circularibus compositus*
>
> The motion of the heavenly bodies is uniform and circular, perpetual, or compounded of circular motions
>
> NICOLAUS COPERNICUS, *On the Revolutions*, I, 4

⁖

The incipient fascination for the spherical figure, emergent from the dawn of Greek philosophy, was adopted by two great Pythagorean thinkers of the fifth and fourth centuries BCE: Philolaus of Croton (*c*.470–*c*.385 BCE) and Plato (*c*.427–*c*.347 BCE). According to Joannes Stobaeus, Philolaus locates:

> … what is controlling [the cosmos] in the central fire, which the demiurgic god set down under the sphere of the whole like a keel [of a ship] (τὸ δὲ ἡγεμονικὸν ἐν τῶι μεσαιτάτωι πυρί, ὅπερ τρόπεως δίκην προϋπεβάλετο τῆς τοῦ παντὸς <σφαίρας> ὁ δημιουργὸς θεός);[1]

and claims that:

> The first thing to be harmonised—the one—in the middle of the sphere is called the hearth (τὸ πρᾶτον ἁρμοσθέν, τὸ ἕν, ἐν τῶι μέσωι τᾶς σφαίρας ἑστία καλεῖται).[2]

---

1 DK 44 A 17; LFP III 159: Joannes Stobaeus (5th c. BCE), *Eclogae physicae, dialecticae et ethicae* (Berlin: Wachsmuth & Hense, 1958), I, 21, 6d; English trans. from C. Proust and J. Steele, eds., *Scholars and Scholarship in Late Babylonian Uruk* (Berlin: Springer, 2019), p. 249.

2 KRS 441; DK 33 B 7; LFP III 161: Joannes Stobaeus, *Eclogae physicae, dialecticae et ethicae*, I, 21, 8.

© GERARDO BOTTERI AND ROBERTO CASAZZA, 2023 | DOI:10.1163/9789004525535_005

Plato, for his part, left at least two relevant cosmological-astronomical accounts: *Republic*, X, 616c–617a and *Timaeus*, 33b–47c. In these he offers complex descriptions of the system of heavens. He describes the universe as limited and spherical, its outer limit being the sphere of the fixed stars, with Earth as its central body. Both texts present the seven wandering stars in the Egyptian order (Moon-Sun-Venus-Mercury-Mars-Jupiter-Saturn), beneath the sphere of the fixed stars.

A group of nodal ideas penetrates Plato's entire work, much as the five transcendentals (*esse, verum, unum, bonum, pulchrum*) permeated medieval thought. Among these "Platonic transcendentals," we find the *primacy of the simple* over the many; the notion of the universe as ruled by a *rational divine principle*; the *logical-eidetic articulation of reality* at all levels; the presence of an *order in nature* (whose symbol is the cyclical nature of celestial motion); the *isomorphic structure* of the cosmic and mental orders; and the *search for logos* in both individual and collective experience. All of these ideas permeate Plato's notion of celestial motions, which he—like his predecessors—conceives within the paradigm of a spherical cosmos.

It is clear that the astronomical account of the *Republic* consists in a propaedeutic presentation of the system of the heavens intended as a literary model, and that the best way to understand it—as has been done by interpreters such as Thompson, Heath and Dreyer—is to analyse the passage in terms of apparent heavenly motions.[3] We will do this briefly now.

In the Myth of Er (614b–621d), the story that concludes the *Republic*, Plato expounds in a mythical manner his doctrine of μετεμψύχωσις (transmigration of souls)—present in other dialogues as well—according to which, as the soul is immortal, all animate beings continue to live after bodily death through successive reincarnations (παλιγγενεσία).[4] The text tells the story of Er, a man from Pamphylia, who dies in battle but is bound by divine will to be resurrected in his same body after a period of twelve days. Upon leaving his body, Er's soul begins its journey, along with many other souls as companions, and

3 Beyond the accounts in the *Republic*, 614b–621d, and the *Timaeus*, 33b–47c, aspects of Platonic astronomy useful to its understanding can be found in a number of additional texts—fully in line with the geometry-based physics of the *Timaeus*—the most significant passages being: *Phaedrus*, 246b–248c; *Republic*, VII, 528d–530c; *Phaedo*, 109a–109e; *Laws*, X, 896b–898e; and *Laws*, XII, 966e–968a.

4 Humans, privileged with a rational soul (even if weakened by its ties to the body), must direct their souls to the purest, most divine objects—that is, science and philosophy. In this way they can elevate their dignity, though not without pain and effort, through successive reincarnations, as described in the Myth of Er. This is how the noble individual attunes themselves with the universal cosmic order, the origin of which is, for Plato, clearly divine.

reaches an attractive meadow with four symmetrically placed openings—two in the ground and two in the sky at each of its right and left edges. Through the openings souls travel up and down, to and from the heavens and the underground, after lengthy spiritual journeys. It is here that Er witnesses the trial of souls. According to the kind of life led in their immediately preceding existence, souls are either required to follow a long underground path in atonement of their faults (which could last up to a thousand years),[5] after which they would be reincarnated;[6] or are elevated to the heavens—a metaphor for the intelligible world—by virtue of lives attuned to philosophy. (An interesting side-note here: paradoxically, and quite in contrast to Christian eschatology, once having completed a journey to the heavens and been reincarnated for a new earthly life, some souls would subsequently follow *less* spiritually edifying paths.) After spending seven days in that beautiful meadow, the souls (including Er's), were obliged to continue their journey for four more days until they reached a summit from where they could see a pure shining light like a pillar extending through the sky,[7] and after one more day of travel they found themselves surrounded by that same light. Using one of his typical literary strategies, at this point in the story Plato dilutes his protagonist's adventure

---

5 In the *Phaedrus*, 248e, it appears that the journey of souls from one incarnation to the next can last even up to ten thousand years.

6 This is a valuable myth as it uses different registers to develop deeply interconnected cosmological, anthropological, and ethical content. In fact, towards the end of the story (617d–621d), Er witnesses the powerful moment when souls choose their new corporeal lives—not always human. Here Plato offers a unique doctrine of freedom: the only instance of freedom is the choice of one's own life before life itself. According to Schopenhauer's interpretation of the myth at the end of his *Preisschrift über die Freiheit des Willens* [*Essay on the Freedom of the Will*, trans. K. Kolenda (New York: Bobbs-Merrill, 1960)], entering space-time (i.e., being born), submits the will to the Law of Causality by which actions are causally bound, denying freedom. However, in a transcendental plane, before incarnation, the soul chose its destiny freely. As individuals do not have knowledge, during their lives, of their own destiny, they are constantly *choosing* without constraint the life that they have *already* chosen in a timeless plane.

7 There are opposing views on this point, involving conflicting hermeneutical frameworks: one considers the shaft of light as a metaphor for the Spindle of Necessity (or the celestial axis, in astronomical terms), thus the pillar of light and the spindle are one and the same; a second interpretation (Boeckh) considers the light to be the Milky Way. This last is based on the passage of the *Phaedrus* 246a–248e, which describes the journey of souls beyond the celestial sphere, where it is said that the Milky Way would be perceived as a pillar. We do not find this interpretation viable, however. Plato indicates that light derives from the poles, and although—according to the myth—the Milky Way draws a circle in the sky, it does not cross the poles, as do the solstitial and equinoctial colures. Our interpretation will therefore be based on the first reading, which takes the shaft of light as metaphor.

in a complex astronomical-cosmological description, which critics have deciphered variously.[8] The following is our own interpretation.

According to the story, at the end points of "a shaft of light, like a pillar" (φῶς εὐθύ, οἷον κίονα)[9]—which traverses the Earth through the spot where Er and his companions are standing and reaches the celestial vault—radiate dense filaments that hold the heavens as if they were cables (δεσμοί), stretching out from one end to the other (τὰ ἄκρα). In the myth, these threads of light are compared to the cables that hold the hull of trireme vessels and can be interpreted as the solstitial and equinoctial colures (Dreyer), which meet at the celestial poles. The pillar of light should be regarded as the celestial axis of positional astronomy (Theon of Smyrna, Stewart, Heath). As the story unfolds—and due to a change of scene, as is common in dreams (Stewart), a standard device in mythical discourse—the column is subtly assimilated to the "Spindle of Necessity" (Ἀνάγκης ἄτρακτον), with eight celestial semi-spheres ("whorls" in most translations) revolving around it. An obscure description of the behaviour of these spheres is the main heuristic challenge of this tale.

Necessity (Ἀνάγκη) is embodied as a goddess who turns the spindle on her lap. The description of the figure alludes to the persistent, cyclical, and irrevocable nature of celestial motion; that is, of Time and its corollaries of order and measurability. Necessity receives help from her three daughters, the Moirai (in the Greek world) or the Parcae (in the Roman world): Lachesis, Clotho and Atropos. More commonly known as the Fates to English readers, these white-robed spinners of human destiny bear, here, a cosmic meaning. According to traditional mythology, Clotho, the spinner, is associated with birth; Lachesis moves the spindle allowing life to unfold; and Atropos, the cutter of the thread, ends the brief singular human live. In the Platonic tale their roles are modified, however. Clotho, who here moves the sphere of the fixed stars westwards with her right hand, is an allegory of the *present*; and Atropos, who sets the inner spheres moving in the opposite direction with her left hand, personifies the *future*; while Lachesis, representing the *past*, alternately touches the spheres with her right and left hand, thereby signalling that the planets also participate in the diurnal motions of the heavens

8 The main interpreters of passage 616c–617a—whose astronomical knowledge is completely reliable—are Theon of Smyrna (fl. 100 CE), Proclus (*c.*440), Boeckh (1858), Martin (1881), Adam (1902), Berger (1903), Thompson (1910), Stewart (1905), Heath (1913) and Dreyer (1953).

9 *The Republic of Plato*, English trans. F. M. Cornford (New York & London: Oxford University Press, 1941) p. 344.

(Dreyer). The motions driven by the Moirai should be understood—from an astronomical point of view—in light of both the *Timaeus*, 38c–40d, and the *Republic*, 617a, which state that the seven wandering stars or planets (Moon-Sun-Venus-Mercury-Mars-Jupiter-Saturn) and the outer circle of the fixed stars move in opposite directions (τοὺς μὲν ἐντὸς ἑπτὰ κύκλους τὴν ἐναντίαν τῷ ὅλῳ ἠρέμα περιφέρεσθαι). This poetic image of the Moirai moving their hands signals the main astronomical point of the Platonic exposition: the clear distinction between the westward motion of the fixed stars and the direct eastward motion completed by the seven wandering stars, each in its own period.[10]

FIGURE 6 Plato's astronomical system as represented in the Myth of Er (*Republic*, X, 616c–617a). This vision of the afterlife, as witnessed by Er of Pamphylia, consists in a system of eight concentric semi-spheres with Earth at the centre. Necessity and her daughters, the three white-robed Moirai (Lachesis, Clotho and Atropos), turn the spindle, which represents the celestial axis.

10 These are "motions a and b," which will be described thoroughly at the beginning of chapter 4 "Eudoxus and Callippus: Planetary Models."

The entire passage celebrates the celestial laws governing the cyclical flow of universal motion. The spindle, as a textile tool, is simply a straight spike with a whorl at the bottom. In the myth, it constitutes a poetic image of the spinner's activity, which consists of spinning the spindle smoothly to twist wool or thread. According to Plato, the axis is made of adamant (or steel, in many translations) implying its great structural strength and the enormous force transmitted by its rotation; it is the engine of all the heavens. In this Plato's view is unlike Aristotle's conception, which denies that such force can be exerted from the celestial axis (*Physics*, VIII, 10, 267a22–267b9 –§ 25–).

From an astronomical point of view, the main difficulty in understanding the model of the *Republic* is found in the passage that describes the spindle as attached and fixed to the celestial sphere—that is, to the North and South celestial poles—and presents the structure and dynamics of the planetary system:

> And from the extremities stretched the Spindle of Necessity, by means of which all the circles revolve [τὰς περιφοράς]. The shaft of the Spindle and the hook were of adamant [ἀδάμαντος], and the whorl [σφόνδυλον] partly of adamant and partly of other substances. The whorl was of this fashion. In shape it was like an ordinary whorl; but from Er's account we must imagine it as a large whorl with the inside completely scooped out, and within it a second smaller whorl, and a third and a fourth and four more, fitting into one another like a nest of bowls. For there were in all eight whorls, set one within another, with their rims showing above as circles [κύκλους ἄνωθεν τὰ χείλη φαίνοντας] and making up the continuous surface of a single whorl round the shaft, which pierces right through the centre of the eighth. The circle forming the rim of the first and outermost whorl is the broadest [πλατύτατον]; next in breadth is the sixth; then the fourth; then the eighth; then the seventh; then the fifth; then the third; and the second is narrowest of all. The rim of the largest whorl was spangled; the seventh brightest [λαμπρότατον]; the eighth coloured by the reflected light of the seventh; the second and fifth like each other and yellower [ξανθότερα]; the third whitest [λευκότατον]; the fourth somewhat ruddy [ὑπέρυθρον]; the sixth second in whiteness. The Spindle revolved as a whole with one motion [ὅλον μὲν τὴν αὐτὴν φοράν]; but, within the whole as it turned, the seven inner circles revolved slowly in the opposite direction [ἐναντίαν]; and of these the eighth moved most swiftly [τάχιστα]; second in speed and all moving together, the seventh, sixth, and fifth; next in speed moved the fourth with what appeared to them to be

> a counter-revolution [ἐπανακυκλούμενον]; next the third, and slowest of all the second.[11]

Here, the guiding astronomical image is a system of eight concentric semi-spheres called σφόνδυλοι ("whorls" or "bowls"), with Earth at the centre. Some interpreters (Adam, Heath, and we as well) agree that the semi-spheres are representations of spheres, while other authors (Burnet, Stewart, Thompson, Dreyer) regard them as rings. Leaving the second interpretation aside for now, the semi-spheres are placed one inside the other. The entire scheme might be considered a didactic tool, meant as a model (Stewart)[12] or other visual aid to describe the celestial sphere and inner planetary spheres.

The quoted passage essentially addresses four topics: 1) sphere order, moving inwards from the outer part towards the centre of the system; 2) the colour and brightness of the celestial bodies associated with each sphere; 3) the angular velocity, as seen from Earth, of the celestial bodies of each sphere; and 4) the order, according to rim breadth (τὰ χείλη), of the spherical rings forming the system (and here arises the discrepancy in interpretations as to whether the whorls are shaped as semi-spheres or rings).

Concerning sphere order and colour, it is possible to deduce the Platonic planetary arrangement directly from the unnamed descriptions in the Greek text—even without resorting to the *Timaeus*, 36b–36d, where he explicitly repeats it[13]—as the colours indicated in the passage match what can be seen through an attentive naked-eye observation of the sky.

11 *The Republic of Plato*, English trans. Cornford, p. 345 (616c–617a). The planet names inserted by Cornford into his translation have here been removed for the sake of our argument.

12 The idea of models of the universe as tools to understand how it functions appears in the *Timaeus*, 40d, in which the demiurge uses a model to fashion its work.

13 *Timaeus*, 36b–c: "This whole fabric, then, he split lengthwise into two halves; and making the two cross one another at their centres in the form of the letter x, he bent each round into a circle and joined it up, making each meet itself and the other at a point opposite to that where they had been brought into contact. He then comprehended them in the motion that is carried round uniformly in the same place, and made the one the outer, the other the inner circle. The outer movement he named the movement of the Same; the inner, the movement of the Different. The movement of the Same he caused to revolve to the right by way of the side; the movement of the Different to the left by way of the diagonal. And he gave the supremacy to the revolution of the Same and uniform; for he left that single and undivided"; English trans. F. M. Cornford, *Plato's Cosmology: The Timaeus of Plato* (Indianapolis: Hackett Publishing Company, 1997 [1935]).

a) Sphere order, moving inwards from the furthest point from the central axis		*b) Colour of the celestial bodies in each sphere*	
first	Sphere of the Fixed Stars	"the first	... was spangled"
second	Saturn	"the second	... and the fifth like each other and yellower"*
third	Jupiter	"the third	... whitest"
fourth	Mars	"the fourth	... somewhat ruddy"
fifth	Mercury	"the fifth	... and the second like each other and yellower"
sixth	Venus	"the sixth	... second in whiteness"
seventh	Sun	"the seventh	... brightest"
eighth	Moon	"the eighth	... coloured by the reflected light of the seventh"

* Mercury and Saturn look very much alike in the sky.

The order proposed by Plato may cause surprise to those who observe the solar system as we know it today. However, one must bear in mind that the pre-Copernican debate in Europe about planetary order presented several alternatives. There was general agreement about the outer position of the fixed stars and the proximity of the Moon to the terrestrial centre, and there were few doubts about the natural order of Saturn, Jupiter and Mars, whose zodiacal periods were agreed as approximately 29.6, 11.9 and 1.9 years, relatively. Therefore, the main problem of the ancient planetary order was the relative position of the Sun, Mercury, and Venus, on the grounds that it takes them a year to complete their zodiacal revolution, a fact pointed out by Plato in the *Timaeus*, 39a. The reason is simple. Since Mercury and Venus are inner planets of the solar system, when seen from Earth they both accompany the Sun's annual revolution. Ptolemaic systems tended to place these three celestial bodies in the following ascending order from the moon: Mercury, Venus, Sun. Such a scheme prevailed in medieval cosmology and was only questioned after the publication of Copernicus's *De revolutionibus* (1543). In the *Timaeus*, 38d, Plato points out that these three stars continually overtake each other in a sort of celestial dance, a phenomenon that can be seen with the naked eye throughout the year.[14]

As regards the angular velocity of each wandering star, Plato simply gathers observational data. The moon can be observed to take 27.23 days to complete a revolution around the ecliptic, it takes a year for the next group (moving

14 That is "motion c," which will be described in chapter 4.

outwards from the moon) to complete their revolutions, and Mars, Jupiter, and Saturn complete well-known zodiacal periods notwithstanding their stations and retrogradations.

c) Planet order according to the angular velocity of apparent eastward motions around the ecliptic from a geocentric point of view

first (fastest)	Moon	27.23 days	"the eighth [sphere] moved most swiftly"
second	Sun	1 year	"second in speed, the *seventh*, sixth, and fifth"
second	Venus	1 year	"second in speed, the seventh, *sixth*, and fifth"
second	Mercury	1 year	"second in speed, the seventh, sixth, and *fifth*"
third	Mars	1.9 years	"next in speed moved the fourth"
fourth	Jupiter	11.9 years	"next the third"
fifth	Saturn	29.6 years	"slowest of all the second"*

* The entire passage considers the sphere of the fixed stars to be the first sphere (or "whorl").

And, finally, as regards rim breadth, there is considerable controversy yet to be resolved.[15] The text cryptically refers to a gradation of breadth:

d) Rim breadth order, from broadest to narrowest

"the rim of the first is the broadest"	i.e., sphere of the fixed stars
"next in breadth is the sixth"	i.e., Venus
"then the fourth"	i.e., Mars
"then the eighth"	i.e., Moon
"then the seventh"	i.e., Sun
"then the fifth"	i.e., Mercury
"then the third"	i.e., Jupiter
"the second is narrowest of all"	i.e., Saturn

Here we find a true interpretation problem since there is not any other clear astronomical account that explains this rim breadth order. There are two main hypotheses that to some extent match Plato's text to astronomical

15 The main proponents on this matter are Adam, Heath, Dreyer, Stewart and Thompson.

phenomena.[16] For the sake of clarity we will call them 1) the lineal hypothesis and 2) the diametrical hypothesis on rim breadth in the *Republic*, 616d. They can be summarised as follows:

1) **Lineal hypothesis.** Developed by J. Adam,[17] and then adopted by T. Heath and J. L. E. Dreyer, it suggests that the rim breadth refers to interplanetary distances.[18] Therefore, the rims would mark the surface occupied by each planetary hoop, as seen from above. The circle traced by each planet would be regarded as the outer circle of its hoop, while the circle followed by the next planet towards the system centre would be regarded as its inner circle. Viewed thus, the text would state that the greatest interplanetary distance is that between the sphere of the fixed stars and Saturn, followed secondly by the distance between Venus and the Sun; thirdly by that of Mars and Mercury; fourthly by that of the Moon and Earth; fifthly the distance between the Sun and the Moon; sixthly the distance from Mercury to Venus; in seventh place the distance between Jupiter and Mars; and, finally, the eighth and smallest distance is that between Saturn and Jupiter. This would not pose any *a priori* problem (the lack of astronomical grounds could derive from assumptions unknown to us). As a consequence of this interpretation, however, the order proposed by the passage does not match the order presented in the *Timaeus*, 35b–36d (as Dreyer has remarked), which is based on a sound, harmonious interpretation influenced by Pythagorean thought. As a matter of fact, in this well-known passage of the *Timaeus*, the inner circle is divided in six sections by double and triple intervals that arise from the square and cube roots of the integers 1, 2 and 3, and that have a musical dimension (36a–36b). The numeric series 1, 2–4–8 and 3–9–27 (the simplest geometrical progressions) are the key to revealing planetary distances as a result of a rational plan through which the divine principle, which becomes manifest in the acts of the demiurge, flows over celestial bodies.

---

16 There is a third hypothesis, developed by philologist, mathematician and biologist D'Arcy Wentworth Thompson in his article "On Plato's 'Theory of the Planets' Republic x. 616 E," *The Classical Review* 24, no. 5 (August 1910), pp. 137–142. This is a truly ingenious and satisfactory hypothesis but as it requires accepting a new set of explanatory principles, we do not find it a feasible option. Thompson suggests that the rim breadth denotes, for Saturn, Jupiter and Mars, their arc of *retrogradation*; for Mercury and Venus, their *greatest elongation from the sun*; and for the Sun and Moon, their *declinations*. If we accept these assumptions (which we do not), Plato's order becomes feasible.

17 *The Republic of Plato*, English trans. J. Adam (Cambridge: Cambridge University Press, 1902).

18 J. L. E. Dreyer, *A History of Astronomy from Thales to Kepler*, revised by W. H. Stahl (Cambridge: Cambridge University Press, 1953), pp. 61–69.

Thus, Plato places the seven known planets in orbits with distances that gently adjust themselves at the interwoven intervals of these two series: 0 (Earth), 1 (Moon), 2 (Sun), 3 (Venus), 4 (Mercury), 8 (Mars), 9 (Jupiter), and 27 (Saturn), with the sphere of the fixed stars at an unspecified (though we suppose high) number. However, this order does not match that proposed in the *Republic* according to the lineal hypothesis, so the problem is yet to be resolved.

2) **Diametrical hypothesis.** This hypothesis was summarised by Proclus (410–485 CE), presented as an old reading of the question by earlier authors. In this interpretation, the rims would not correspond to interplanetary distances but to the breadth of celestial spheres determined by the size of the planets (Adam, Stewart). The rim of each sphere would replicate, proportionally, the diameter of each body. According to this idea, Plato apparently supposed that the larger the celestial body, the greater would be the space needed for the corresponding sphere, such that the planetary order in the *Republic*, 616d would be an order of "planetary diameters." Therefore, in descending order—of both planetary diameter and sphere rim width, as seen from above—we find: the sphere of the fixed stars, Venus, Mars, the Moon, the Sun, Mercury, Jupiter, and Saturn. This hypothesis does not have any textual or astronomical basis, and we reject it, but the fact that it was put forward by a late Neoplatonist, reveals, at least, that this passage was also the focus of questions and hypotheses within the ancient Academy.[19]

The description of the cosmic scheme and celestial motions developed in the *Timaeus*, 39c–40d, is in many ways less problematic. There, Plato describes how the celestial sphere is manufactured by the demiurge, and elaborates on the intersection between the circle of the Same and the circle of the Different. The circles clearly refer to the celestial equator and the ecliptic, respectively. Through this image, Plato associates the smooth regular westward motion of the sky with the intelligible domain (the Same, the world of Ideas), and the slight anomalies in planetary motions—pauses, retrogradations, etc.—with the sensible domain (the Different, the phenomenal world subject to change and irregularities).

Interesting as they are, the celestial myths and descriptions in the *Republic* and the *Timaeus* were not Plato's most influential contribution to the Sphairopoietic cosmology. His greatest legacy was undoubtedly "the Platonic mandate," by which historians of astronomy refer to a certain heuristic demand,

19 T. Heath, *Aristarchus of Samos, the Ancient Copernicus* (Oxford: Clarendon Press, 1913), p. 156.

supposedly proposed by Plato, that heavily influenced European thought through many centuries.

Reaching us through Simplicius—who tells us it was recorded by Sosigenes (fl. *c.*45 BCE), one of the last astronomers from late Antiquity, who in turn learned it through Eudemus (*c.*350–*c.*290 BCE)—the legend tells of Plato asking one of his disciples,[20] the mathematician and geometer Eudoxus of Cnidus, to find an explanation for the visible motions of the sky using only simple circular trajectories.[21] This legend builds on a crucial epistemological attribute, as it contains the basis for the homocentric astronomical systems of Eudoxus, Callippus, and Aristotle, which will be described in the following pages. Their proposals provided a matrix for the development of other such systems down to the dawn of modernity.

All of these early Greek cosmological-astronomical models assume a final, limiting sphere, but it was not until Aristotle that a systematic argument for the *need* of an outer spherical limit was postulated. Based on a sound philosophical structure, the Aristotelian argumentation in favour of the unicity and sphericity of the cosmos is the key chapter in the array of ideas produced by Greek thinkers on the matter. In several passages, mainly in *On the Heavens, Physics* and *Metaphysics*, Aristotle speaks at length about the unengendered, incorruptible nature of the world, the perfection of heavens, the finitude of the universe, and the need for its unicity; he lists arguments in favour of the sphericity of the cosmos and presents a number of kinetic aporiae derived from claims that the universe is infinite.

These ideas can be summarised in two arguments; one addresses the superlunary domain while the other focuses on the sublunary domain. The first argument reasons that given the circular motion of the Whole, if it were infinite there would be bodies infinitely apart from the centre of rotation and moving in circular trajectories with infinite lengths. Therefore, it would take the bodies an infinite amount of time to complete their trajectories. Such motion, however, would be impossible as there is not any infinite distance that can be covered either in a finite or infinite time. In fact, and contrary to the previous hypothesis, the time needed by the universe as a Whole to complete a revolution is finite (twenty-four hours), which indicates that the universe is itself finite.[22] The second argument is a variant of the first but applied to the

20 It does not, in the end, matter much whether the legend is true or not. Plato's thought possessed all the necessary elements for the so-called "Platonic mandate" to become a heuristic imperative for studies that grew out of this intellectual foundation.

21 *In Aristotelis De caelo commentaria*, 492, 25 ff (corresponding to *On the Heavens*, II, 12, 291b22–293a15).

22 *On the Heavens*, I, 5, 271b27–272a5.

sublunary domain. It posits that if the universe were infinite, given that the elements are finite in number, the quantity of each element would be infinite and, therefore, have no limit in terms of its natural place. As a result, for instance, the Earth would infinitely move downwards to *no specific place*, which is inconsistent with Aristotelian teleology: "It is impossible therefore even to be moving to a place where nothing can ever in its movement actually arrive" (ἀδύνατον ἄρα καὶ φέρεσθαι ἐκεῖ οὗ μηθὲν δυνατὸν ἀφικέσθαι φερόμενον).[23] The universe is, then, finite. Aristotle's arguments are based on his understanding of the finitude of space and motion and on his belief that the superlunary domain is the natural limit to both, since "the revolution of the heaven is the measure of all motions, because it alone is continuous and unvarying and eternal" (ἔτι δ' εἰ τῶν μὲν κινήσεων τὸ μέτρον ἡ τοῦ οὐρανοῦ φορὰ διὰ τὸ εἶναι μόνη συνεχὴς καὶ ὁμαλὴς καὶ ἀΐδιος) and "therefore the motion of the heaven must clearly be the quickest of all motions" (δῆλον ὅτι ταχίστη ἂν εἴη πασῶν τῶν κινήσεων ἡ τοῦ οὐρανοῦ κίνησις), referring to the outermost sphere.[24]

Aristotle stands out from his predecessors—who shared his conviction regarding the sphericity of the cosmos—because of the refined and systematic way in which he sets definite limits to the sphere and establishes the first rational principles postulating its necessity. Another passage in *On the Heavens*, II, 4, 286b10–26, provides a sublime synthesis of the idealistic imposition of the ideal on the real, which characterised classical Greek thought at its highest expression (§§ 5–6):

> The shape of the heaven must be spherical [σχῆμα δ' ἀνάγκη σφαιροειδὲς ἔχειν τὸν οὐρανόν]. That is most suitable to its substance, and is the primary shape in nature [καὶ τῇ φύσει πρῶτον]. But let us discuss the question of what is the primary shape, both in plane surfaces and in solids. Every plane figure is bounded either by straight lines or by a circumference; the rectilinear is bounded by several lines, the circular by one only. Thus since in every genus the one is by nature prior to the many, and the simple to the composite [πρότερον (τῇ φύσει) ἐν ἑκάστῳ γένει τὸ ἓν τῶν πολλῶν καὶ τὸ ἁπλοῦν τῶν συνθέτων], the circle [κύκλος] must be the primary plane figure. Also, if the term "perfect" is applied, according to our previous definition, to that outside which no part of itself can be found, and addition to a straight line is always possible, to a circle never, the circumference of the circle must be a perfect [τέλειος] line: granted therefore that the perfect is prior to the imperfect [εἰ τὸ τέλειον πρότερον

23 *On the Heavens*, I, 7, 274b17; English trans. Guthrie, p. 57.

24 *On the Heavens*, II, 4, 287a25–27; English trans. Guthrie, p. 159.

τοῦ ἀτελοῦς], this argument too demonstrates the priority of the circle to other figures [πρότερον ἂν εἴη τῶν σχημάτων ὁ κύκλος]. By the same reasoning the sphere is the primary solid [δὲ καὶ ἡ σφαῖρα τῶν στερεῶν], for it alone is bounded by a single surface, rectilinear solids by several. The place of the sphere among solids is the same as that of the circle among plane figures.[25]

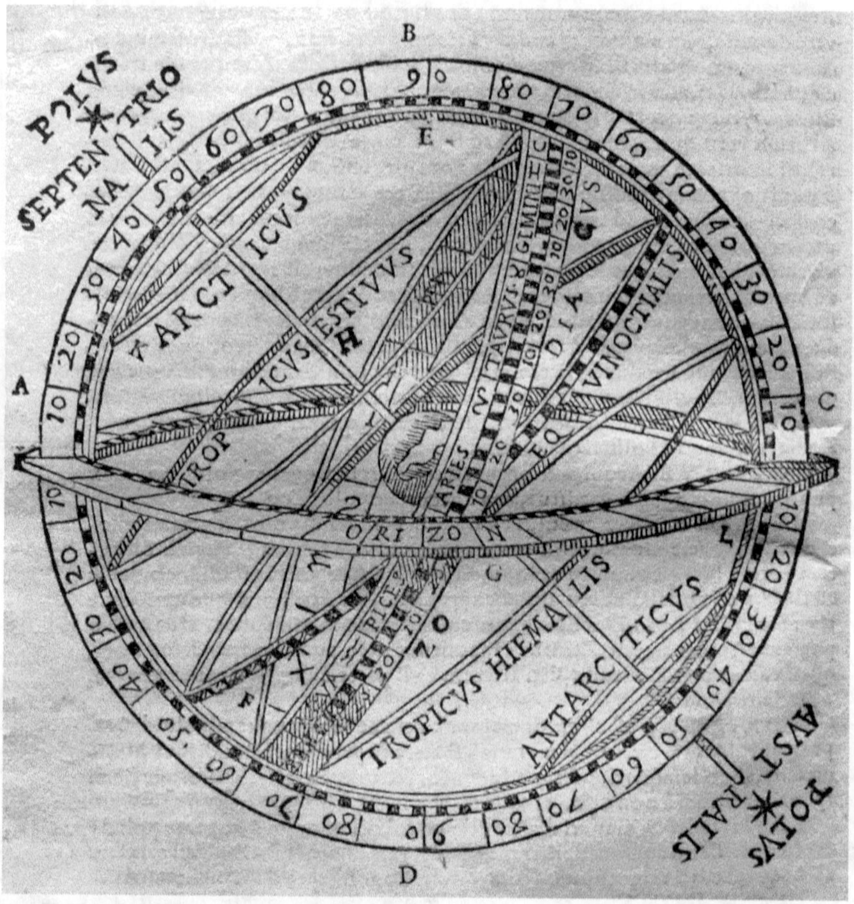

FIGURE 7 The coordinates of the celestial sphere (poles, celestial axis, equator, ecliptic, tropics, polar circles, prime meridian, horizon) in an engraving from Christophorus Clavius's *In sphaeram Ioannis de Sacro Bosco commentarius*, Rome, 1585. Sala del Tesoro, Biblioteca Nacional.

25 English trans. Guthrie, pp. 155–157.

Once the priority of the spherical figure over the remaining solids is established, the Stagirite presents it as the most appropriate figure for all strata of the superlunary domain (*On the Heavens*, II, 4, 287a3–10):

> ... but the primary figure belongs to the primary body, and the primary body is that which is at the farthest circumference, hence it, the body which revolves in a circle, must be spherical in shape. The same must be true of the body which is contiguous to it, for what is contiguous to the spherical is spherical, and also of those bodies which lie nearer the centre, for bodies which are surrounded by the spherical and touch it at all points must themselves be spherical, and the lower bodies are in contact with the sphere above. It is, then, spherical through and through, seeing that everything in it is in continuous contact with the spheres.[26]

This goes hand in hand with the idea that the spherical is perfect, its attributes being immutability, beauty, and order. It is a final order that places humankind as the central observer of that which is Becoming. Aristotle celebrates this perfection in *On the Heavens*, II, 4, 287b15–21, where he states that:

> Our arguments have clearly shown that the universe is spherical, and so accurately turned that nothing made by man, nor anything visible to us on the earth, can be compared to it. For of the elements of which it is composed, none is capable of taking such a smooth and accurate finish as the nature of the body which encompasses the rest.[27]

With Aristotle, the sphere is perfectly delimited, understood and legitimised to the extent that, thereafter, it would be no easy task for classical or post-classical thinkers to dent its claim or deny it. More interesting even than his arguments are the presuppositions and axiological considerations that led Aristotle to postulate the sphere as the natural shape of the cosmos. *On the Heavens* is full of considerations—appearing also in his other works—on "what is most divine," "what is best," "simpler," and "prior," expressions that organise longstanding elementary notions forged at the dawn of Greek philosophy. The following testimony of Iamblichus about Pythagoras (*c.*582–*c.* 507 BCE) is an instructive example of the primacy of that which is simple and prior in Greek thought:

26 English trans. Guthrie, pp. 157, 159.
27 English trans. Guthrie, p. 163.

FIGURE 8 God as geometer, from the frontispiece of the Codex Vindobonensis 2554 (Österreichische Nationalbibliothek), a mid-thirteenth century French Bible. The primordial shapeless mass is being fashioned into the form of a sphere by the Maker.

> A few days also after this [his arrival to Croton], he entered the Gymnasium, and being surrounded with a crowd of young men, he is said to have delivered an oration to them, in which he incited them to pay attention to their elders, evincing that in the world, in life, in cities, and in nature, that which has a precedency is more honorable than that which is consequent in time. As for instance, that the east is more honorable than the west; the morning than the evening; the beginning than the end; and generation than corruption. In a similar manner he observed, that natives were more honorable than strangers, and the leaders of colonies than the builders of cities: and universally Gods than daemons; daemons than demigods; and heroes than men. Of these likewise he observed, that the authors of generation are more honorable than their progeny.[28]

Notable in Pythagoras's testimony is that axiological considerations had a leading role in establishing the basis for an emerging philosophy that, in a sense, reached its culmination in the Aristotelian system. It is clear that Pythagoras's astronomical system is based on a set of foundational values that we could call "pre-cosmological" and that can be summarised as an exaltation of simplicity and ideality. This attitude towards the real explains, for instance, that celestial motions constituted a kinetic model for the entire φύσις, or that the simplest two-dimensional and three-dimensional figures—namely the circle and the sphere—formed the primary eidetic substratum of the real, which ultimately depends on the divine and unchanging.

---

28 LFP I 256: Iamblichus of Chalcis (4th c. CE), *De vita Pythagorica liber*, ed. L. Deubner, Bibliotheca Scriptorum Graecorum et Romanorum Teubneriana (Leipzig: Teubner, 1937; revised ed. by U. Klein, Stuttgart: Teubner, 1975), VIII, 37–45; English trans. from T. Taylor, *Iamblichus' Life of Pythagoras* (London: J. M. Watkins, 1818), p. 17.

FIGURE 9 *Per monstra ad sphaeram* (By way of monsters to the sphere) is the motto on the *ex libris* of astronomy historian Franz Boll's (1805–1875) personal collection. The motto, chosen from Boll's works by Aby Warburg when adding his books to the Kulturwissenschaftliche Bibliothek Warburg, summarises—according to Warburg—the journey of Western rationality, which had to resort to monstrous myths in order to understand its surroundings and finally reach a simplified order composed of a beautiful cosmos limited by the celestial sphere. Thus, Boll's motto illustrates the complex alliance between the rational and irrational in Euro-American culture.

CHAPTER 4

Eudoxus and Callippus: Planetary Models

Democritus quoque, subtilissimus antiquorum omnium, suspicari se ait plures stellas esse quae currant, sed nec numerum illarum posuit nec nomina, nondum comprehensis quinque siderum cursibus. Eudoxus primus ab Aegypto hos motus in Graeciam transtulit.

Democritus too, the most acute of all the ancients, said he suspected that there are more moving stars, but he did not give either their number or their names, for the motions of the five stars were not yet understood. Eudoxus first introduced knowledge of their orbits to Greece from Egypt.

SENECA, *Natural Questions*, VII, 3

∴

As we have seen in the preceding chapters, the ancient Greek cosmos is spherical and limited. The outermost sphere contains the fixed stars, so named for their invariant relative positions, and encloses the totality of existing things. It is the constancy of the sphere of fixed stars that makes it possible to identify groups of stars as constellations or asterisms. Through the ages, the most significant constellations for observers of the sky have always been those of the zodiac. This "circle of animals" (ζῴδιον) refers to the animal-like shapes that crowd the zodiac belt—along with the Moon, the Sun, and the five planets visible to the naked eye: Mercury, Venus, Mars, Jupiter and Saturn—and move in cyclic annual trajectories. For Aristotle's cosmology the twelve zodiac constellations are extremely important, as the Sun, given its unusual annual trajectory, represents the cosmic force responsible, through the succession of seasons, for generation and corruption in the sublunary realm (§§ 29–30).[1] The regular changes in the sky, registered anew by each generation (as Aristotle himself points out),[2] greatly influenced the minds of those who, accepting the incessant repetition of the same, supposed that Being, too, was ongoing and

1 *On Generation and Corruption*, II, 10, 336b3–10 –§§ 29, 30–.

2 *On the Heavens*, I, 3, 270b12–16 –§ 4–.

© GERARDO BOTTERI AND ROBERTO CASAZZA, 2023 | DOI:10.1163/9789004525535_006

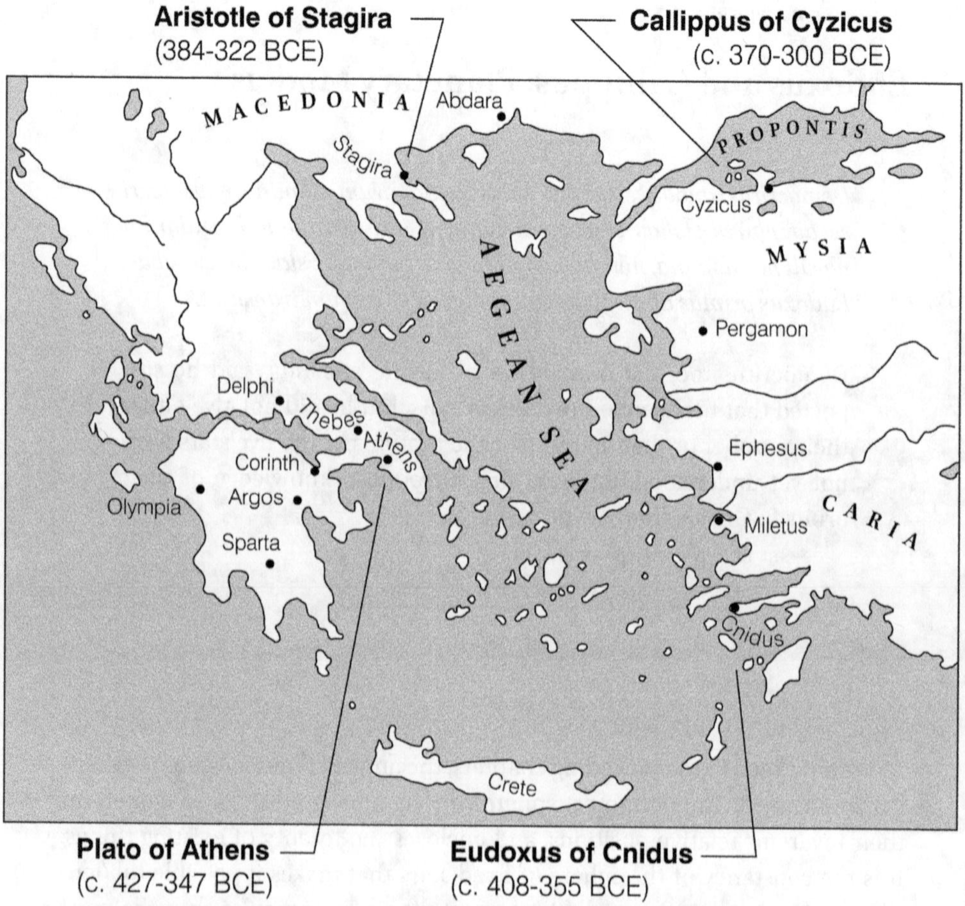

FIGURE 10 Greek cosmologists from the 4th c. BCE: Plato, Eudoxus, Callippus, Aristotle.

sought an explanation for its dynamic. They were convinced that the order of nature rested on the intelligibility of reality, which originated from the divine.

The smooth flow of the heavens held, however, an inconvenient enigma for the geometers, astronomers, and philosophers gathered at the Academy. These observers noticed a phenomenon difficult to comprehend: amidst the regular motion of the sphere of the fixed stars (ἡ ἀπλανὴς σφαῖρα)—also known as the "sphere of heaven" (ὁ κύκλος τοῦ οὐρανοῦ) or the "celestial vault" (τὸ κύτος ἀστέριον)—they noticed a wandering motion of the planets. An erratic element was observed mainly in the motions of Mercury, Venus, and Mars but to a lesser extent also in the paths of Jupiter and Saturn. Today we know that these motions are apparent, but at the time they were believed to be real. These observations led to explanatory models that minimised the issue of wandering

motion, relating it to λόγος, thus translating complex planetary trajectories into idealised, rational schemes suitable to the discovery of proportions and harmonies in the observable. Direct open-sky observation was the main method at their disposal for studying these phenomena, and a proto-trigonometry was the language of interpretation. Their observations can be repeated by us, and it is easy to imagine how the serenity of celestial motions led ancient observers to inquire into the causes of such regularity, and might predispose them to a view of the celestial order as κόσμος (the ornamented), born of a divine principle of intelligibility. This is, at least, how these Platonic thinkers reviewed here (Aristotle included) conceived what they saw in the sky.

Plato, Eudoxus, Callippus, and Aristotle, all of whom followed the tradition of observational astronomy—a relatively old approach by the mid-fourth century BCE—made a clear distinction among the three main observable motions of the heavens. Hereafter, we will call them "motion a," "motion b," and "motion c." At that time, just as now, these motions could be seen through patient and systematic naked-eye observation of the night sky, and as we will see, these motions are the guiding thread for our analysis of Aristotle's astronomical system. In fact, the introduction of numerous Unmoved Movers in *Metaphysics*, Λ, 8 is fundamentally related to an explanation of such celestial motions, as we shall show in this section.

These observers recognised three essential elements. First, they recognised a *uniform rotational motion* (motion a), the most evident of the motions and that which moves all visible bodies westward, completing a revolution in approximately twenty-four hours while maintaining the relative position of the so-called fixed stars. The terrestrial observer perceives this motion (the rotation of the celestial sphere in its entirety) as if the totality of the heavens rotated on a single axis, the *celestial axis*, attached to the celestial poles. One of these poles is always visible to any given observer while the other is always hidden (for instance our Greek philosophers could only see the North Pole, whereas from South America one can only see the South Pole).

The centre of the celestial sphere was naturally placed at the point of observation, consistent with the ancient adoption of a geocentric (or even topocentric) understanding of the universe. This dominant and most evident of celestial motions is the only motion attributed to the sphere of the fixed stars by the Greeks.

Secondly, they also distinguished the *circular motion of the planets in the opposite direction (eastward) and along independent orbits, all contained by the band of the zodiac and each having its particular period of revolution*. This is known as *direct* planetary motion (motion b), and is superimposed on the diurnal motion (motion a). It is noticed primarily because of the slower

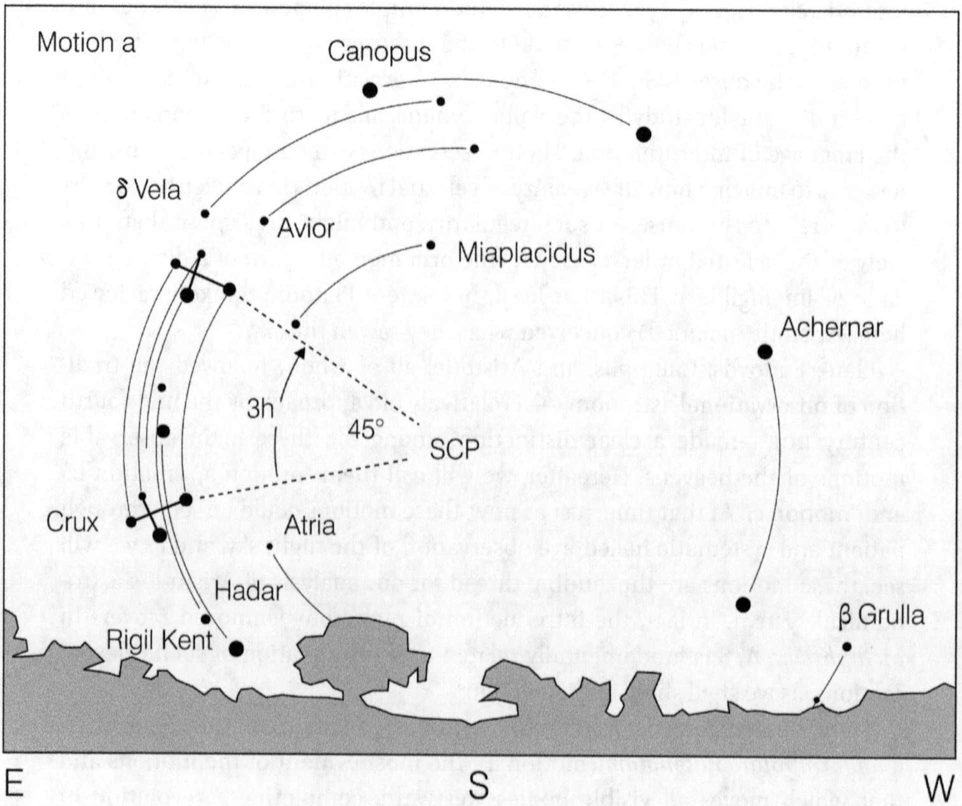

FIGURE 11 Motion a: A uniform rotational motion, the most evident of all, which moves all visible bodies westward, completing a revolution approximately every 24 hrs, and maintaining the relative position of the fixed stars. The terrestrial observer perceives this motion (the rotation of the celestial sphere in its entirety) as if the whole of heaven rotated on a single axis. In positional astronomy, this is called the celestial axis, which is attached to the celestial poles.

relative motion of the planets in relation to the fixed stars as they move westward along the band of the zodiac. The corresponding path of the Sun draws a line through the middle of the zodiac, known as the ecliptic, and the completion of its trajectory is what we call the year.

And thirdly, they were able to perceive *the wandering motion of the visible planets* (*Mercury, Venus, Mars, Jupiter, Saturn*) *along their orbits* (motion c). Indeed, a thorough and prolonged observation of the sky reveals that planets occasionally exhibit "stations" (they seem to be at a halt against the background of the fixed stars) and "retrogradations" (they seem to move in the same direction as the fixed stars (i.e., westward), during several days or weeks). These motions are actually only apparent, as Copernicus conclusively demonstrated.

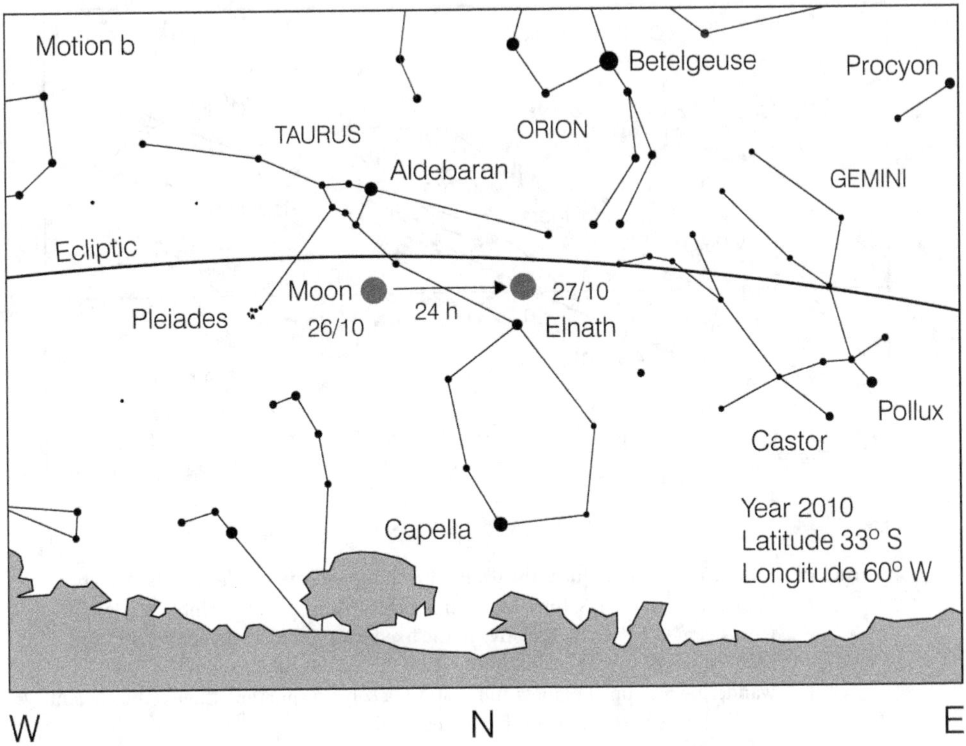

FIGURE 12 Motion b: For the terrestrial observer, the planets and the sun exhibit a circular motion, opposite to the diurnal motion (a), i.e., eastward. They have independent orbits and specific periods of revolution. All of these motions are slower than the diurnal motion and are contained within the zodiac band.

Ancient astronomers believed them to be real, however, because they were convinced that the Earth was at rest. The astronomical systems of Eudoxus, Callippus, and Aristotle attempted to account for all of these motions simultaneously. Each made contributions, but it was Eudoxus who first designed the geometric scheme upon which they were built.

The basic structure of the Greek cosmos has the Earth at rest in its centre, with the seven planets known at that time situated between the Earth and the fixed stars. As regards known celestial bodies, Eudoxus, Callippus, and Aristotle maintained the number and structure established by Plato in *The Republic*, 616c–e, and *Timaeus*, 38d–39a. Following the Egyptian arrangement, Plato orders the planets from the centre to the periphery: 1) ☾ Moon (Σελήνη), 2) ☉ Sun (Ἥλιος), 3) ♀ Venus (Ἀφροδίτη, Ἑωσφόρος), 4) ☿ Mercury (Ἑρμῆς), 5) ♂ Mars (Ἄρης), 6) ♃ Jupiter (Διός), and 7) Saturn (Κρόνος). The last five exhibit this wandering behaviour and can be properly called πλανήτης (this word

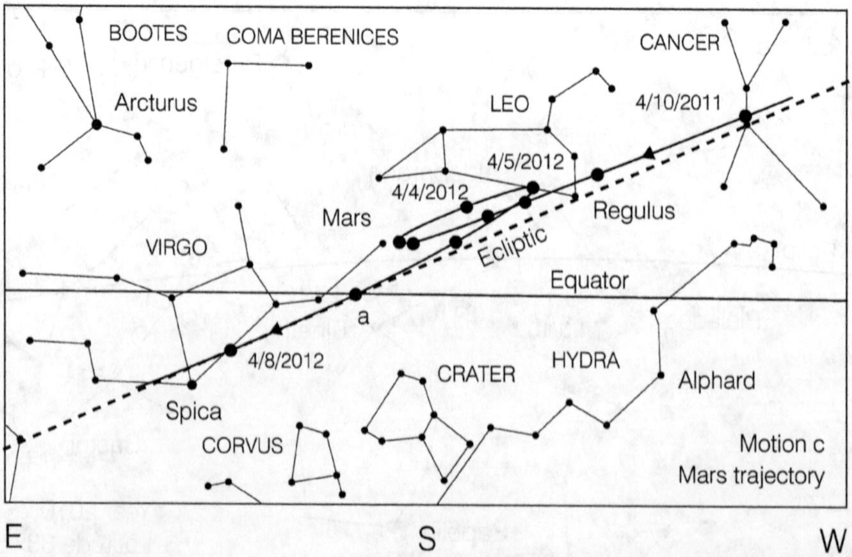

FIGURE 13 Motion c: This reproduces the wandering motion of the visible planets (Mercury, Venus, Mars, Jupiter, Saturn) along their orbits. These planets exhibit "stations" (they seem to be at a halt relative to the fixed stars) and "retrogradations" (they seem to move faster than, but in the same direction as, the fixed stars). The wandering motion of planets and motions a & b are apparent, though the ancient Greeks believed them to be effective and real.

derives from the verb πλανάω, meaning *wander* or *roam*), that is to say, *wandering star*. The search for a rational explanation for the wandering motion of bodies in the sky drove astronomical enquiry until the Copernican shift from the geocentric to the heliocentric paradigm in the modern era and the further developments of Kepler and Newton that definitively resolved the issue.

4.1 The Heavens and the Compass

Chapter Λ, 8 of Aristotle's *Metaphysics* (1073a14–1074a34 –§ 15–) addresses his famous disquisition on the Prime Mover, whose physical-astronomical necessity has been frequently overlooked at the expense of its metaphysical necessity. In this passage, Aristotle presents an astronomical system that, based on intelligible immaterial principles (the Unmoved Movers), aims to explain observable celestial motions in an integrated manner, to show how they come to operate as a coordinated Whole. In order to create his own system, Aristotle uses the geometrical systems of concentric or homocentric spheres—that is,

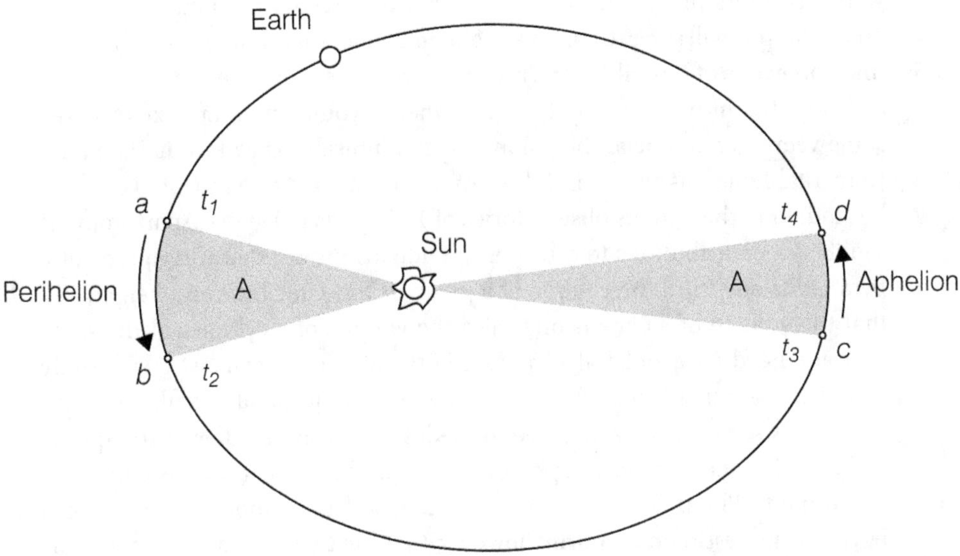

FIGURE 14 We owe to Kepler the major discovery of the elliptical orbits in planetary motions (First Law), which marked the end of the conception of the heavens as spherical and made possible the discovery of the explanation we now accept for the irregularities in the orbital speed of planets. In his *Astronomia nova* (part 3, chapter 40), Kepler expresses: "My first error was to suppose that the path of the planet is a perfect circle, a supposition that was all the more noxious a thief of time the more it was endowed with the authority of all philosophers, and the more convenient it was for metaphysics in particular"; trans. Donahue, p. 419. With the sun located at one of the foci of the planet's elliptical orbit, the line joining the two bodies carves out equal areas in equal intervals of time (Second Law), so if the shaded areas are equal, it takes the planet the same time $(t_2 - t_1) = (t_4 - t_3)$ to cover different distances (ab is greater than cd). As a consequence, the speed at aphelion $V_A = cd/(t_4 - t_3)$ is slower than the speed at perihelion $V_P = ab/(t_2 - t_1)$. The Second Law of Kepler shows that the Earth, as well as any other planet, moves along its orbit at a variable speed. This is why, from the point of view of a terrestrial observer, the apparent speed of the sun is also variable.

with the *same* (ὅμοιον) *centre* (κέντρον)—developed by Eudoxus and Callippus at the Academy. Drawing from them, he describes the specific motions of each star. The result is a remarkable system of numerous concentric spheres that succeeds in providing a general account of apparent celestial motions even if some aspects of its operation could not be fully explained.

Our interpretation of *Metaphysics*, Λ, 8, based essentially on apparent celestial motions, suggests understanding the Aristotelian proposal as a *sui generis mechanical integration* of Callippus's system which, following Eudoxus, highlights the *kinematic* comprehension of celestial motions (that is, a study concerned with the description of phenomena) rather than their *causes*, which

were Aristotle's main concern. An analysis of the contributions of Eudoxus and Callippus will serve, then, as a useful introduction towards developing our main objective of describing the astronomical system of Aristotle.

Instructed in medicine by Philistus when a young man, Eudoxus worked some years as a physician but soon devoted himself instead to mathematics (remarkably, he was also a disciple of Archytas) and astronomy (it is likely that he studied in the famous observatories of Heliopolis in Egypt). Apart from his significant contributions to astronomy, Eudoxus proved that the volume of a pyramid is one-third the volume of a prism of the same base and height, and that the volume of a cone is one-third the volume of a cylinder of the same base and height. He took full advantage of the method of exhaustion (the only complex mathematical tool available before infinitesimal calculation) and his studies served as a basis for Archimedes's extensive studies of the sphere and cylinder, and of conoids, spheroids and spirals. So great was his fame that, according to Plutarch, Plato gave him a cold welcome upon his return from Egypt. Other testimonies affirm, however, that the two were on friendly terms (and this is the version we prefer to believe). Unfortunately, Eudoxus's *On Speeds*, which describes his astronomical system, has been lost, but his great influence is clear.[3] In fact, Eudoxus directed the Academy while Plato was in Syracuse and during that time Aristotle was his student.

To Plato's question (how is it possible to account for celestial motions resorting only to simple and regular circular motions?) Eudoxus found a clever geometrical answer. The solution was simply to interpret observed motions as the result of superimposed uniform circular motions created by a series of nested homocentric spheres. The device he used—a thought experiment—reveals his remarkable acuity and sagacity. For each sphere he laid out the corresponding axis of rotation, tilted at a fixed angle to the others, and constructed the overall structure by mounting each axis upon that of the next immediately following sphere. As a result, each would rotate on its own axis while also being dragged by the rotation of the outer spheres. Thinking like a geometer, he isolated the motions of each planet in order to treat them independently and was able to understand celestial motions by studying separately the spheres carrying each planet. In each group of planetary spheres, the planet, fixed at a point on the equator of the innermost sphere, receives the compound motion of the (two or three) upper spheres. This composition roughly reproduces the apparent motion of each planet (motions a, b and c).

We know little about the Eudoxian system from direct and reliable sources, and the same is true of the modifications to it that were introduced by

3 Simplicius, *In Aristotelis de Caelo commentaria*, 2, 492.31–494.12.

Callippus. The fragment of *Metaphysics*, Λ, 8 (1073b4–31 –§ 15–) is a rare exception. There, Aristotle aims to provide his own integrated system of the heavens by bringing together the individual planetary subsystems of Eudoxus—as drawn by Callippus—into a single integrated model, mechanically speaking. Since Aristotle conceived of the heavens as a non-fragmented totality, he was driven to find a comprehensive explanation for the motion of every known star by means of a single harmonious unit, just as the heavens appear before an observer. In an obscure passage of *Metaphysics*, Aristotle emphasises this idea of the heavens as a unified whole:

> But it is necessary, if all the spheres [of Callippus's planetary subsystems] combined are to explain the phenomena, that for each of the planets there should be other spheres (one fewer than those hitherto assigned) which counteract those already mentioned and bring back to the same position the first sphere of the star which in each case is situated below the star in question; for only thus can all the forces at work produce the motion of the planets. Since, then, the spheres by which the planets themselves are moved are eight [Callippus's spheres for the upper stars: Saturn (4) and Jupiter (4)] and twenty-five [Callippus's spheres for the lower stars: Mars (5), Mercury (5), Venus (5), sun (5) and moon (5)], and of these only those by which the lowest-situated planet [the moon] is moved need not be counteracted, the spheres which counteract those of the first two planets will be six in number [two upper stars each having three rewinding spheres (2 × 3 = 6)], and the spheres which counteract those of the next four planets will be sixteen [the lower stars, except for the moon, each having four rewinding spheres (4 × 4 = 16)], and the number of all the spheres—those which move the planets [8 + 25 = 33] and those which counteract these [6 + 16 = 22]—will be fifty-five [33 + 22 = 55]. And if one were not to add to the moon and to the sun the movements we mentioned, all the spheres will be forty-nine in number.[4]

The Aristotelian model described by this short passage is based on Callippus's planetary subsystems, which were built upon those of Eudoxus. In order to understand how these "mechanisms" work, we will first analyse the simpler subsystems developed by Eudoxus. We will then see how Callippus supplemented them to improve their description of phenomena. According to the brief description of these geometrical representations offered in that passage

4 *Metaphysics*, Λ, 8, 1073b36–1074a14 –§ 15–; English trans. W. D. Ross in Barnes, *Complete Works of Aristotle*, vol. 2, p. 1697.

of the *Metaphysics*—a text crucial to a reconstruction of the models of Eudoxus and Callippus—Eudoxus accounted for the motions of the Sun and the Moon by way of qualitatively similar subsystems, each made of three homocentric spheres.[5] As many nineteenth- and twentieth-century authors have agreed, based on Aristotle's description as well as the commentary of Simplicius on *On the Heavens*,[6] it appears that, in the case of the Moon, the arrangement, rotation speeds, and function of the spheres (starting from the outermost) was as follows: the first sphere reproduces the westward rotation of the sphere of the fixed stars, with a period of approximately 24 hrs. The axis of the second sphere is inclined at an angle of approximately 23.5° to the first sphere, so its equator coincides with *the middle line of the zodiac* (the ecliptic). This sphere produces the retrogradations of the lunar nodes, with a period of (westward) rotation of 223 lunations (synodic months), that is approximately 18 yrs, 11 days and 8 hrs.[7] The third sphere holds the Moon fixed to its equator, revolving about an axis

5 *Metaphysics*, Λ, 8, 1073b17–22 –§ 15–.

6 The subsystems of Eudoxus and Callippus, introduced by Aristotle in *Metaphysics*, Λ, 8 and further discussed by Simplicius in *In Aristotelis de Caelo commentaria*, have been successfully explained by Giovanni Schiaparelli (1835–1910) in his collection of classical works, *Scritti sulla storia della astronomia antica*, Prima Parte, *Scritti Editi* (Bologna: N. Zanichelli, 1926). For the topics discussed in this book, see Schiaparelli's volume 2. Schiaparelli's foundational work has been enriched by the comments of John Louis Emil Dreyer (1852–1926), D'Arcy Wentworth Thompson (1860–1948), Thomas Little Heath (1861–1940), William David Ross (1877–1971), Joseph Owens (1908–2005), Philip Merlan (1897–1968), Norwood Russell Hanson (1924–1967), Geoffrey Ernest Richard Lloyd (1933–), and more recently, István Bodnár, Theokritos Kouremenos, and Ido Yavetz, as listed in the Bibliography section of this volume.

7 The values provided here are the current values of the astronomical parameters for each star, unless otherwise stated. We have chosen this methodology for two reasons: first, in many cases, there is no accurate record of the values assigned to these parameters by ancient Greek astronomers; second, the ultimate goal of this work is to determine whether these models were an appropriate reproduction of phenomena, which can only be done by adjusting them to modern measurements. We consider that Greek geometrical-mathematical thought focused more on procedure or the path to results rather than on the results themselves (in this case, the correspondence between the model and the phenomena), as suggested by Julio Rey Pastor and José Babini, *Historia de las matemáticas* (Barcelona: Gedisa, 2000), vol. 1, p. 66. Moreover, as regards the empirical nature of ancient astronomy, except for some astronomers (Callippus included) who did achieve accurate measurements (e.g., the duration of seasons or the remarkably accurate Callippic cycle to calculate the lunar cycle), in general, as pointed out by Schiaparelli in reference to Eudoxus's planetary periods, Greek astronomers were searching for approximate values (*Scritti*, vol. 2, pp. 66–67). In addition, we cannot disregard the poor tools of measurement available in the fourth century BCE and the fact that these issues were largely studied without the benefit of a large or interconnected community. As regards measurements, especially, the benefit of collective efforts such as are normal today is of course enormous. The fact that the philosophical and scientific works addressed in this book have been rescued from oblivion and are today very much valued can make us

inclined at an average of 5° 09′ to the axis of the second sphere, and completes its eastward revolution in approximately 27.21 days. This period corresponds to the draconic month (motion b).[8] The figure below shows a schema derived from this interpretation. The arrangement and rotation periods shown in the figure are based on Schiaparelli's interpretation of Simplicius's description in his commentary on *On the Heavens*.[9]

Schiaparelli rightly interpreted, we believe, that Simplicius seems to have mistaken the correct position of the two inner spheres of the moon. He places the slower sphere third with the purpose of reproducing the slow retrogradation of the nodes,[10] whereas it is in fact achieved by placing it second, as shown. Simplicius does correctly indicate that the retrogradation is revealed by the fact that the moon does not return to the same position, relative to the constellations of the zodiac, each time it passes through the maximum latitudinal digression in the course of its slow westward motion. However, if Simplicius's description was correct, the moon should go across the ecliptic approximately every nine years, rather than every two weeks, as observed. The maximum latitudinal digression characteristic of the lunar orbit, like the nodes, are found again almost at the same points relative to the zodiac only after 223 lunations (or synodic months). Even though Simplicius does not make any reference to this period (also known as the *Saros cycle* due to its relevant role in eclipse prediction), Schiaparelli has grounds to assume that Eudoxus was aware of it.[11]

---

forget that they were most likely unknown, or did not interest to their contemporaries and remained the intellectual effort of a very few.

8 The draconic month is the time it takes the Moon to cross the same orbital node two consecutive times.

9 Schiaparelli, *Scritti*, vol. 2, pp. 20–23.

10 Simplicius, *In Aristotelis de Caelo commentaria*, 494, 23; 495, 16. For an English translation see A. C. Bowen, *Simplicius on the Planets and Their Motions* (Boston: Brill, 2013).

11 The Saros is the period between any eclipse and the time it takes the moon to return to the same phase and position, relative to its orbital node, in which it was located during that inicial eclipse. This period marks the recurrence of eclipses following the same order, as a consequence of the complex but regular motion of the bodies involved: moon-earth-sun. Taking as reference, for example, a total solar eclipse, it is easy to establish the length of the Saros, as the conjunction of the sun and the moon takes place while the centres of both stars are aligned to one of the lunar nodes. Considering that the moon passes through any given node of its orbit every 27.21 days (*draconic month*) and repeats the same phase approximately every 29.53 days (*synodic month*), while the sun passes through any given lunar node every 346.62 days (*draconic year*), there will be another total eclipse every time the beginning of these three periods coincides. The Saros is therefore determined by the fact that 242 draconic months (6585.36 days) equal approximately 223 synodic months (6585.32 days = 18 years 11 days 7 hours 42 minutes) or 19 draconic years (6585.78 days). It was believed that the Babylonians used the Saros to predict eclipses, but everything points to the conclusion that this was not possible until the second or

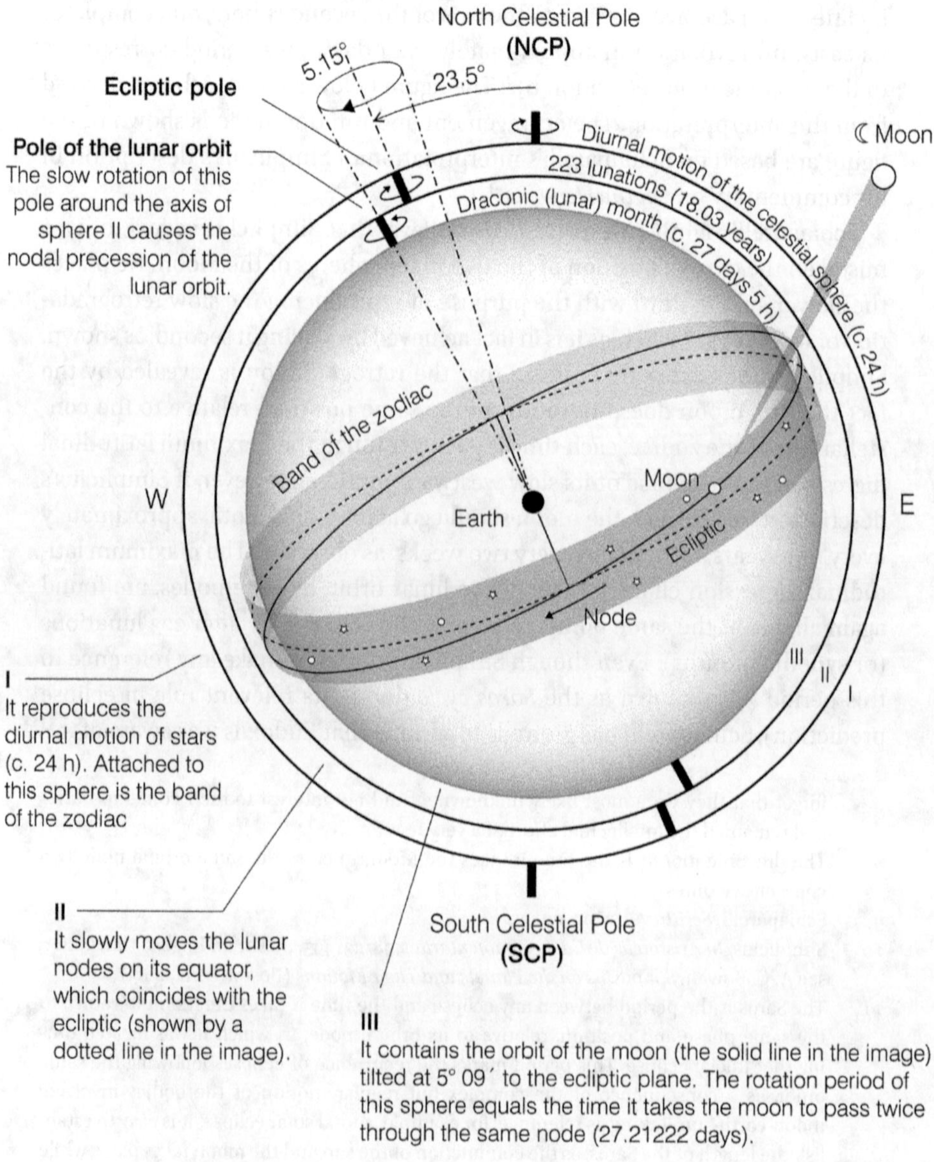

FIGURE 15 Geometrical subsystem created by Eudoxus to explain the motion of the Moon. The angles and rotation periods correspond to current values.

This interpretation, which seems to us appropriate, leads to a system in which the layout and rotation periods of the three lunar spheres is just as described. Such a system merely corrects what was undoubtedly a mistake on the part of Simplicius regarding the position of the spheres and adds the periods that were neglected to describe the phenomena addressed in the source text.

The Aristotelian description of Eudoxus's solar subsystem is similar: the first sphere reproduces the motion of the fixed stars; the second sphere—with the equator at the middle line of the zodiac—moves the Sun eastward and is the slowest of the three; the third sphere—with an axis less inclined than the second sphere and at a smaller angle than in the case of the Moon—moves the star in a direct way, in the same direction as the second sphere, completing one revolution in a year. The reason why Eudoxus introduces three spheres into the solar subsystem is not yet well explained. A mutation of the solar orbit analogous to that imposed on the Moon by the second sphere of the lunar subsystem is not necessary.

The rotation periods and directions identified for the spheres of the sun (as well as for the moon) are those suggested by nineteenth- and twentieth-century interpreters who built on Simplicius's description. Strictly speaking, according to Simplicius, the third sphere is that which turns at the slowest pace, and the second sphere rotates with the annual solar period of revolution.[12] And just as with the lunar subsystem, in this description the third sphere reproduces a purported retrogradation (or direct motion if we maintain the rotational direction of the second and third sphere as given by Simplicius) of a change—also

third century BCE, during the Seleucid Empire, after the encounter of the Babylonian and Greek cultures; see O. Neugebauer, *The Exact Sciences in Antiquity* (New York: Dover, 1969), pp. 141–143. Schiaparelli, who acknowledges the ability of ancient Babylonians to predict eclipses, also doubts they were able to determine the length of the Saros accurately during the Sargonid period, between 2340–2150 BCE (*Scritti*, vol. 1, p. 75). For more information on eclipse prediction in Babylonia in more recent periods, see John M. Steele, "Eclipse Prediction in Mesopotamia," *Archive for the History of Exact Sciences* 54 (2000), pp. 421–454. For a detailed account of the Saros and eclipses in general during Antiquity, see D. H. Kelley, E. F. Milone, *Exploring Ancient Skies: A survey of Ancient and Cultural Astronomy* (New York: Springer, 2011). The paper "La explicación de los eclipses en la Antigüedad grecolatina" by Roberto Casazza and Alejandro Gangui, *Revista de Estudios Clásicos* 39, (Mendoza, Universidad Nacional de Cuyo, 2012), pp. 79–103, addresses the scope of knowledge on the Saros and its use to predict eclipses during ancient times, and reviews some eclipses (and reactions to them) as documented in ancient and late antique literary sources.

12 Simplicius, *In Aristotelis de Caelo commentaria*, 494, 1–13.

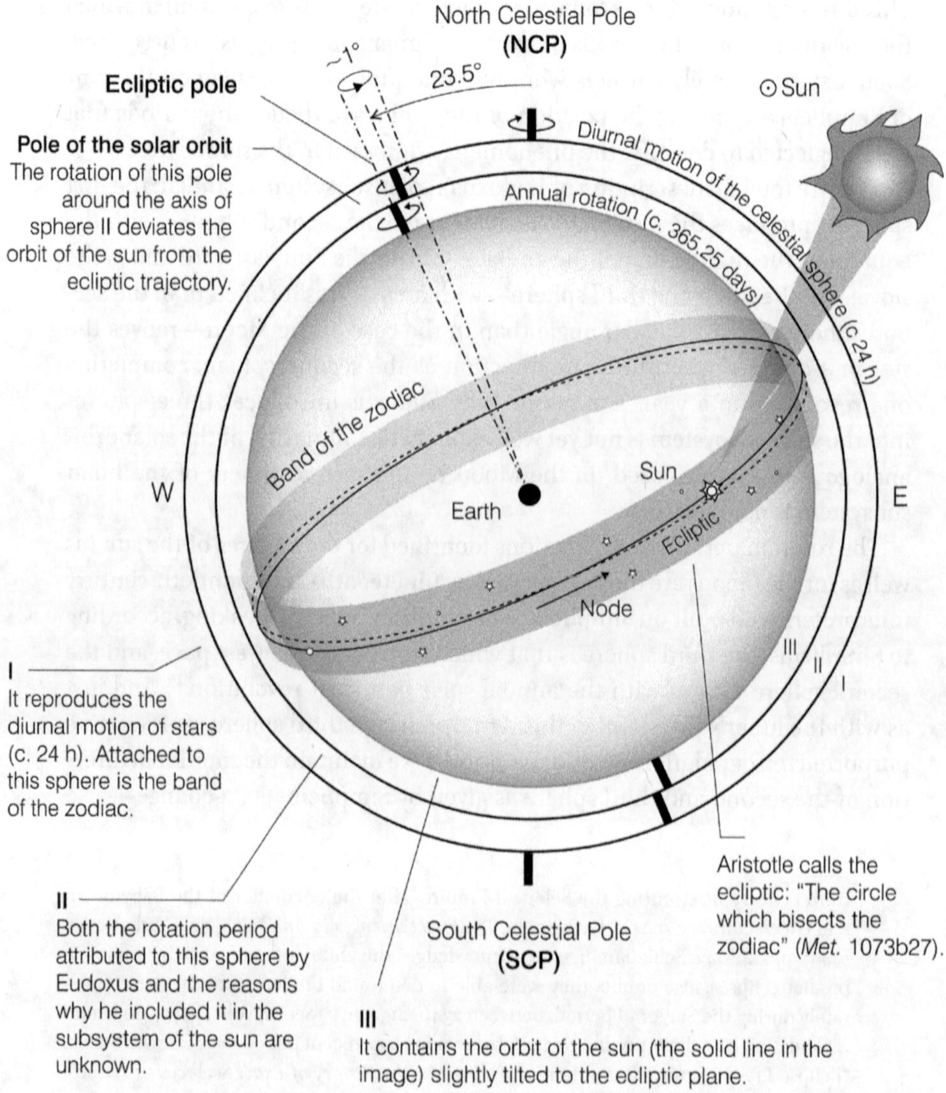

FIGURE 16 Most likely layout of Eudoxian solar subsystem. The rotation period and role of the second sphere, the slowest of this subsystem, remains unknown. It is probable that, as Schiaparelli suggests (*Scritti*, vol. 2, pp. 26–27), the precession of the solar orbit relative to the ecliptic caused by the second sphere was an attempt to reproduce a presumed variation in orbital latitude (postulated from observational errors and by analogy with the latitudinal variation of the Moon).

apparent—in the latitude of the star, signalled by the fact that the sun does not always rise at the same place on the horizon during the summer and winter solstices.[13]

Ross[14] agrees with Schiaparelli[15] (and so do we) that Simplicius must have wrongly interpreted the Eudoxian scheme. If one would uphold Simplicius's description, the sun would have to remain almost in the same latitude for years and this is not the case. It is more reasonable to assume that academic astronomers introduced the slowest sphere of the solar subsystem—thus a change of latitude for the sun—by analogy with the recurring change in latitude exhibited by the planets and the moon.[16] This also leads to the assessment, as several interpreters following Schiaparelli found,[17] that the second sphere described by Simplicius, apart from being the slowest, must move in a retrograde direction, rather than eastward, as he suggested, therefore producing an actual retrogradation of the orbital node similar to that observed in the case of the moon. As regards the tilt of the third sphere relative to the ecliptic axis, Pliny (23–79 CE) considered it to be at 1°,[18] whereas Theon of Smyrna (*c.*70–*c.*135 CE), following Adrastus of Aphrodisias (fl. beginning of 2nd c. CE), calculated it at 0.5°.[19]

There has been an attempt to correlate the specific layout and motion of the three solar spheres with the phenomena caused by the *precession of the equinoxes*. However, as far as we know, the phenomenon responsible for the variation in the duration of the seasons was only discovered by Hipparchus in the second century BCE.[20] As Schiaparelli indicates, if Eudoxus had wanted to describe a phenomenon of this kind, being himself a geometer, he would not have missed the fact that, to explain the precession of the equatorial axis around the ecliptic axis, he needed only two spheres: the first sphere, that carrying the celestial equator, would have a slightly faster motion than that of the fixed stars, and the second sphere, with its axis inclined at an angle of 23.5°, would have the direct annual motion of the sun. Nevertheless, with this structure of homocentric spheres, it is not possible to make this configuration of

13 Simplicius, *In Aristotelis de Caelo commentaria*, 493, 15.

14 *Aristotle's Metaphysics. A Revised Text with Introduction and Commentary*, 2 vols, ed. W. D. Ross (Oxford: Clarendon Press, 1924, reprint 1997), p. 387.

15 *Scritti*, vol. 2, pp. 21 and 24–25.

16 *Scritti*, vol. 2, p. 27.

17 *Scritti*, vol. 2, p. 32.

18 Pliny the Elder, *Natural History*, II, 16.

19 Theon of Smyrna, *Platonici Liber de Astronomia*, ed. Th. H. Martin (Paris: Reipublicæ Typographeo, 1849; reprinted Groningen: Bouma, 1971), p. 174.

20 Clagett, M., *Greek Science in Antiquity* (New York: Abelard-Schuman, 1955), p. 96; Heath, *Aristarchus of Samos*, p. 200.

two spheres compatible with another that, bearing the fixed stars, would provide the necessary background to produce the phenomenon as understood by Hipparchus and as it is known today.[21] The astronomical phenomenon of the precession of equinoxes cannot be reproduced through a system of three concentric spheres, which would require that such a system maintain the orbit of the sun fixed at the middle line of the zodiac (ecliptic) while at the same time causing a precession of the celestial poles around the ecliptic axis, resulting in a displacement of the celestial axis relative to the fixed stars.

The mathematical tools available to Eudoxus were limited, but this is not the primary reason to dismiss the possibility that his model was an attempt to reproduce motions that were too complex for the geometry of homocentric spheres. Instead, as Dreyer points out,[22] this is more likely explained by the fact that ancient astronomy, as practiced by Eudoxus and Callippus, and despite being identified as a science—to the extent that it seeks to explain phenomena—was strongly conditioned by aprioristic metaphysical principles which its practitioners neither aimed to change nor were able to. In this sense, it is difficult to reconcile the systems of Eudoxus, Callippus, and especially that of Aristotle, with the idea of a displacement of the celestial axis, because according to such metaphysical principles the axis was the immutable support of the entire universe, which is eternally unmoved. In fact, we do not know the origin of this "myth about the mutation of the solar orb," as Schiaparelli has called it. Despite Schiaparelli's efforts to trace the phenomena or the imaginings that inspired the myth, he recognises the lack of any concrete lead to solve the enigma. Nonetheless, Schiaparelli suggests some values for the rotation period of the third sphere of the sun. He deduces a period of 2,922 years from the descriptions of Adrastus, and a period of 7,200 years from Jean Sylvain Bailly (*Histoire de l'astronomie ancienne*, Paris, 1781). Karl Richard Lepsius (*Chronologie der Alter Aegypter*, Berlin, 1849), also quoted by Schiaparelli, assigned a period of 3,600 years to this sphere.[23]

In order to explain the erratic planetary movements, Eudoxus assigns four spheres to each planet, the two outer spheres being identical in disposition to the first and third spheres of his lunar and solar subsystems.[24] Consequently, the first (outer) sphere of each planetary subsystem moves the planet westward and parallel to the fixed stars, completing its revolution once a day (motion a);

21 We should bear in mind that in Simplicius's description, the first sphere of the sun is not that containing the fixed stars. This is why he and Theophrastus called this and the others planetary spheres ἀνάστρους (Simplicius, *In Aristotelis de Caelo commentaria*, 493, 18).

22 Dreyer, *A History of Astronomy from Thales to Kepler*, p. 167.

23 *Scritti*, vol. 2, pp. 28–40.

24 *Metaphysics*, Λ, 8, 1073b22–32 –§ 15–.

the second sphere moves the planet eastward in the opposite direction from the fixed stars, along the middle line of the zodiac—the ecliptic—and in a direct way (motion b), with a rotation period that, for the outer planets, matches their corresponding sidereal periods (1.9 years for Mars; 11.9 years for Jupiter; 29.5 years for Saturn), and, for the inner planets (Mercury and Venus) that accompany the motion of the Sun at least in appearance, constitutes a year.[25]

Eudoxus introduced the two remaining spheres to explain variations in latitude, retrogradations, and stations of planetary trajectories, which, as we know now, are only apparent (motion c).[26] The axis of the third sphere is fixed to the equator of the second sphere, such that their poles remain on the ecliptic. The fourth sphere, which carries the planet at its equator, is inclined at a *small angle* with respect to the third sphere. The two inner spheres (this being the decisive point of the model) revolve in opposite directions but have the same revolution period.[27] Given that in the Eudoxian subsystems the motions reproduced by the spheres depend on the position of the planet relative to the sun (elongation), their revolution time corresponds to their synodic period, at least in the case of Venus and Mercury.[28] Due to a small angular difference between the axes of the third and fourth spheres, the trajectory drawn by their combined motion resembles a horizontal figure eight (∞), named *hippopede* (ἱπποπέδη) by the Greek mathematicians. This figure reproduces quite accurately the apparent back-and-forth movement of the planets against the background of the fixed stars.[29]

---

25 *On the Cosmos*, traditionally attributed to Aristotle but considered apocryphal by most philologists on account of its style and its way of introducing peripatetic philosophy, presents the following periods for the direct planetary orbits: 1 year for Venus and Mercury, 2 years for Mars, 12 years for Jupiter and 30 years for Saturn (*On the Cosmos*, 399a8–11). These periods match those supposedly known to Eudoxus, according to Simplicius (*In Aristotelis de Caelo commentaria*, 495, 26–29).

26 The apparent wandering motion of planets is explained—in the heliocentric model—by the fact that the Earth is not at the centre of the system but orbits like the other planets around the Sun, and that the Sun and the planets have different orbital speeds.

27 See Plato's description of the celestial dance performed by the sun, Venus, and Mercury throughout the year in *Timaeus*, 38d–39a.

28 *Synodic period* or *synodic revolution* is the time elapsed between two consecutive oppositions or conjunctions of the planet (in this case Venus or Mercury) with the sun. Of course, this is a measurement of apparent motions from a geocentric point of view.

29 The "Hippopede" or "horse shackle" is a curve similar to the feint made by the foot of a horse during certain riding exercises. Today it is known as "lemniscate." We should remember that in the case of planets, this sort of flat Moebius strip traced by the mobile point never closes itself or returns to the original point. This is because the planets are at the same time being dragged by the annual motion of the star.

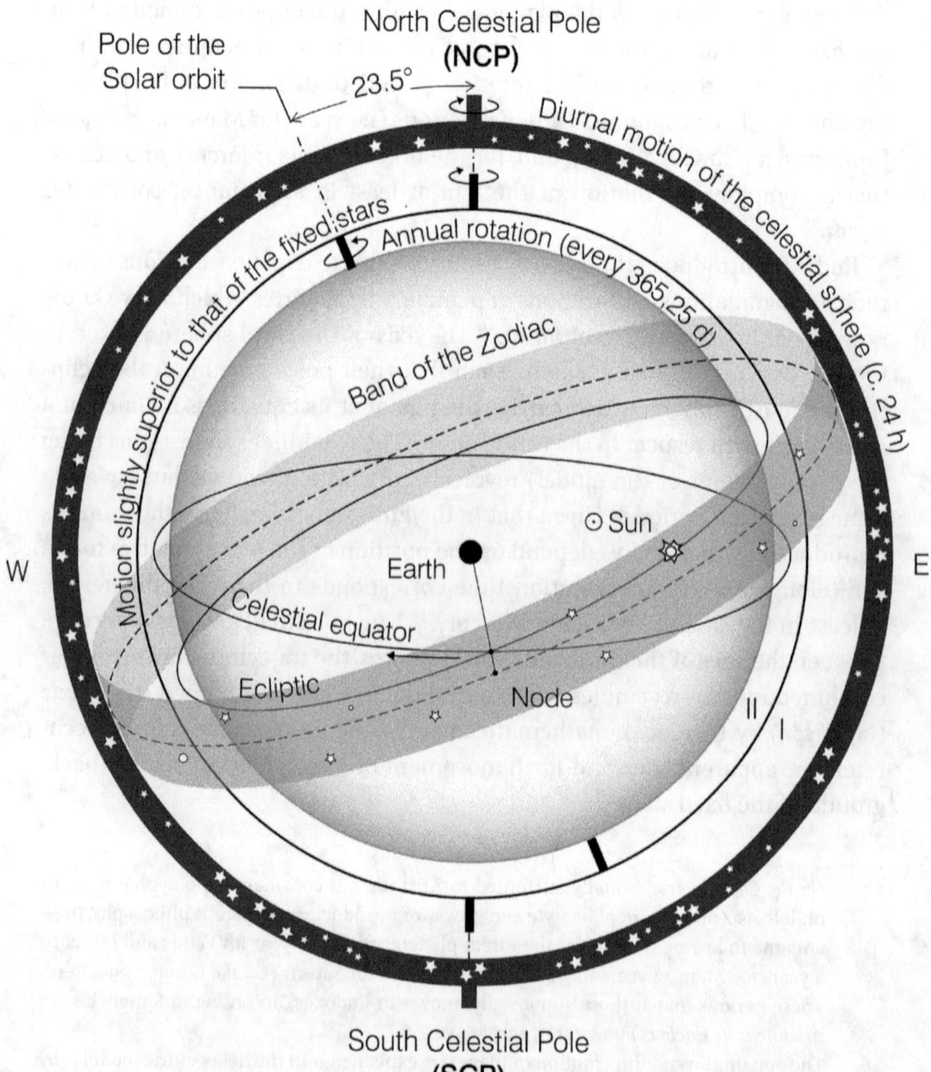

FIGURE 17 Schiaparelli (*Scritti*, vol. 2, p. 41) believes that, if Eudoxus would have wanted to reproduce a motion similar to the precession of the equinoxes, he would have replaced (as shown in the figure) the first two spheres of his solar subsystem with a single sphere revolving in the same direction as the sphere of the fixed stars, at a slightly superior speed than the daily revolution (or slightly inferior, if the intention would have been to reproduce a displacement of the equinoxes in a direct way, as happens in a real precession). Thus, the axis of this sphere, perpendicular to its equator, would describe a precession relative to the ecliptic axis. Even then, the observable phenomena of the precession of the equinoxes would not have been achieved since the solar orbit does not remain in the middle line of the zodiac. There has to be a third sphere rotating with a 24-hr period and containing the sphere of the fixed stars.

The joint motion of these two spheres (motion c), plus the motion of the second sphere (motion b) and the first (motion a), reproduces annual (b) and diurnal (a) motion, while accounting as well for planetary loops (motion c) as seen by the terrestrial observer.

The complete system of Eudoxus is made up of twenty-six spheres, and his greatest achievement is having imagined a conceptual mechanism able to translate the irregularities into regularities; that is, to comprehend the heavens as a sublime manifestation of the intelligible order. However, his observations were bettered by Callippus, who realised that a small number of additional spheres could be added to the model in order to account for some phenomena overlooked by Eudoxus.

Trained alongside Eudoxus and Polemarchus, the astronomer Callippus of Cyzicus (fl. 330 BCE), who also spent a season studying with Aristotle in Athens,[30] improved upon Eudoxus's model by taking into account the different duration of the seasons. Moreover, he devised a calendar made of 940 lunar months or 76 tropical years that proved to be an improvement on the Metonic cycle[31] and synchronised lunar and solar calculations. According to Aristotle (*Metaphysics*, Λ, 8, 1073b32–1074a6 –§ 15–), Callippus added to Eudoxus's independent planetary subsystems, achieving a better description of phenomena. In fact, it was this homocentric model, based on that of Eudoxus, that was adopted by Aristotle to develop his own vision of the heavens. Callippus did not make any changes to Eudoxus's subsystems for Jupiter and Saturn (which described the motions of these planets accurately). However, he added one extra sphere to each of the subsystems of Mars, Mercury, and Venus, and two each to the subsystems of the Sun and Moon.

30 In *Commentary on Aristotle's Metaphysics*, Bk. 12, Lesson 10, Aquinas states, following Simplicius, that Aristotle, with Callippus's help, corrected and improved (*corrigens et supplens*) Eudoxus's system (*quae ab Eudoxo inventa fuerant*) when the two were working together in Athens. This relationship must be called into question since the Aristotelian system of fifty-five spheres is quite different from the Callippic system of thirty-three spheres. The short passage of *Metaphysics*, Λ, 8 does not confirm the existence of this so-called friendship or scientific collaboration between the two philosophers. Ross (*Aristotle's Metaphysics*, p. 142), however, accepts the idea that Aristotle and Callippus were friends.

31 The Metonic cycle is named after Meton, a 5th c. Greek astronomer who accurately established, for example, the summer solstice on June 27, 432 BCE. The Metonic cycle establishes the (almost exact) equivalence between 6,940 days, 235 lunar months and 19 tropical years (there is a difference of only two hours between lunar months and tropical years). This equivalence is very useful for eclipse prediction: as moon phases repeat themselves in every cycle, it is possible to estimate future syzygies through these cycles.

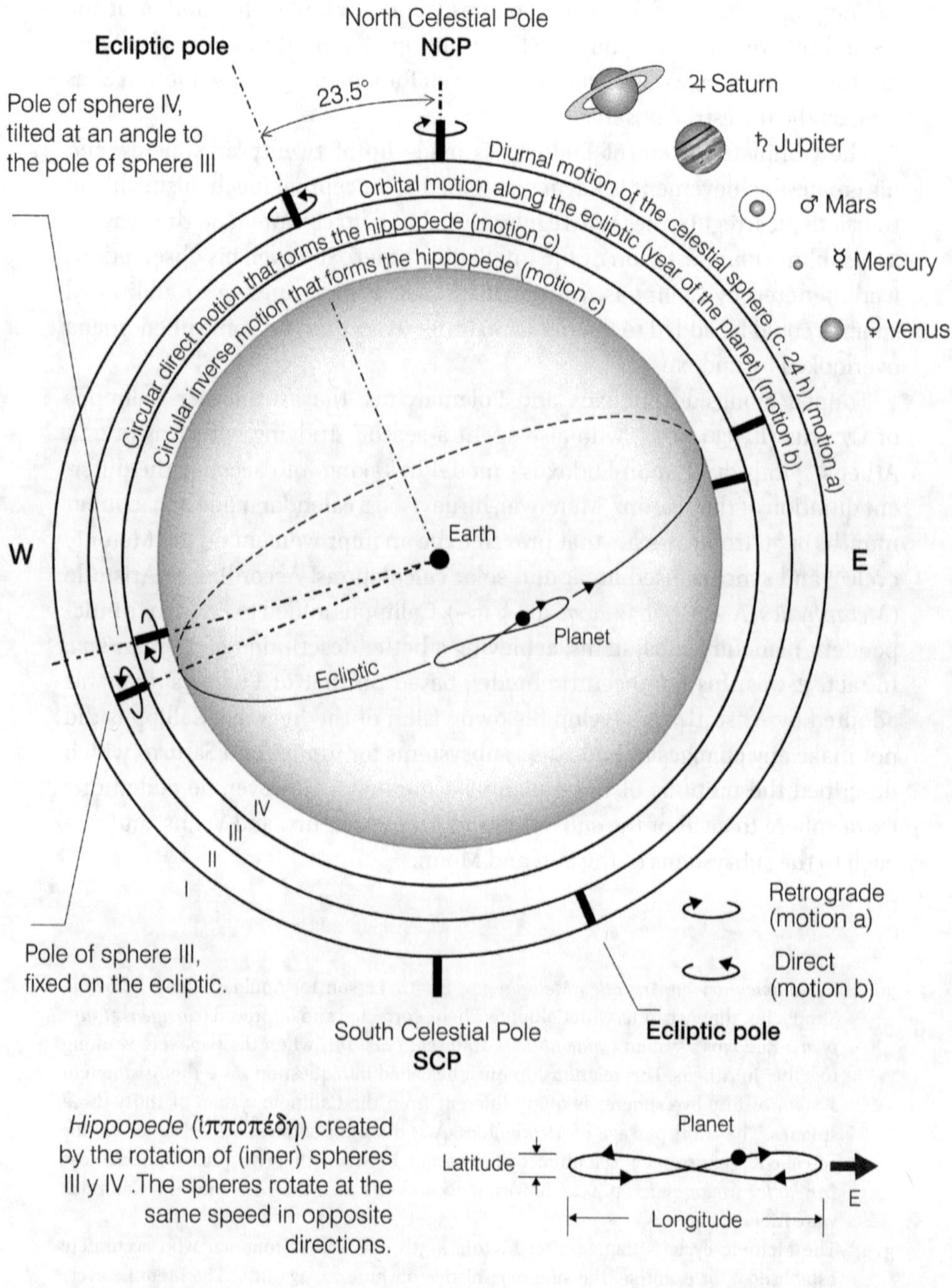

FIGURE 18 Cluster of spheres used by Eudoxus to describe the motions of Saturn, Jupiter, Mars, Mercury, and Venus.

The two spheres added to the solar subsystem behave in the same way as the third and fourth spheres with respect to the Eudoxian planetary subsystems generally. Thus, the axis of the fourth sphere of the sun is positioned as a diameter of the equator of the third sphere, while the axis of the fifth sphere is inclined at a small angle to the fourth sphere. It is possible that through this addition Callippus accounted for the longitudinal irregularities exhibited by the apparent orbit of the Sun in the Eudoxian subsystem. In so doing, he explained the unequal duration of seasons, ignored by his predecessor, and measured them with great accuracy. He established (by 330 BCE) the following values from the vernal equinox (Aries), for an observer in the Northern Hemisphere: spring, 94 days; summer, 92 days; autumn, 89 days; and winter, 90 days.[32] Thomas Little Heath, who analysed the Callippic system in detail, has pointed out that this representation of the motion of the Sun was as accurate as that obtained years later through the Ptolemaic systems, formed using eccentric circles and epicycles.[33] As regards the addition of two extra spheres for the Moon, similar to the two extra solar spheres, they appear to have been introduced to correct the observed irregularities in longitude of the lunar orbit. Such a discovery was apparently made by comparing the time of certain lunar eclipses with the corresponding lunar longitudes. Eudoxus's ἱπποπέδη also helped to explain these phenomena even when it failed to account for evection.[34]

The addition of an extra sphere to the subsystems of Venus, Mercury, and Mars was probably due to the failure of the Eudoxian system to accurately reproduce the apparent orbits of these planets, especially that of Mars. Schiaparelli has proven that the addition of an extra sphere to the two that produce the ἱπποπέδη in the Eudoxian subsystem for Mars gives that planet a higher direct retrograde speed without changing the latitude of its motion. Thus, they better reflect the apparent motions as long as the correct synodic periods (corresponding to the rotation period of the third and fourth sphere)

---

32 It is strange that Eudoxus shows no knowledge of the measurements of Meton and Euctemon—it may have been a purposeful omission to simplify his system. These two Athenians had already noticed a few decades earlier some differences among the seasons. When Callippus acknowledged these differences, he was forced to introduce new spheres (Ross, *Aristotle's Metaphysics*, vol. 2, p. 391).

33 Heath, *Aristarchus*, p. 216.

34 Heath, *Aristarchus*, p. 216; Ross, *Aristotle's Metaphysics*, vol. 2, p. 391. Evection is a variation that regularly affects lunar motion by virtue of the solar attraction on the earth-moon system. Its main effects are a shift in the perigee of the orbit and a variation in orbital eccentricity (between 0.045 and 0.065). The evection period lasts 31 days, 19 hours and 26 minutes. Its amplitude, positive or negative, is $1^{\circ}\ 16'$. In the 2nd century CE, Claudius Ptolemy noticed this phenomenon for the first time.

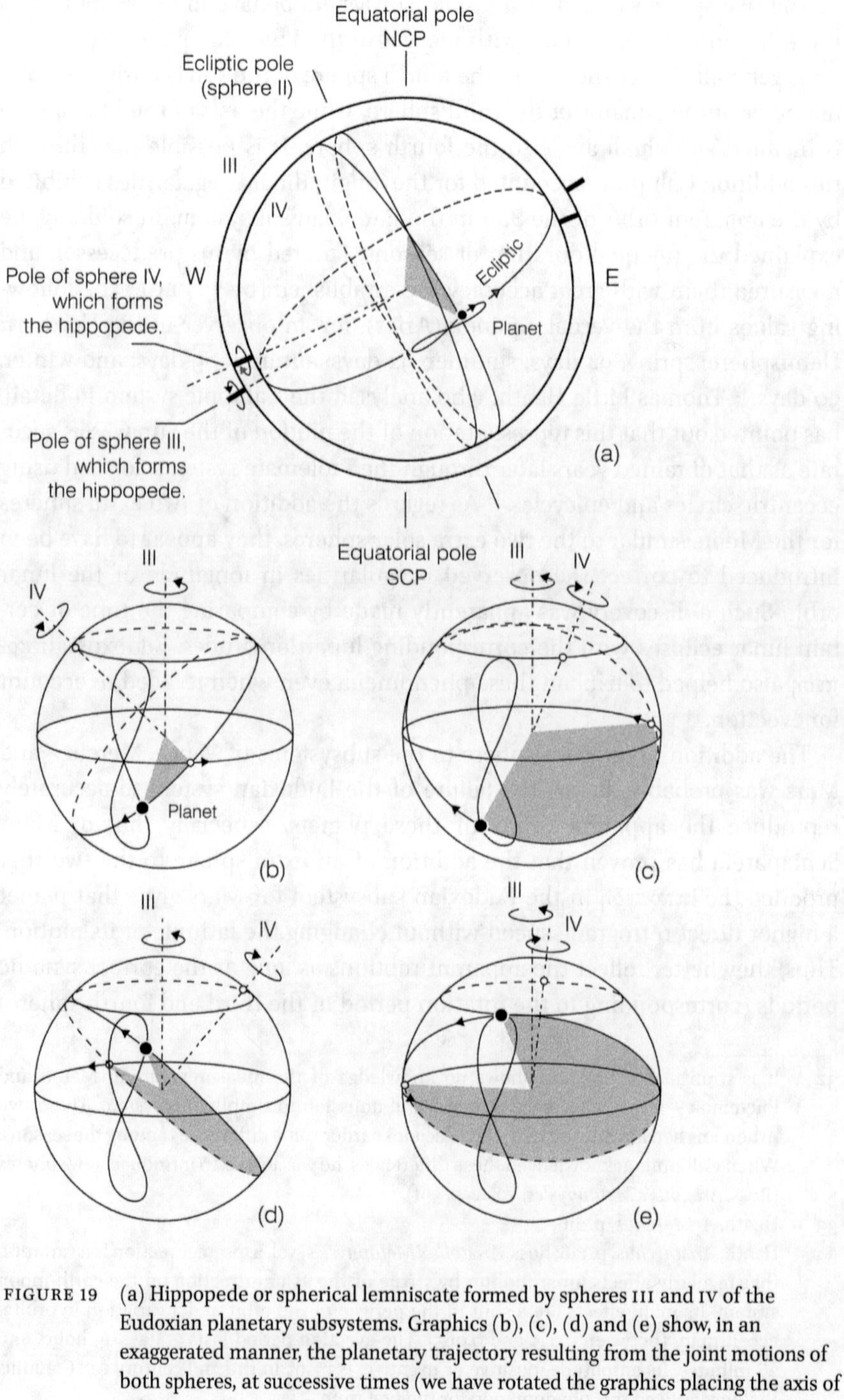

FIGURE 19 (a) Hippopede or spherical lemniscate formed by spheres III and IV of the Eudoxian planetary subsystems. Graphics (b), (c), (d) and (e) show, in an exaggerated manner, the planetary trajectory resulting from the joint motions of both spheres, at successive times (we have rotated the graphics placing the axis of sphere III in an upright position so as to clearly appreciate the formation of the hippopede).

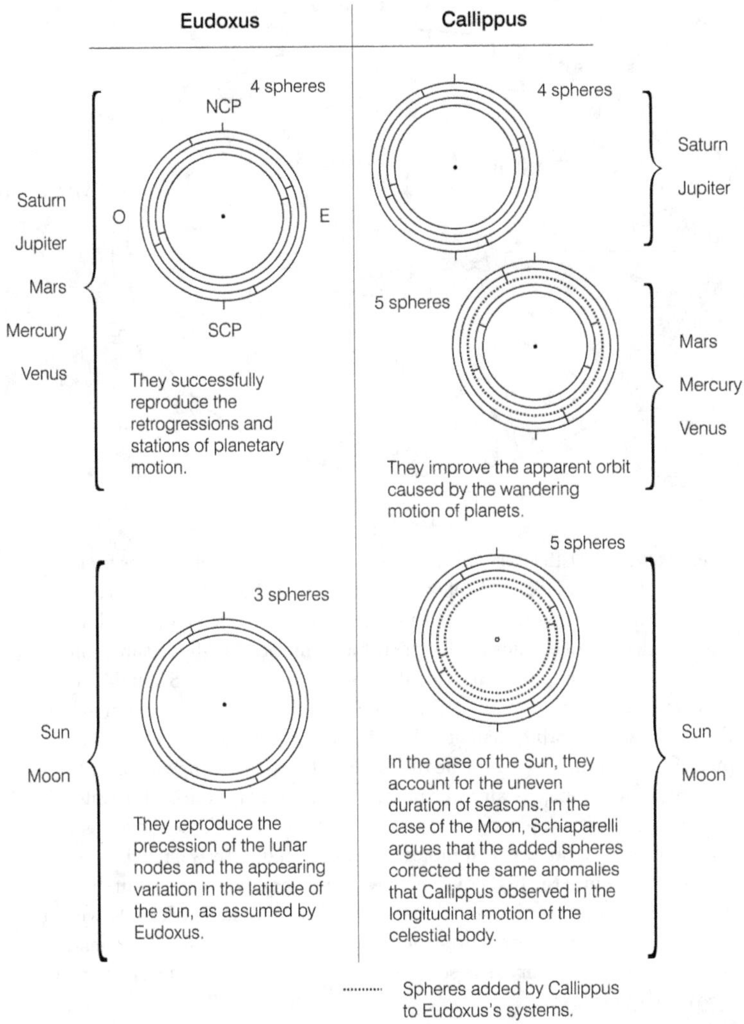

FIGURE 20 Difference in the number of spheres between the planetary models of Eudoxus and Callippus. One of the main improvements introduced by Callippus is the uneven duration of seasons in the solar subsystem. Callippus seemed to have measured the seasons with great accuracy as he assigned them very similar values to those currently used. Another remarkable fact is that the indirect determination of the lunar cycle (or synodic period: the time elapsed between two successive conjunctions of the moon and the sun) based on Callippus's 27.759-day cycle—corresponding to 940 lunations—gives, for the lunar cycle, a value that only differs by approximately 10 seconds from the current value of 29 days, 12 hours, 44 minutes and 12 seconds (Schiaparelli, *Scritti*, vol. 2, p. 85). In the representation of Callippus's five-sphere subsystems for Mars, Mercury, and Venus, we have placed the three inner spheres according to the orientation of the poles, as estimated by Schiaparelli (pp. 79–81).

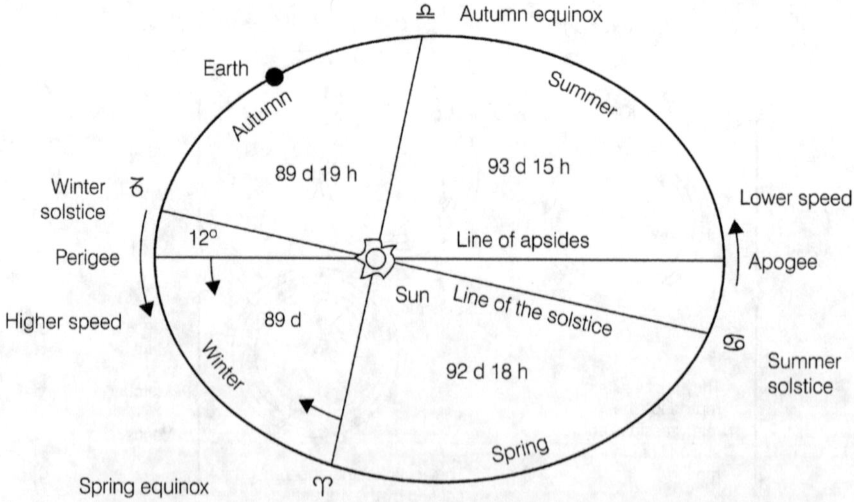

FIGURE 21 Equinoxes and solstices according to the current duration of seasons for the Northern Hemisphere. As is well known, "season" is the name given to the period of time it takes the Earth to go from a solstice to an equinox and from an equinox to a solstice. Since the Earth's orbit is elliptical (with the sun in one of its foci), it has a non-uniform speed throughout its trajectory (Second Law of Kepler) thus, the duration of seasons is uneven. If the apsis line (major axis of the Earth's elliptic orbit) matched the solstice line, the duration of the seasons on each side of this line would be the same (this actually happened in 1280 CE, so winter = autumn, and spring = summer). At present (as during the time of Eudoxus, Callippus, and Aristotle), the lines do not coincide, so the four seasons have uneven durations. The duration of seasons is constantly changing, at a very slow pace, because the first point in Aries (vernal equinox) retreats in the zodiac 50″ (seconds of arc) each year due to the precession of the Earth's axis, while the perigee advances 12″ during the same period. Therefore, one point approaches the other by almost 62″ every year; see Martín Asín, *Astronomía*, (Madrid: Paraninfo, 1979), pp. 250–251.

are applied (in place of the incorrect periods applied by Eudoxus); that is to say, the 260-day period utilised by Eudoxus must be replaced by a 780-day period. Ross adds that a similar improvement in the description of phenomena is seemingly achieved with the addition of a fifth sphere for Venus and Mercury.[35]

We do not know whether these astronomers were in fact able to make predictive calculations using these cunning geometrical models. This was probably not their main motivation. It seems, rather, that their energies were

35 Schiaparelli, *Scritti*, vol. 2, pp. 79–80.

directed towards discovering the intelligible background of celestial phenomena, which were considered divine as well as a manifestation of the order and permanence of reality. Regardless, if we apply current values to the parameters of the planetary models of Eudoxus and Callippus as adopted by Aristotle, we will be able to reproduce the main characteristics of planetary orbits with remarkable accuracy (at least for short astronomical periods). Today, thanks to the rotation matrices based on Euler angles, we can easily calculate planetary trajectories from Eudoxian and Callippian models and compare them with the exact orbits obtained applying current astronomical knowledge.

The results obtained are striking and show that these systems are more than mere philosophical perspectives or primitive scientific ideas. They are the historical origin of the way we do science and understand the cosmos. Indeed, the distance between our heuristic matrices and those used by the Greek masters is, in a sense, only apparent.

It has been often said that Aristotle's approach is distant from modern thinking because his doctrines on nature disregarded mathematics, the basis of current physics. However, the knowledge and use of geometrical systems in relation to the cosmos demonstrated in *Metaphysics*, Λ, 8 (to name but one example) shows that his mathematical knowledge and skills equalled that of his fellow astronomers and geometers (we know in fact that he not only knew Eudoxus but admired him as much as he did his master, Plato).

The Aristotelian contribution to the astronomy of Eudoxus and Callippus is quite significant. It aims to improve on their vision by means of a single integrated system. In order to achieve this, Aristotle introduces so-called counteracting (or rewinding) spheres and immaterial movers for all spheres of the system of planetary motion described in *Metaphysics*, Λ, 8, 1073b39–1074a14, thus adding further complexity to the scheme and showing a clear comprehension of the geometry and dynamics implicit in the previous systems. Aristotle's astronomical approach goes beyond motions a + b + c to account for their *causes*. He identifies them with separate intelligible substances, whose ontological status has been extensively discussed by the classical, late antique, and medieval philosophical traditions (one should bear in mind that the Judaeo-Christian tradition identified these movers with the angelic powers distinguishing the first and most eminent of them, the Prime Mover, as the pagan expression of the God of Scriptures). Taking into account the complexity of Aristotle's reformulation of this issue in astronomical as well as ontological/metaphysical terms, we will pause in our analysis of celestial motions to focus in the next chapter on the general premises of Aristotelian thought, without which it is impossible to comprehend his system of the heavens. We will resume the issue of the specific celestial motions as introduced by the

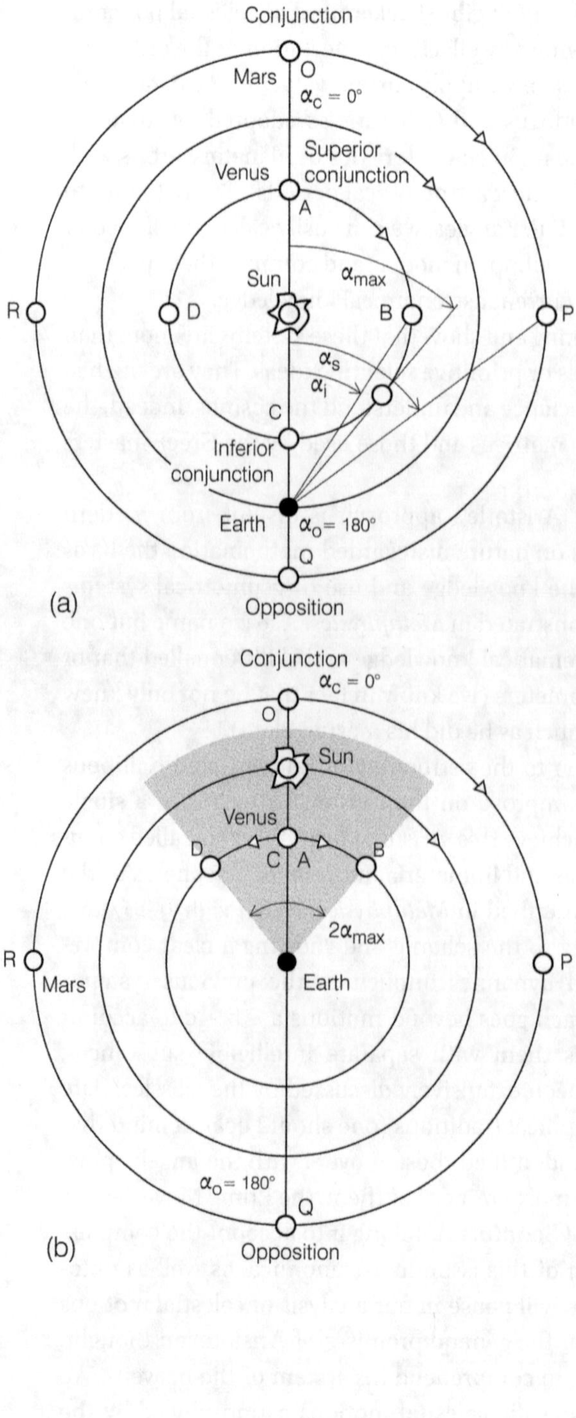

FIGURE 22
Conjunctions and oppositions caused by the direct motion of stars (motion b) from a heliocentric (a) and geocentric (b) perspective. In the heliocentric system (image a), the upper planets show conjunctions and oppositions with elongations (angular distance between the sun and the planet, as seen from the Earth, angle α in the figure) that range from 0° (conjunction) to 180° (opposition) on both sides of the sun. Lower planets only exhibit upper and lower conjunctions with elongations that range from 0° to a maximum value set between 18° and 28° for Mercury, and between 45° and 47° for Venus (these variations are caused by the orbital ellipticity). The graphic shows the upper and lower elongations of Venus (inner planet) and the conjunction and opposition of Mars (outer planet). In Eudoxus's geo-homocentric system (image b), the variations in elongation indicate the autonomous motion of the upper planets, while the motions of the inner planets seem to be bound to the sun (Mercury and Venus), as stated in *Timaeus*, 39a. In image b, the area where the inner planet moves relative to the sun is represented by a shaded cone.

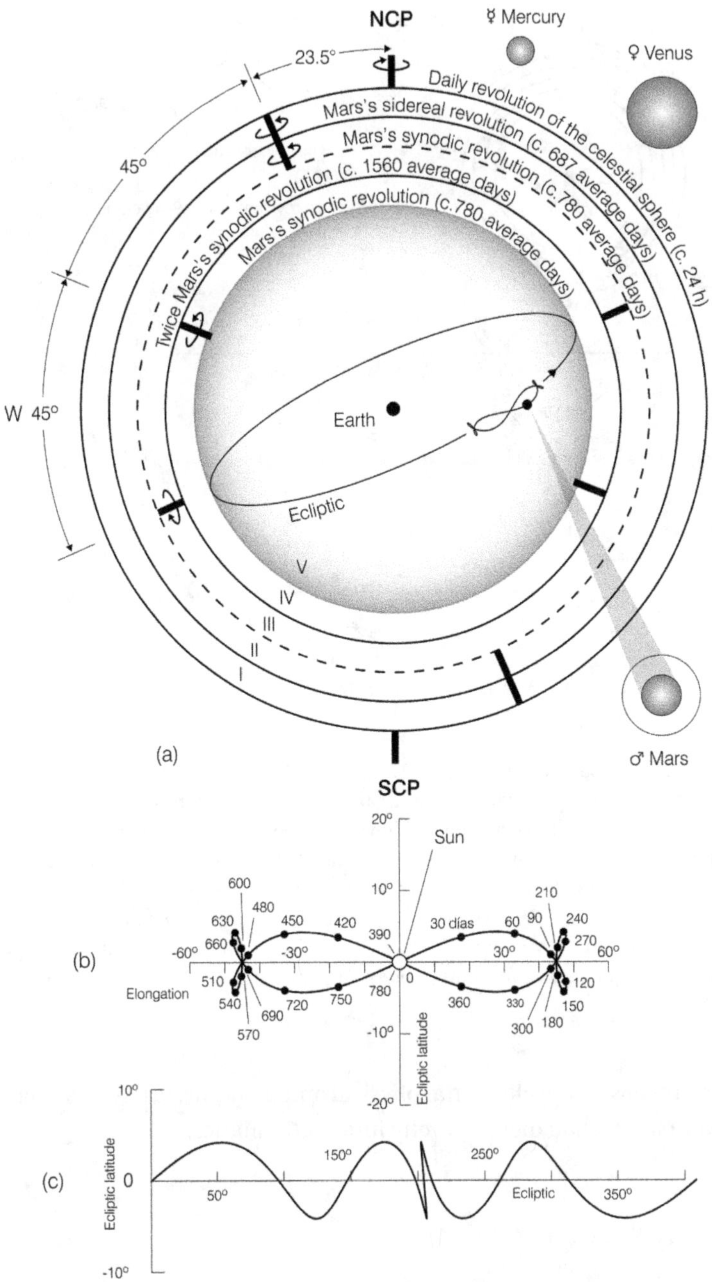

FIGURE 23 Trajectory of Mars. (a) Callippic model for Mars. We have placed the spheres according to Schiaparelli's interpretation. The revolution periods taken into consideration for the numerical calculation of trajectories correspond to current values. (b) Figure generated by Callippic spheres III, IV, and V. (c) The direct motion of Mars along the ecliptic, generated by Callippic spheres II, III, IV, and V.

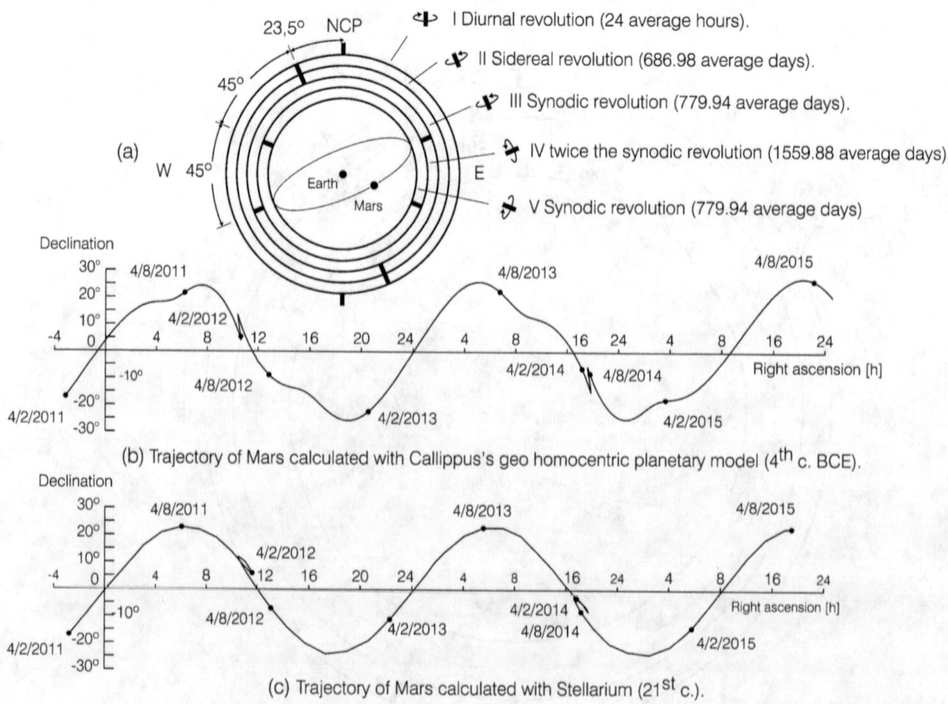

FIGURE 24 Comparison of the apparent trajectories of Mars, calculated (a) with Callippus's geo-homocentric planetary model (4th c. BCE) and (b) with Stellarium software (21st c.). The values assigned to Callippic parameters (revolution periods of the spheres) correspond to current values, i.e., for sphere I, an average of 24 hours; for sphere II, an average of 686.98 days (sidereal revolution); for sphere III, an average of 779.94 days (synodic revolution); for sphere IV, an average of 1559.88 days (twice a synodic revolution); for sphere V, an average of 779.94 (synodic revolution). We should bear in mind that the remarkable similarity between the orbits described holds only for short periods. For long periods, the curves tend to stray considerably.

Stagirite after a review of the elemental physical and metaphysical Aristotelian doctrines. But first, we shall digress briefly into mathematics.

4.2 Planetary Trajectories

Using current mathematical and computational resources we can quite easily calculate the planetary trajectories produced by geometrical astronomical systems of homocentric spheres such as those developed by Eudoxus and Callippus. These systems are based on the supposition that complex planetary trajectories are a consequence of simple superimposed circular motions

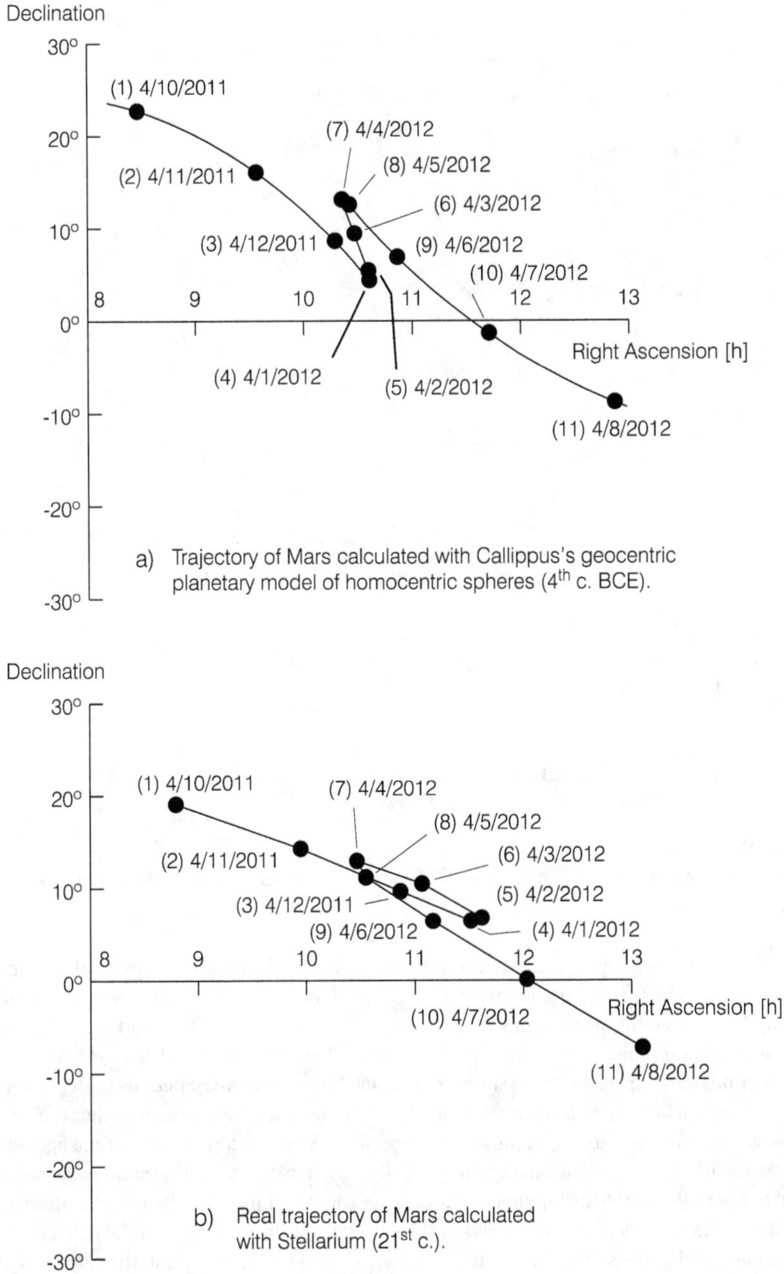

FIGURE 25 The apparent trajectories of Mars, calculated with Callippus's geo-homocentric planetary model (4th c. BCE) and with Stellarium software (21st c.) This graphic is an enlarged image of the retrogradation area described in the previous figure, using the same angles and periods. The values assigned to the parameters of Callippus's model are not those used by Eudoxus; rather they are current values, like those used in the previous graphic.

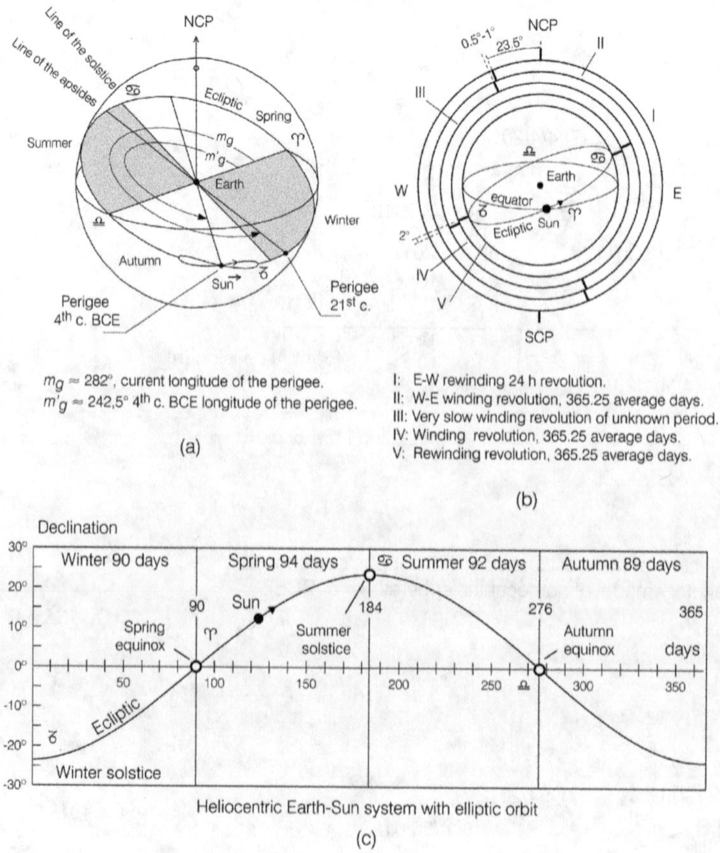

FIGURE 26 The trajectory of the Sun and the length of the seasons, calculated with the callippic solar system.

Image (c) shows the sun's trajectory in equatorial coordinates, calculated with the Callippic solar system, represented in image (b). The first three spheres—which follow Eudoxus's solar system—are represented according to Simplicius's description, though we have disregarded the motion of the third sphere, as its period is unknown. Such simplification does not produce major alterations to the calculations as they cover very short periods in comparison to the supposedly much longer revolution periods of this sphere. Assuming that the hippopede formed by spheres IV and V aims at reproducing the uneven duration of seasons, the first step of our calculations, as shown in image (a), was to match the centre of the figure to what was the position of the perigee back in the time of Callippus. In his system, direct and retrograde motions related to the trajectory of the hippopede produce a variable ecliptic speed for the sun during its annual revolution. This equals the effect caused by the ellipticity of the solar orbit in the heliocentric system. The variable speed of the sun relative to the position of the apsis (major axis of the elliptic solar orbit) is responsible for the uneven duration of seasons, whose beginning and end are determined by the orthogonal solstice and equinox lines. Because of the angular separation between the solstice and apsis line (longitude of the perigee), the sun does not present the same speed during the intervals of each season, resulting in uneven lengths. The seasons only have approximately the same duration if considered in pairs (winter = autumn; spring = summer) and only when these two lines coincide (longitude of the perigee = 0°). Surprisingly, the duration of seasons properly calculated in the Callippic model, image (c), coincides almost exactly with the Callippic values. We do not know, however, which method of calculation he used.

effected by concentric spheres conveniently positioned and fitted one inside the other. The first and most original of these systems was that created by Eudoxus to describe the motion of Saturn, Jupiter, Mars, Mercury, and Venus. As we have mentioned above, by framing each planet within four spheres and setting the correct orientation and speed of rotation for each sphere, Eudoxus succeeded in reproducing planetary stations and retrogradations with remarkable accuracy, as modern calculations now reveal.

In order to describe the motion of the Moon and the Sun, Eudoxus designed simpler systems of just three spheres. Callippus maintained the same basic principles of functioning as his master Eudoxus, introducing minor changes to the system's structure. He added, in particular, an extra sphere to each of the subsystems of Mars, Mercury, and Venus, and two extra spheres for the Sun and the Moon, as described above.

The astronomical model of homocentric spheres gained in diversity and complexity through Callippus's improvements and the Aristotelian integration of previous models into a single system comprising all celestial motions, but the key element in the solution to the thorny question of irregular planetary trajectories was Eudoxus's system of four spheres. In this system, the two inner spheres (III and IV) turn with the same revolution period but in opposite directions, and their axes form a small angle that we will call θ. Under these motions, the star located at the equator of the innermost sphere (IV) draws a distinctive trajectory that Eudoxus called the *hippopede*. Sphere III, for its part, is articulated at the equator of sphere II, which is on the ecliptic plane. The rotational motion of Sphere II moves the two inner spheres, untying, so to speak, the hippopede. Given the correct orientation and speed for each sphere, the set reproduces with considerable accuracy the direct motion of each star with their corresponding stations and retrogradations.

The motion of a star within these astronomical systems, produced by these two spheres revolving simultaneously relative to a third sphere, is analogous to the rotational motion of a rigid body, which can be described using Euler angles θ, φ, ψ. By means of these angles, the motion of any given point (in this case, the position of the star) on a rigid body with a pure rotational motion can be broken down into the three successive rotational motions that cause it. The rotated point is expressed by the coordinates of the reference system relative to which each rotational motion is produced, by means of transformation matrices B, C, D.

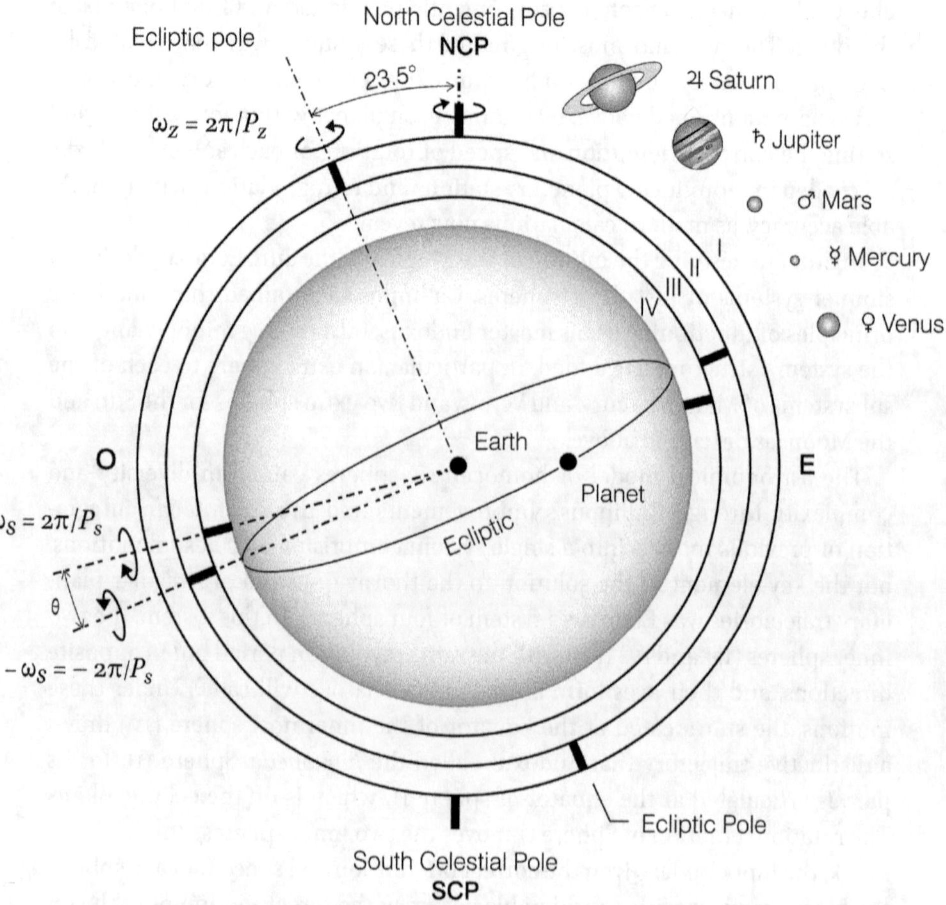

FIGURE 27 Cluster of planetary spheres designed by Eudoxus for Saturn, Jupiter, Mars, Mercury, and Venus. P_z and P_s are the zodiac (or sidereal) and synodic revolution periods, respectively, while ω_z and ω_s are the angular speeds of the spheres for the same revolution periods, expressed in radians per unit of time.

Each of the matrices restores the rotated point to the coordinates of the non-rotated system.

$$B=\begin{pmatrix} \cos\varphi & -\text{sen}\,\varphi & 0 \\ \text{sen}\,\varphi & \cos\varphi & 0 \\ 0 & 0 & 1 \end{pmatrix},\ C=\begin{pmatrix} 1 & 0 & 0 \\ 0 & \cos\theta & -\text{sen}\,\theta \\ 0 & \text{sen}\,\theta & \cos\theta \end{pmatrix},\ D=\begin{pmatrix} \cos\psi & -\text{sen}\,\psi & 0 \\ \text{sen}\,\psi & \cos\psi & 0 \\ 0 & 0 & 1 \end{pmatrix}$$

$$X = B.\eta$$
$$\eta = C.\zeta$$
$$\zeta = D.X'$$

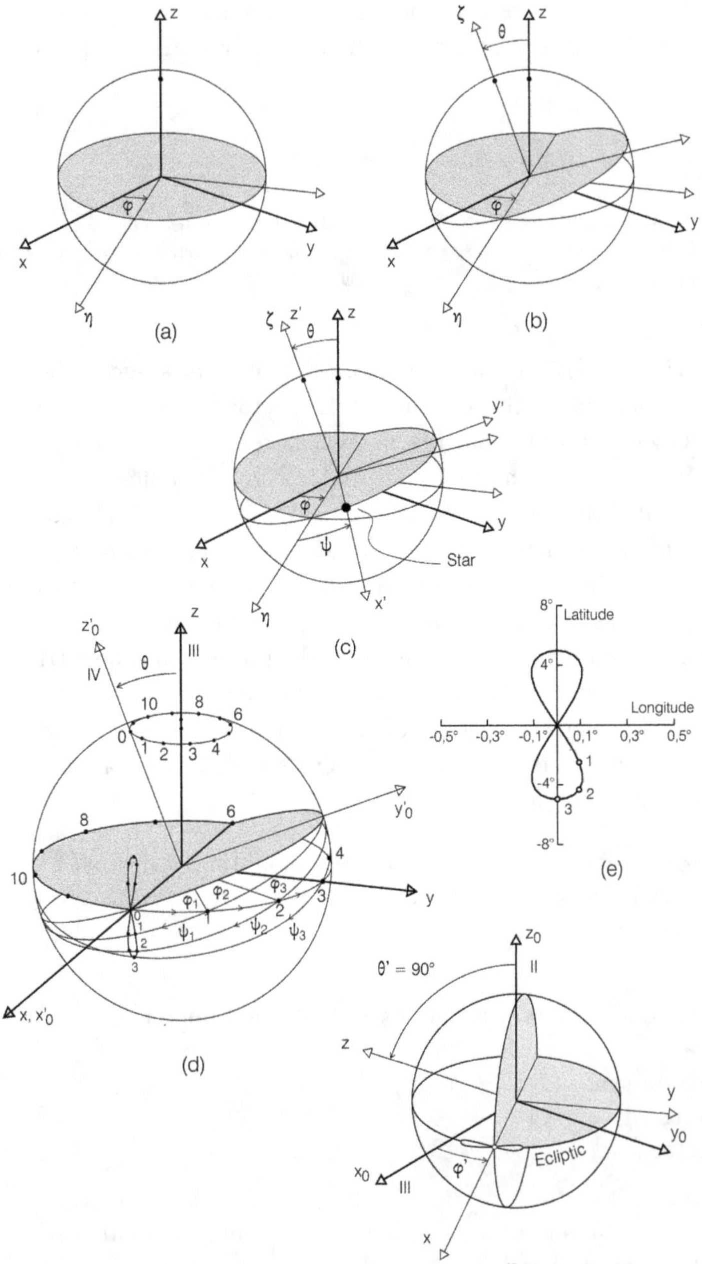

FIGURE 28 (a), (b) and (c) are representations of the three interdependent rotations of a solid in space, as defined by Euler angles. (d) Rotations caused by spheres IV and III of Eudoxus's planetary subsystems, based on Euler angles. As the image shows, the rotation of Eudoxus's two inner spheres is enough for the star, fixed on the equator of sphere IV, to draw the hippopede. This typical figure of ancient astronomy is also exhibited with more precision in image (e), represented in ecliptic coordinates. (f) Direct ecliptic or zodiacal rotation of the star caused by the rotation of the Eudoxian sphere II, also based on Euler angles.

An arbitrary rotation of any point in space can be expressed in the coordinates of a reference system at rest applying the three previous transformations simultaneously.

$$X = BCD.X' = A.X' =$$

$$= \begin{pmatrix} \cos\psi\cos\varphi - \cos\theta\ \text{sen}\,\varphi\,\text{sen}\,\psi & -\text{sen}\,\psi\cos\varphi - \cos\theta\ \text{sen}\,\varphi\ \cos\psi & \text{sen}\,\theta\ \text{sen}\,\varphi \\ \cos\psi\ \text{sen}\,\varphi + \cos\theta\ \cos\varphi\,\text{sen}\,\psi & -\text{sen}\,\psi\ \text{sen}\,\varphi + \cos\theta\cos\varphi\ \cos\psi & -\text{sen}\,\theta\ \cos\varphi \\ \text{sen}\,\theta\ \text{sen}\,\psi & \text{sen}\,\theta\ \cos\psi & \cos\theta \end{pmatrix} \cdot \begin{pmatrix} x' \\ y' \\ z' \end{pmatrix}$$

As we can see from image (d) of the previous figure, the rotation caused by the Eudoxian spheres III and IV relative to sphere II is exactly equivalent to the rotation described by means of the Euler angles, which results from superimposing the rotations exhibited in images (a), (b) and (c), properly adjusted and combined. We can calculate, for each point in time, the position of the star placed in system *X'* fixed to sphere IV, relative to system *X* fixed to sphere II, as long as we agree that: 1) the star is at rest relative to sphere IV in coordinates $(x', y', z') = (1, 0, 0)$; 2) constant angle θ equals the angle formed by the axes of rotation of spheres III and VI; 3) angle φ describes the rotation of sphere III at angular speed $\omega = 2\pi/P_S$ (where P_S is the period of synodic revolution of the star), thus $\varphi = \omega.t$; and 4) angle ψ describes the rotation of sphere IV at speed $-\omega$, thus $\psi = -\omega.t$. If we assign these values to the Euler angles, we obtain the following transformation equation:

$$\begin{pmatrix} x \\ y \\ z \end{pmatrix} = \begin{pmatrix} \cos^2\omega t + \cos\theta\ \text{sen}^2\,\omega t & (1-\cos\theta)\cos\omega t\ \text{sen}\,\omega t & \text{sen}\,\theta\ \text{sen}\,\omega t \\ (1-\cos\theta)\cos\omega t\ \text{sen}\,\omega t & \text{sen}^2\,\omega t + \cos\theta\ \cos^2\omega t & -\text{sen}\,\theta\ \cos\omega t \\ \text{sen}\,\theta\ \text{sen}\,\omega t & \text{sen}\,\theta\ \cos\omega t & \cos\theta \end{pmatrix} \cdot \begin{pmatrix} 1 \\ 0 \\ 0 \end{pmatrix}$$

From this, we obtain the parametric equations for Eudoxus's hippopede:

$$\begin{pmatrix} x \\ y \\ z \end{pmatrix} = \begin{pmatrix} \cos^2\omega t + \cos\theta\ \text{sen}^2\,\omega t \\ (1-\cos\theta)\cos\omega t\ \text{sen}\,\omega t \\ \text{sen}\,\theta\ \text{sen}\,\omega t \end{pmatrix}$$

By rotating some points in succession, we have represented the motion of the star that results from the rotation of the spheres III and IV, as illustrated in image (d). Image (e) shows the hippopede or lemniscate produced by this motion, based on latitude and longitude coordinates. In order to calculate the trajectory of the planet relative to the fixed stars or, in other words, its direct motion around the ecliptic, we need to apply a new transformation to the coordinates (x, y, z) obtained.

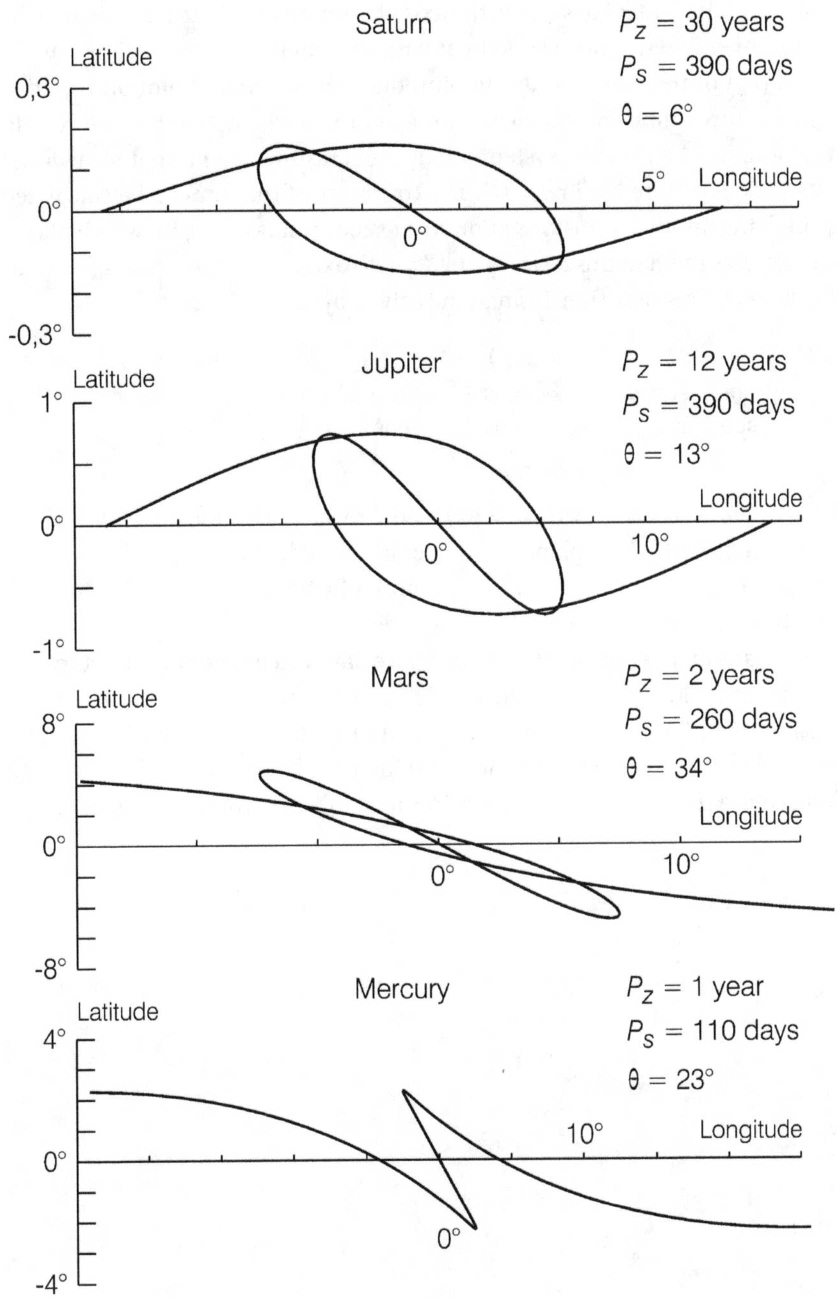

FIGURE 29 The trajectories of some planets produced by the four-sphere subsystems of Eudoxus and calculated through coordinate transformations based on Euler angles.

Since, in the Eudoxian system, the axis of rotation of sphere III is fixed to the equator of sphere II and carried by it with the sidereal or zodiacal revolution period (P_z) of the star, in order to reproduce the rotational motion resulting from the three innermost spheres, we need to consider the previous coordinates as part of a rotation system relative to a resting system that we will call system X_o, illustrated in image (f). The trajectory of the star can be calculated by applying another transformation to the coordinates (x, y, z), which allows us to express them in this new system X_o, with axis z in the direction of the axis of sphere II. This new transformation is given by:

$$\begin{pmatrix} x_o \\ y_o \\ z_o \end{pmatrix} = \begin{pmatrix} \cos\psi' & -\mathrm{sen}\,\theta & 0 \\ \cos\theta'\,\mathrm{sen}\,\psi' & \cos\theta'\cos\psi' & -\mathrm{sen}\,\psi' \\ \mathrm{sen}\,\theta'\,\mathrm{sen}\,\psi' & \mathrm{sen}\,\theta'\cos\psi' & \cos\theta \end{pmatrix} \begin{pmatrix} x \\ y \\ z \end{pmatrix}$$

If $\theta' = 90^\circ$ and $\psi' = \omega_z.t$, with $\omega_z = 2\pi/P_z$, the result is the images above, where we have represented the planetary trajectories by adjusting the transformation parameters according to the values proposed by Schiaparelli for the different planets in Eudoxus's astronomical models.[36]

By means of an equivalent procedure, we have calculated the planetary trajectories of Callippus's subsystems for Mars and the Sun, presented in the previous section. These geometrical models do not differ in essence from those designed by Eudoxus. Callippus, however, increased the number of spheres and changed the rotation speeds and orientations to adjust them to phenomena.

36 *Scritti*, vol. 2, pp. 66–77.

CHAPTER 5

Aristotle's Astronomical System

Caelum quoque primum inter cetera corpora habet armoniam …
In motu quoque, si mobile est, maximam habet uniformitatem,
si vero quietum est, maximam habet quietis unitatem.

And the first heaven is the most harmonious of all bodies …
For in motion, it has the greatest uniformity of motion,
And if at rest, it has the greatest unity in repose.

ROBERT GROSSETESTE, *De operationibus solis* (*On the Operations of the Sun*), 1

Clearly, Aristotle's astronomical system is founded on a set of values and ideas that the systems of Eudoxus and Callippus did not make explicit, namely: 1) the perfect (divine, unchanging) is prior to the imperfect (changing); 2) celestial motions belong to the divine domain, thus must be regular, eternal, and rational; 3) the circular motions of celestial spheres have neither end nor beginning, therefore, they must be caused by Unmoved Movers, whose action is unceasing, considered as ἐντελέχειαι full of life and happiness; 4) the unicity of the universe demands a Prime Mover; 5) the eternal motion of the universe is ensured by the pure actuality of this Prime Mover, which Aristotle calls "god" (θεός). Furthermore, this Unmoved Mover is thinking itself, and also self-thinking-thought (νόησις νοήσεως; 1075a34 –§ 15–); that is, about the most beautiful possible object of thinking, which is none other than itself. It moves the heavens perpetually by its ontological plenitude—which is to say by the Good, Beauty, and Truth (these are not Aristotle's words but this is how we understand his meaning)—that replicates the plenitude of its divine reality. The astronomical system of Aristotle rests not only on physical grounds, then, but also, and mainly, on metaphysical assumptions. This is important to keep in mind when analysing the particularities of the Aristotelian understanding of celestial motions.

© GERARDO BOTTERI AND ROBERTO CASAZZA, 2023 | DOI:10.1163/9789004525535_007

Building on Eudoxus's astronomical model (more specifically, the version improved by Callippus), Aristotle composed a unified celestial system consistent with his doctrine of the Prime Mover. The Aristotelian worldview regards this first cause—the root of any change—as that producing the first eternal motion, unique in its kind,[1] and he identifies such motion with the simple movement (rotation) of the universe as a Whole (τὴν τοῦ παντὸς τὴν ἁπλῆν φοράν; 1073a29 –§ 15–). Because the first motion is directly caused by the first οὐσία, it corresponds specifically to the motion of the fixed stars, yet also indirectly determines the conditions for planetary motions. Its action therefore engages the entire superlunary and sublunary universe.

The introduction of an immaterial intelligible principle is not, however, the main difficulty when it comes to adjusting the Eudoxian and Callippic model to that of Aristotle: as Eudoxus and Callippus did not advance an ultimate cause of motion, it was a simple matter for Aristotle to locate it in another plane, so to speak. A greater challenge lies in the introduction, in *Metaphysics*, Λ, 8, of numerous immaterial movers—substances that are by nature motionless and lack magnitude (τοσαύτας τε οὐσίας ἀναγκαῖον εἶναι τήν τε φύσιν ἀΐδίους καὶ ἀκινήτους καθ' αὑτάς, καὶ ἄνευ μεγέθους; 1073a37 –§ 15–)—to account for the different planetary motions that had already been noticed in the Academy and which had been accounted for, in the systems of Eudoxus and Callippus, by the presence of multiple spheres.

In order to justify the specific motions of the wandering stars, attributed by Eudoxus and Callippus to the rotation of a series of superimposed concentric spheres, Aristotle—abiding by the principles of his cosmology—introduces the same number of Unmoved Movers as there are simple circular motions or rotational spheres required to describe these motions. From a physical point of view, the difficulty is that, given that each Unmoved Mover causes just one simple rotation, there are—in addition to the eternal simple motion of the heaven as a Whole (motion a), which could ultimately be explained by means of a single Unmoved Mover—several simultaneous eternal rotations (motions b and c) that are needed to explain the motions of the wandering stars (ἄλλας φορὰς οὔσας τὰς τῶν πλανήτων ἀϊδίους; 1073a32 –§ 15–). The multiple rotations postulated in the models of Eudoxus and Callippus also have different directions, speeds, and orientations with respect to the motion of the totality of the universe, thus creating the need for many different movers to explain visible phenomena.

1 *Metaphysics*, Λ, 8, 1073a23–26.

As these "Unmoved Movers of Planetary Spheres" also move the sensible eternal οὐσίαι directly—that is, the celestial spheres (and the stars attached to some of them)—they are necessarily οὐσίαι, since, according to Aristotle, a substance can only be moved by another substance (1073a33 –§ 15–). In addition, like the Prime Mover, they are pure actuality, necessarily moving in a superphysical manner; that is, *without physical contact*. Evidently, these substances are immaterial and eternal but inferior and subordinate to the Prime Mover (Aristotle is unclear about the precise characteristics of this relation).[2] This inferiority can be interpreted in part as rooted in the notion that the bodies moved by each of the inferior movers are at the same time moved by the general rotational motion of the entire universe, whose ultimate origin is the Prime Mover. In other words, the Unmoved Movers, hierarchically arranged below the Prime Mover and in successive order in relation to planetary movements (1073b1–3), are but partial causes of the motion experienced by their spheres; the motions they produce are mounted, so to speak, on the unique motion of the Whole. Therefore, they cannot be considered ultimate causes, in a strict sense. That is an attribute of the Prime Mover alone. Despite their inferiority, however, because they are intelligible principles and pure actuality, they share of the ontological nobility and dignity of the Prime Mover.[3]

---

2 *Metaphysics*, Λ, 8, 1073a36–1073b3 –§ 15–.

3 In regard to book Λ, chapter 8, Guthrie ("The Development of Aristotle's Theology—II", *The Classical Quarterly* 28, no. 2, (April 1934, pp. 90–98) considers that, faced with the novelty of this text in the context of *Metaphysics*, Λ, the main question that comes to mind is not how these Unmoved Movers, being pure forms, can be distinguished from one another (an idea that was challenged by Plotinus –§ 22–), but rather how they are subordinated to the Prime Mover. This question, we believe, can only be answered by incorporating the astronomical dimension (that is, its link to physics). If we agree that the Unmoved Movers cause celestial motions and that these motions have different directions and magnitudes, the systematic conclusion would be a hierarchical variety of Unmoved Movers. For his part, Jaeger considers that the relation between the Prime Mover and the many movers is completely obscure (*Aristotle*, p. 351). Owens ("The Reality of the Aristotelian Separate Movers," *Review of Metaphysics* 3, 1949/1950: 333) sustains that all forms moving the celestial bodies must be different from one another and do not belong to any genus; Merlan ("Aristotle's Unmoved Movers," *Traditio* 4, 1946: 1–30) denies that each Unmoved Mover is a species within a genus, and draws a parallel with Plato's Ideal Numbers: each form a series such that the main relation is one of priority and posteriority (i.e., the former is necessary to the existence of the next). Following Merlan's argument part way, Harry Wolfson ("The Plurality of Immovable Movers in Aristotle and Averroes," *Harvard Studies in Classical Philology* 63, 1958: 242–245) argues that the inferior movers are one in species but (though they have no matter) not in number, differing from one another in terms of their priority and posteriority within the system, and considers unfeasible the Neoplatonic interpretation that postulates the existence of emanations between Unmoved Movers. At this point, and given our own explanation of

That said, how many movers, wonders Aristotle, do in fact draw the celestial spheres in rotation? He believes that it is through the discipline of astronomy, being the mathematical science most akin to philosophy (1073b6–8 –§ 15–), that this number can be discovered. Moreover, it is evident for him, following Eudoxus and Callippus, that the number of eternal movements is greater than the number of eternal entities—namely, the planets—that are moved (πλείους τῶν φερομένων αἱ φοραί … πλείους γὰρ ἕκαστον φέρεται μιᾶς τῶν πλανωμένων ἄστρων; 1073b8–11 –§ 15–).

The hierarchical inferiority of the Unmoved Movers with respect to the Prime Mover (and also of the innermost movers with respect to the nobler outer movers) is clearly stated by Aristotle in *On the Heavens*, II, 12, 291b29–293a12 –§19–. There, he conceives the multiplicity of spheres and motions corresponding to each celestial body as the consequence of the planet's desire to reach the state of harmony and plenitude of the perfect motion of the fixed stars. Here, the Prime Mover is conceived as a kind of *attractor*, so to speak, of all celestial movements. In this lies a distinctive feature of the Aristotelian system: its hierarchical teleology, in which the end of the motion of celestial spheres depends on the end *par excellence*: the Prime Mover, the first cause of all change.

Having considered this, let us see now how Aristotle builds his unified view of the heavens, successfully integrating Callippus's astronomical systems into a single scheme. The key idea consists in linking the Callippic planetary subsystems to one another while maintaining the isolated motions, as Callippus conceived them, that each system transmits to its associated planet. It is evident that if these systems were simply linked to one another following the established celestial order, from Saturn to the Moon, they would each—except for Saturn—cease to function properly. For example, linking the first sphere of Jupiter's system directly to the innermost sphere of Saturn's system would cause the transmission of the motions of Saturn to the orbit of Jupiter, diverting that body from its proper trajectory as imagined by Callippus, and which Aristotle wished to preserve. If the integration was to maintain (from a formal point of view) the operational independence of Callippic subsystems, the specific motions of each planet had to be counteracted so as to preserve their shared diurnal motion, which is also that of the fixed stars, and which

sphere order (see the section "Two Celestial Systems"), we should raise the issue of whether the hierarchy is continuous from sphere two to sphere fifty-five, that is, from the outermost to the innermost sphere, or whether the first spheres of Jupiter, Mars, Mercury, etc. should be designated as the second, third and fourth noblest. The interpretation becomes even more problematic if we try to arrange the fifty-four spheres according to the primacy of their motions.

is produced by the first sphere of each Callippic subsystem. Aristotle realised that in order to achieve the joint operation of these planetary subsystems as conceived by Callippus, he simply had to introduce between each subsystem a number of counteracting or rewinding spheres equal to one less than the number of forward-revolving spheres in the upper subsystems. This we can read in the only remaining Aristotelian fragment available to help reconstruct his celestial system:

> But it is necessary, if all the spheres combined [i.e. the spheres of Callippus's planetary subsystems] are to explain the phenomena, that for each of the planets there should be other spheres (one fewer than those hitherto assigned) which counteract those already mentioned and bring back to the same position the first sphere of the star which in each case is situated below the star in question; for only thus can all the forces at work produce the motion of the planets.[4]

According to this description, the added rewinding spheres interposed between the Callippic subsystems—which mirror the corresponding forward-revolving spheres (henceforth simply "revolving spheres") of the upper celestial body—rewind and cancel the motions of the revolving spheres one by one until the first sphere of the lower subsystem is restored to its original position. And this restored position must be the exact position and orientation of each of the first spheres of the Callippic subsystems when viewed in isolation. This position is determined by the orientation of the axis of rotation belonging to the sphere of the fixed stars. In the Aristotelian conception, this sphere is identified exclusively with the first sphere of Saturn and, as a result of his integration, the first spheres of each of Callippus's subsystems are also aligned with the orientation of the first sphere of Saturn. The following figure—generally regarded by ancient and modern interpreters as the most appropriate representation of *Metaphysics* Λ, 8—shows the system that most reasonably complies with Aristotle's description.

If conceptually simple enough, an analysis of the Aristotelian system of the heavens nevertheless poses several problems. First, it raises the following questions about the systems of Eudoxus and Callippus: is the first sphere of Saturn alone to be identified with the sphere of the fixed stars or, rather, is each of the first spheres one and the same with the sphere of the fixed stars, replicated out of necessity for the proper composition of each of the subsystems?

4 *Metaphysics*, Λ, 8, 1073b38–1074a5 –§ 15–; English trans. Ross in Barnes, *Complete Works of Aristotle*, vol. 2, p. 1697.

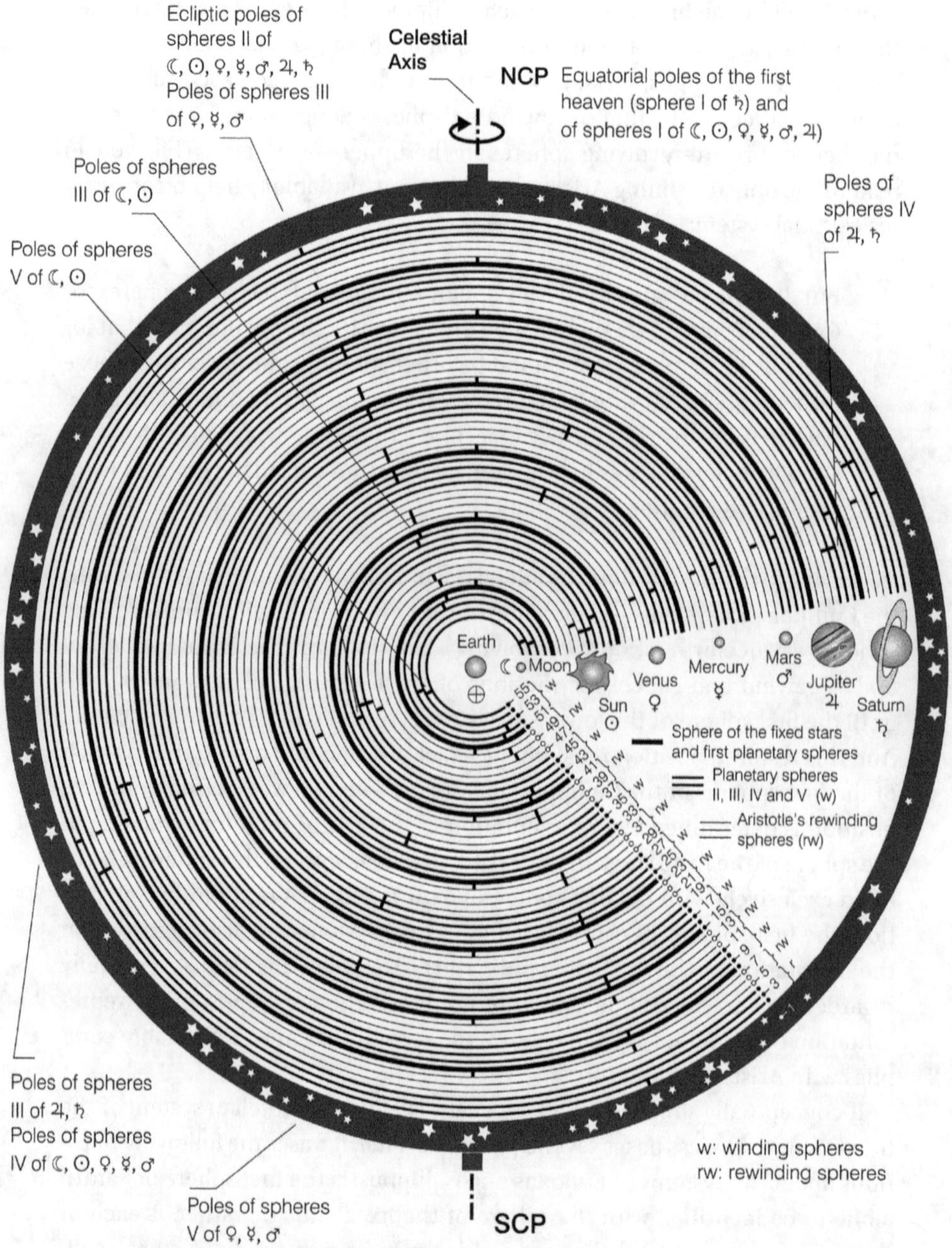

FIGURE 30 Aristotle's astronomical system. The system brings together the subsystems of Eudoxus and Callippus by interposing rewinding spheres that counteract the motion of the upper planetary spheres, thus preserving the motion of the first sphere of each subsystem, which will have same revolution period as the sphere of the fixed stars. This system of fifty-five spheres adequately reproduces the apparent celestial motions (motions a, b, and c).

Second, a specific query arises regarding the integrated system: What constitutes, according to Aristotle, the link between the sphere of the fixed stars (the first sphere of Saturn's subsystem) and the first spheres of each of the remaining subsystems, from Jupiter to the Moon? And finally: Is it feasible, this composition of a single system, constructed by integrating Callippus's geometrical subsystems and then adding the precise number of rewinding spheres indicated by Aristotle? Some interpreters (Norwood Russell Hanson stands out as a representative of these) consider that the system supposedly imagined by Aristotle would not function as he intended.[5]

In the section "The Integration of Planetary Spheres" later in this chapter, we discuss in more detail the integration of spheres in the Aristotelian system and offer a schematic chart showing, according to different interpretive approaches to the questions raised above, possible variants with respect to the total number of spheres corresponding to the systems of Eudoxus, Callippus, and Aristotle. Each of these models, developed from the core of Λ, 8, seek to shed light on this highly hermetic text. The text itself presents internal difficulties but also some seeming contradictions with respect to Λ, 7 and Λ, 9–10, allowing for—and even demanding—imaginative interpretations.

Our attempt to deal with the difficulties presented by the Aristotelian system of the heavens demands a clarification of the principles of Aristotle's physics and metaphysics. This is necessary in order to understand both the extent and the basis of his astronomical ideas as manifested in Λ, 8. In the ensuing sections, we will cover these aspects and gradually approach the reconstruction of a multi-faceted system that so far we have only presented in a structural manner. In so doing, we shall discuss the main objections to the system as regards not only its physical-astronomical aspects, but also its metaphysical elements, and will try to show, finally, its feasibility and coherence with the central ideas of Aristotelian thought.

5.1 The Prime Mover and Unmoved Movers

Aristotle's supposed affirmation of the existence of a single Unmoved Mover (Λ, 7, 9–10) that is, at the same time, several (Λ, 6 and Λ, 8), and that causes the motion of all things, turns out to be particularly troublesome for the argumentative coherence of *Metaphysics*, Λ. This ambiguity had already been challenged by Neoplatonic thought and, especially, by Christianity, which aimed to emphasise an idea of the single Prime Mover that could be assimilated to

5 Hanson, *Constellations and Conjectures*. Edited by Willard C. Humphreys, Jr. Dordrecht/Boston: D. Reidel, 1973.

the Judeo-Christian God and would ensure the intelligible harmony of nature. Moreover, other interpretations of *Metaphysics*, Λ have tended to minimise the *plurality* of Unmoved Movers, clearly postulated by Aristotle in Λ, 6 and especially in Λ, 8, therefore overlooking some central *physical* and *astronomical* aspects of Aristotelian thought. The analysis of these aspects, however, reveals that it is possible to reconcile the Aristotelian need for a unique Prime Mover (hierarchically and ontologically pre-eminent) with the existence of other movers of a similar nature in order to account for the organised complexity of φύσις, as well as for the phenomenal expressions of it. Furthermore, this affirmation rests on a subtle argument that we have only just touched on thus far, namely, that a multiplicity of natural entities requires the existence of at least two Unmoved Movers to rupture the *absolute sameness* that would result if just one intelligible principle were to cause all motion.

The main purpose of this section is to stress the *need* for several movers in the economy of Aristotle's physical system. Indeed, the *plurality* of Unmoved Movers (whose number will be fixed at fifty-five for the purpose of this exposition, though the number has been extensively debated and was even presented in tentative terms by Aristotle) is far from being the product of an extemporaneous introduction of concepts foreign to book Λ, as Jaeger hypothesises. Rather, it is a core element of Aristotelian metaphysics, as the passages reviewed in the following pages will show.[6]

As we know, the number of Unmoved Movers is intimately related to the account of planetary motions. Aristotle's worldview considers that the Prime Mover—the origin of all change—is the direct cause of the first eternal motion of the sphere of the fixed stars, which consists in a simple rotation, and the indirect (mediated) cause of all other motions in the universe. The first simple motion determines planetary motions and is the ultimate cause of all change in the sensible realm, as its action affects the entire superlunary and sublunary domains.

Many are the issues raised by Λ, 8, the most significant being that Aristotle would seem to affirm, simultaneously, the existence of a single Unmoved Mover (Λ, 7 and 9–10), but also several (Λ, 6 and 8) Unmoved Movers. The affirmation of multiple Unmoved Movers poses the further difficulty (especially from the viewpoint of a systematic understanding of Aristotelian metaphysics) of what would be, then, the principle of individuation among these immaterial beings, given that Aristotle explicitly states that the only principle of individuation

6 *Metaphysics*, Λ, 2, 1069b22–28 and Λ, 6, 1072a4–17.

for formal or specific things is *matter*.[7] Jaeger attributes this apparent inconsistency to the fact that Λ, 8 was written later than the rest of *Metaphysics*, Λ. There could, therefore, have been a turn in Aristotelian thought in the time that passed between the composition of these chapters of Λ (at least as they are in the current canonical edition).[8] Another line of interpretation, whose representative figure is Düring, disagrees with Jaeger and believes that book Λ is a perfectly coherent unit written together at one time, probably while Aristotle was still at the Academy, and composed in part to challenge the cosmology of the *Timaeus*.[9]

As regards the problem of adding several Unmoved Movers, Jaeger says:

> Either all these intelligible essences must arise from the first, and must, just as the spheres which they move fit into the outermost sphere and are governed by it, be contained in the highest *Nous* as its objects, which would give an intelligible world like Plato's, or each of them must be an independent principle, and if so there is no order or structure among them, and they cannot explain the symphony of the cosmos.[10]

These difficulties had already been pointed out by Theophrastus (*On First Principles*, 1, 4a1–8, 5a21 –§ 20–), and later by Plotinus (*Enneads*, V, 1, (8), 15–(9), 27 –§ 22–), who added a further difficulty to the individuation of transcendent movers: if a stratified or ordinal relation among the intelligible principles were to be established, there would be no alternative but to arrive at a Neoplatonic-like hierarchy, as indicated in *Enneads*, V, I, 9, 15–27 –§ 22–:

---

7 We understand here by "systematic" the intention, repeated again and again in the Aristotelian tradition, to seek integral coherence in Aristotle's thought. This concept does not allude here to the fact that Aristotle was a systematic thinker (there is no doubt that he was such to a very high degree!). It seeks, rather, to point out those approaches to his work that privilege, over other possible perspectives of analysis, the search for a fully systematic system.

8 Jaeger, *Aristotle: Fundamentals of the History of His Development*. Translated by Richard Robinson. Oxford: Clarendon Press, 1934. William Guthrie shares this view, though with some reservations in "The Development of Aristotle's Theology—I," *The Classical Quarterly* 27, no. 3/4 (July–October 1933), pp. 162–171.

9 Ingemar Düring, *Aristoteles: Darstellung und Interpretation seines Denkens* (Heidelberg: C. Winter, 1966); on the timeline of *Metaphysics* Λ, pp. 189–194; on the challenge to Plato's *Timaeus*, pp. 231–235; Philip Merlan also believes that *Metaphysics*, Λ was written all together, at once ("Aristotle's Unmoved Movers," *Traditio* 4, 1946, pp. 1–30).

10 Jaeger, *Aristotle*, p. 351.

> And one might enquire whether Aristotle thinks that the many intelligibles derive from one, the first, or whether there are many primary principles in the intelligible world; and if they derive from one, the situation will clearly be analogous to that of the heavenly spheres in the sense-world, where each contains the other, and one, the outermost, dominates; so that there too the first would contain the others and there will be an intelligible universe; and, just as here in the sense-world the spheres are not empty, but the first is full of heavenly bodies [the fixed stars] and the others have heavenly bodies [the planets] in them, so there also the moving principles [the Unmoved Movers] will have many realities in them, and the realities there will be truer. But if each is primary principle, the primary principles will be a random assembly; and why will they be a community and in agreement on one work, the harmony of the whole universe? And how can the perceptible beings in heaven be equal in number to the intelligible movers? And how can the intelligible even be many, when they are incorporeal, as they are, and matter does not divide them?[11]

As suggested by Jaeger, if the principle of individuation is matter (a doctrine often reiterated by Aristotle), then either the movers of spheres cannot be immaterial, because they would compose a plurality of specimens of the same genus, or Aristotle contradicts himself when he maintains the immateriality of movers, which would *eo ipso* render their multiplicity unfeasible.[12] Despite Jaeger's assertions, the multiplicity of immaterial movers is indisputable, because these immaterial principles constitute the specific character of Aristotle's physics and metaphysics, as the reasons provided below will demonstrate. In addition, the plurality of immaterial entities is hardly an inconsistency caused by a turn in Aristotle's thought during his final years, as Jaeger believes. The idea of an eternal, unchanging, uncreated heaven made of several motions that can be accounted for by a set of superimposed homocentric spheres—thus resulting in the need for several Unmoved Movers as stated in Λ, 8—is also present in other treatises: *On the Heavens*, II, 12, 292b4–12 –§ 19–, and *On Generation and Corruption*, II, 10, 337a17–22 –§ 18–, which Jaeger himself regards as early Aristotelian works.[13]

---

11 Plotinus, *Enneads*, V, 1–9; English trans. A. H. Armstrong, p. 43.

12 Jaeger, *Aristotle*, p. 352.

13 Jaeger places the writing of *On the Heavens* and its appendix, *On Generation and Corruption*, shortly after Plato's death, at the time Aristotle started to develop his own physical-cosmological system, freeing himself from the mythical pre-Socratic notions of nature and challenging the mainstream Platonic cosmology within the Academy (Jaeger,

We should remember that Aristotle does not arrive at the conclusion of multiple Unmoved Movers exclusively by way of metaphysical exploration; rather it is the conclusion demanded by his own cosmology. Far from being based on speculative grounds, Aristotelian cosmology assimilates the astronomical knowledge of his predecessors, which, according to the Stagirite, is based on principles derived from experience: observation shows that celestial motions are both unceasing and multiple.[14] The need for several Unmoved Movers derives exclusively from the multiplicity of eternal motions. According to Aristotle's worldview, this results in separate substances of the same condition—immaterial, eternal, unchanging—that operate on different strata of the superlunary world producing, by means of celestial spheres, the celestial motions that astronomy explains through the study of phenomena.[15]

The argument that there can be no principle of individuation allowing for a multiplicity of immaterial principles—that is, based on Aristotle's belief that the principle of individuation is material—does not deny in and of itself the multiplicity of immaterial principles as a means for explaining celestial motions. That Aristotle was not explicit about the principle of individuation for the Unmoved Movers—unlike in his explanation of physical entities—does not mean that we should deny the existence of such a principle, nor that we should extrapolate a notion developed in a different context (that is, matter as a principle of individuation) to a context or plane in which it does not apply.

Aristotle, pp. 306–308). In contrast, Düring considers them earlier works written—together with books I–VII of *Physics* and IV of *Meteorology*—between 355 and 347 BCE, the year of Plato's death (Düring, *Aristoteles*, pp. 218–220). Although Aristotle probably revised and slightly modified *On the Heavens* and *On Generation and Corruption* near the end of his second Athenian period, as suggested by Düring (*Aristoteles*, pp. 348–349), in *On the Heavens*—which offers, as in *Metaphysics*, Λ, 8, clear evidence that Aristotle, following the Eudoxian systems, considered celestial motions to be the result of circular motions caused by homocentric spheres—we find a *terminus post quem*. It appears in one of the few passages in which Aristotle narrates his personal experience (*On the Heavens* II, 12, 292a5 –§ 19–), and even if it cannot prove the early production of that section of the treatise, it validates its possibility: Aristotle here claims to have seen the concealment of Mars behind the Moon, a phenomena that—Kepler tells us—would have occurred for the last time during Aristotle's lifetime in 357 BCE, when Aristotle was 28 (Düring, *Aristoteles*, pp. 346–347).

14 *Prior Analytics*, I, 30, 46a17–21.

15 Joseph Owens ("The Reality of the Aristotelian Separate Movers," pp. 322 and 328–329) notes the possibility—though he deems the idea unfeasible—that Unmoved Movers could simply be thoughts (of another thinking substance). Would such thoughts, he wonders, be the product of a celestial soul taking part in some kind of activity (something like the Platonic World Soul)? Or perhaps of the Prime Mover itself? Notwithstanding these unresolved questions, the notion that Unmoved Movers are thoughts of a different substance opens several lines of interpretation, and for that reason is worth keeping in mind.

Unless we can affirm that there is just one Unmoved Mover, which would render the principle redundant, the question remains open.

In addition, we should regard the hierarchic and distinctive characterisation of the movers as an original Aristotelian development, which, even if undeniably influenced by the ideas of Plato, is in no way a mere continuity of such thought, as some of the interpretations and criticisms reviewed seem to suggest.

We believe that for Aristotle it is the physical manifestation of these movers that reveals their own specificity, as it is within the context of astronomy that we find both the proof of their existence and Aristotle's metaphysical inquiry into them. The Unmoved Movers relate to one another and manifest in the sensible world in a concrete way, much more *physical* than the Platonic Forms. The "mover" action performed by these immaterial entities, different for each, expresses their essence and place in the hierarchy. The very first of all, the Prime Mover, is the most powerful, causing the fastest motion[16] and carrying the most bodies, while the motions produced by the remaining Unmoved Movers have different speeds and orientations. The spheres moved by these movers and the celestial bodies moved in turn by superimposed motions are spatially ordered according to the hierarchical place of the mover.[17] The identification of each mover and its place in the cosmic ranking derives from astronomical study, as the particular and single motion caused by each is the sensible expression of its specificity.

In order to understand the physical-metaphysical problematic from which arises the need to introduce multiple Unmoved Movers in the account of celestial motions, we first have to bear in mind that, for Aristotle, there are three kinds of entities or substances (Λ, 1069a30–34, 1071b4–5): a sensible eternal entity (celestial bodies); a sensible entity subject to generation and corruption (μία μὲν αἰσθητή –ῆς ἡ μὲν ἀΐδιος ἡ δὲ φθαρτή); and an unmoved entity (ἄλλη δὲ ἀκίνητος). The first two have matter, though the eternal entity has a kind of matter that is transferable but not generable (ἀλλ' οὐ γενητὴν [ὕλην] ἀλλὰ ποθὲν ποί; Λ, 1069b25–27). These are both studied by physics (Λ, 1069b1), while inquiry into the unmoved entity is undertaken through the "first philosophy," later called metaphysics. This classification of stars and the celestial spheres as eternal entities places astronomy—despite its being, properly speaking, a physical-mathematical discipline[18]—between the first philosophy and physics, constituting a link between the two: celestial bodies share with the object

16 *On the Heavens*, II, 12, 292b27–293a4 –§ 19–.

17 *Metaphysics*, Λ, 8, 1073a36–1073b4 –§ 15–.

18 *Physics*, II, 2, 194a7–13.

of metaphysics the condition of the eternal, and share with the object of physics the aspect of sensible matter. It is not surprising, then, to find a twofold problem at the heart of Aristotelian astronomy: on the one hand, the justification of the *eternity of motion* and, on the other hand, the basis for the *diversity of entities* in the universe, also everlasting. As we will see, these problems are directly associated with the need for numerous Unmoved Movers.

We should remember that this need emerges as a corollary of physical investigation, as revealed in key passages of the *Physics* and *Metaphysics* as well as in the Aristotelian treatises on heavenly bodies, meteorology, the movement of animals, and generation and corruption, in which he analyses motion at large, particularly heavenly motions. To understand the indissoluble relationship between physics and metaphysics exhibited in Aristotelian cosmology, we must also comprehend the scope of his notion of nature (φύσις), which defines the realm of physics as a particular science. According to Aristotelian thought, the concept of "nature" entails, in each entity, an inclination (τέλος) towards the full realisation of its form (μορφή, εἶδος), which is identified with its essence (τὸ τί ἦν εἶναι): the ultimate and intimate metaphysical principle of specific entities. This natural inclination is a principle of both motion (κίνησις) and rest (στάσις).[19] For Aristotle motion (κίνησις) is essentially the passage from potentiality (δύναμις) to actuality (ἐνέργεια, ἐντελέχεια), which involves not only change of location (displacement) but also change of quantity (increase and decrease), change of quality (e.g., of colour) and especially change in substance (generation and corruption). For change in substance to happen, for instance the generation of a child, the *potential* child must come to be by virtue of an *active* principle, in this case, the parents' act of generation and the gestation. In a similar way, change of location requires a mover that, working as an active principle, moves—due to its actuality—the thing moved. That said, Aristotle distinguishes the change of place from all other relative changes and considers it a primary motion (πρώτη).[20] He regards it as ontologically prior to any other motion (ἄλλαι κινήσεις ὕστεραι τῆς κατὰ τόπον; Λ, 8, 1073a12 –§ 15–).

The ontological priority of local motion (motion in a "where") is the physical theoretical basis that, seeking to explain the *cause* of the unceasing rotational motion of the Whole, demands the existence of a Prime Mover as the origin and kinetic principle of any other local motion and, therefore, of all change, generally speaking. The priority of local motion is stressed in *Physics*, VIII, 7, 260b30–261a17, where Aristotle affirms that the *temporal* priority of translational motion over other motions or changes, as well as in *On Generation*

19 *Physics*, II, 1, 192b8–24.

20 *Physics*, VIII, 9, 265b17–266a4.

and Corruption, II, 10, 336a19–23, where he postulates the temporal priority of translational motion over generation.

We will look now at the main reasons why local motion is viewed as the first and most important kind of change. The fact that φύσις is motion, broadly speaking, and that it has always existed and always will, is not particularly challenged by Aristotle. Instead, he uses this statement as a starting point for his physical-metaphysical reasoning.[21] The key problem is not to prove the existence and eternity of change but to explain *how* the eternal motion responsible for an everlasting nature is even *possible*. That said, in *Physics* Aristotle proves that, of all motions involved in change (with the exception of the first motion),[22] only the circular motion[23] of the stars can be continuous and eternal without creating contradictions. Therefore, such motion should be the cause and mover of all other particular change and, consequently, change generally speaking is also eternal. Even absolute change in substance (generation and corruption), which begins by virtue of local motion (as do quantitative and qualitative changes),[24] mirrors circular motion as it is cyclical.[25] In addition, the number of the elements (five in all, four belonging to sublunary physics and one to the superlunary) is determined by this condition.[26] As a result, translational motion, and therefore circular motion, due to its eternity, has ontological pre-eminence because it is the only motion that can cause the persistence of all other change. The Aristotelian notion of φύσις rests, then, on the affirmation that there is a local, continuous, and eternal motion that can only be realised in the stars, a motion that is confirmed through the empirical evidence of celestial motions. In a certain sense, the Stagirite identifies this motion with time.[27]

Astronomy is, then, the science to which corresponds the problem of motion when understood as local, circular, uniform, and eternal but not when

21 Concerning motion, Aristotle says: "the *existence* of motion is asserted by all who have anything to say about nature" (εἶναι μὲν οὖν κίνησιν πάντες φασὶν οἱ περὶ φύσεώς τι λέγοντες; *Physics*, VIII, 1, 250b15–16; English trans. R. P. Hardie and R. K. Gaye in Barnes, *Complete Works of Aristotle*, vol. 1, p. 418. Further on he adds that: "there never was a time when there was not motion, and never will be a time when there will not be motion" (ὅτι μὲν οὖν οὐδεὶς ἦν χρόνος οὐδ' ἔσται ὅτε κίνησις οὐκ ἦν ἢ οὐκ ἔσται, εἰρήσθω τοσαῦτα; *Physics*, VIII, 1, 252b5–6; trans. ibid., vol. 1, p. 420).

22 *Physics*, VIII, 8 and 9.

23 For a unique argument justifying the primacy of curved things in the universe, see *On the Heavens*, II, 4, 286b10–26 – § 6–.

24 *Physics*, VIII, 7, 260b7–14.

25 *On Generation and Corruption*, I, 3, 319a23–28.

26 *On Generation and Corruption*, II, 4, 331b2–3; II, 5, 332a30–332b4.

27 *Metaphysics*, Λ, 6, 1071b7–12.

considered in its broader meaning, identified with the notion of φύσις, whose fullest expression is found in the sublunary domain. Having established the phenomenal evidence and the metaphysical need for a continuous eternal motion identified with the motion of the stars, *Metaphysics*, Λ, 6 and 7 raise a specific issue: How is such motion possible on the basis of the act-potency scheme? Aristotle does not solve this difficulty in the above-mentioned texts but he discusses it at length in his physical treatises, particularly in *Physics*, VII and VIII.

Because Aristotle refuses to accept remote non-mediated action, every motion must result from the action of a mover in immediate *contact* with the object moved (VII, 2, 243a32–35 –§ 34–). He reasons as follows: either the mover moves by itself or is moved by another (VII, 2, 243a11); if moved by another, this necessarily leads to the postulation of an indefinite number of movers and moved. Aristotle provides the example of a stone moved by a stick, which is moved by a hand, which is moved by a person (VIII, 5, 256a3–8), which is moved by desire but also by the conditions of the surrounding medium (VII, 6, 259a32–259b14). As each term of such a mover-moved series involves a finite spatial motion that is distinct from all others, and as every motion must be simultaneous (so that the entire motion happens over the same period as that of any term in the series), if the series is infinite, the sum of all the motions—Aristotle concludes—would result in infinite displacement over a finite period, which is impossible (VII, 1, 242a49–242b53). Consequently, *all the series* must necessarily *share* an original mover to account for their motions and this mover must necessarily move without being moved, being therefore an unmoved mover.

Because, as we have seen, the first motion has to be unceasing, and because this is only possible in the *circular* motion of the heavens, Aristotle locates its first physical manifestation in the "limit of the universe,"[28] from where all things are embraced and dominated by the regular eternal motion of the first heaven. The *first object moved* or *primum mobile* (moved by the Prime Mover),

28 Given that Aristotle necessarily regards the limit as the end of something and the beginning of something else, we cannot assign physical meaning to this "limit of the universe," as what is limited here is the totality of being. Concerning the limit, Aristotle states: "No determinate divisible thing has a single termination, whether it is continuously extended in one or in more than one dimension" (*Physics*, IV, 10, 218a23; English trans. R. P. Hardie and R. K. Gaye in Barnes, *Complete Works of Aristotle*, vol. 1, p. 370). If we apply the double limit to a spherical body limited by a single surface, such surface, in so far as it works as a limit, should be reachable by two paths (hence the double limit), which are precisely inside and outside the body. Clearly, this would not be possible for Aristotle with respect to the "limit of the universe" as there is no being outside the universe.

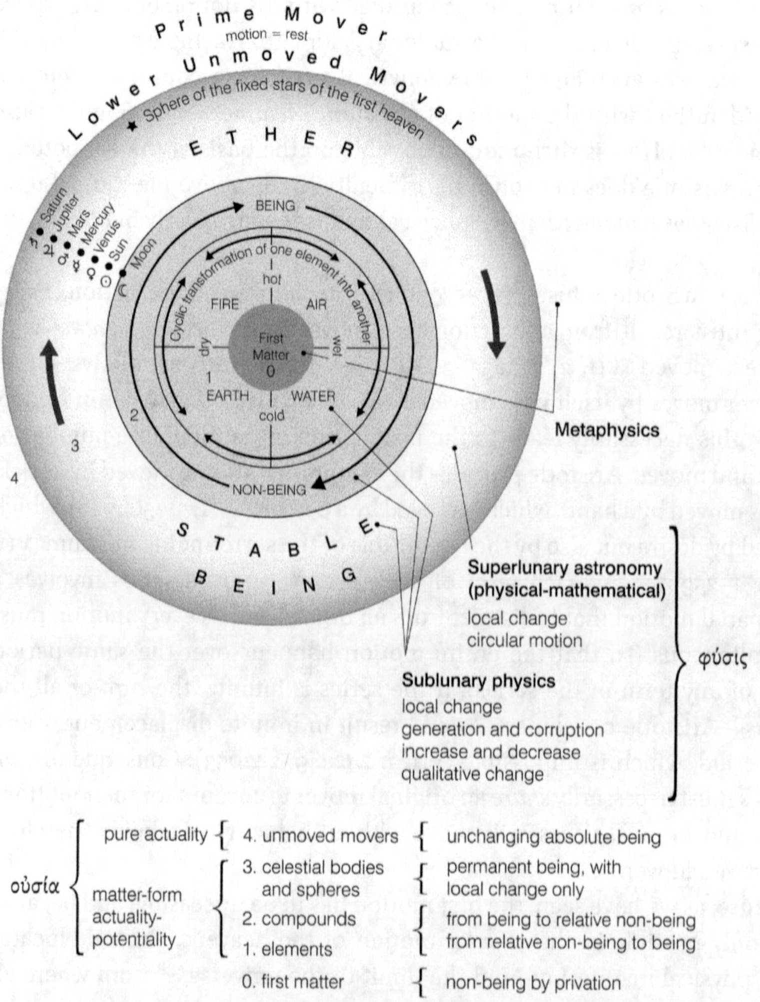

FIGURE 31 Aristotle's circular cosmology. The Φύσις, in all its manifestations, exhibits a circular motion that mirrors the first eternal motion of heavens. In the superlunary domain, celestial bodies are changeless in their being, but change their position due to the superimposition of circular motions in the surrounding ether. In the sublunary domain, entities change from being to relative non-being, and from relative non-being to being. The most fundamental expression of this dynamic is the cyclical transformation of the elements, though Aristotle considers that the cross-transformation of elements (i.e., air-earth and fire-water) is also possible, if more difficult to realise. We have introduced the notion of "first matter" in our representation of the Aristotelian cosmology because, though unknowable, it is a metaphysical reality—allowing of no separation in ontological terms—that penetrates the intimate structure of being and is, for Aristotelian thought, the ultimate limit of intelligible reality.

which Aristotle identifies with the sphere of the fixed stars, belongs to the sensible order (though with the highest degree of perfection) and cannot be the original mover, strictly speaking, because it is a material entity.[29] The Prime Mover—being the cause of continuous eternal motion—must move forever and in the same fashion, thus lacking the potentiality that is proper to matter. That which has potentiality cannot remain forever the same,[30] so, for Aristotle, the Prime Mover needs to be pure immaterial actuality.

A main principle of Aristotelian thought is that the indefinite persistence of motion—a change that is the very essence of nature and, therefore, of all entities—rests primarily on the local motion of the first heaven, which is continuous and circular, a motion resulting from the attraction exerted by the ontological plenitude of the Prime Mover (*Metaphysics*, Λ, 8, 1074a38 –§ 15–). Aristotle defines the Prime Mover (πρῶτον κινοῦν) as an eternal (ἀΐδιος) unmoved (ἀκίνητος) entity (οὐσία) separated from sensible things (κεχωρισμένη τῶν αἰσθητῶν), impassive (ἀπαθὲς), unalterable (ἀναλλοίωτον; 1073a7–11 –§ 15–), and ultimately pure actuality or ἐντελέχεια as it is completely immaterial (οὐκ ἔχει ὕλην; 1074a36–37 –§ 15–).[31]

The existence of a *unique* entity that, without being moved, moves as that which is loved, as an object of love, (κινεῖ δὴ ὡς ἐρώμενον; 1072b4) does not ensure the *plurality* of the real. Only the eternity and continuity of the first heaven's circular motion (the necessary cause of natural motions) has been guaranteed thus far, but this is not enough to explain the *diversity of entities*, which, according to Aristotle, is equally imperishable. Nature cannot be fully justified by the notion of the Prime Mover alone, which is characterised by the principles of motion and rest. In Aristotelian teleology every motion has an end that it eventually fulfils, even if it does so imperfectly due to its material burden; the principle of rest manifests itself in the variety of entities that have form (οὐσία) in each present, just as motion reveals itself when one entity becomes another.

We will briefly interrupt our discussion of this aspect of Aristotelian thought to illustrate, by analogy, the (physical-metaphysical) problem created for Aristotle by the idea of the existence of a *single Unmoved Mover*. Classical

29 Even though ether becomes matter only for local changes, it is not without potentiality.

30 *Metaphysics*, Λ, 6, 1071b17–21.

31 Because change is the "fulfilment of a potentiality," if there ever was a *first change*, then there should had been something prior capable of changing and something else before that capable of producing such change. If the causes of change are to be finite, as Aristotle intended, the *first cause* capable of changing other entities cannot be subject to the process of becoming, thus lacking potentiality. Otherwise, one would have to postulate a prior cause *ad infinitum* (that the Prime Mover has potentiality is, for Aristotle, unfeasible).

physics once faced a similar problem in the field of thermodynamics, before the development of theories on the evolution of complex systems far from thermodynamic balance. Thermodynamics had previously rested on two essential principles, namely the first and second principles of thermodynamics, summarised in Clausius's famous cosmological statements (1865): the first being that "the energy of the universe is constant" (*Die Energie der Welt ist konstant*) and the second, that "the entropy of the universe tends to a maximum" (*Die Entropie der Welt strebt einem Maximum zu*). The first is a principle of *conservation* that can be equated to the "finitude of the universe," in Aristotle's physics. The second indicates that all things *evolve in a single direction* (expressed by a continuous increase of entropy in the universe), which can be equated to the tentative notion (ultimately dismissed by Aristotle) of a single Prime Mover as final cause. This tendency, whether expressed in terms of the existence of a single Unmoved Mover or ever-increasing entropy, would necessarily lead to the extinction of variety or, in the field of thermodynamics, to what is called the "heat death of the universe," a state in which the Whole becomes homogeneous and isotropic and, if being a spherical continuous Whole, a perfect sphere whose rotation would eventually be identical to its state of rest.[32]

Aristotle is aware of the difficulty brought about by a *single active principle*, which naturally leads to a *single tendency* in nature, as shown in his critique of Anaxagoras's physics. According to the Stagirite, allowing this would mean that a representation of the origin of the cosmos would be identical to that of its end: an ultimate state of disorder comparable to an original chaos similar to Anaxagoras's mixture of homoeomerous substances in which—both Aristotelian and nineteenth-century physics agree on this point—the Whole is unable to generate variety by itself.

While commenting on the limitations of Anaxagoras's thought, Aristotle argues for a necessary plurality of movers. As we know, he draws an analogy between his hylomorphic conception of entities and Anaxagoras's physics,[33] and gives the philosopher credit for having envisaged the Aristotelian theory of matter (ὕλη). He even identifies Anaxagoras's Reason (Νοῦς) with the original pure act: that which moves generation and is ultimately responsible for the actuality of the form (μορφή, εἶδος) of specific entities, the act required for the actualisation of potentiality in material things. That is why Anaxagoras's impassive (ἀπαθῆ) unmixed (ἀμιγῆ) Reason—being the ultimate cause that accounts

32 As all points equidistant in every direction from the axis of rotation of a perfect, homogeneous, and continuous sphere are identical, the sphere is invariant under axial rotation, whereby its resting state is indistinguishable from its rotation.

33 *Metaphysics*, A, 3, 989b31–21; Λ, 2, 169b15–23.

for reality—can be equated with the Prime Mover.[34] Aristotle's criticism of his predecessor is that if Reason is one, it cannot explain the *multiplicity* of being realised in matter. In Aristotelian terms, this would mean that two principles (Reason, equated with the Prime Mover; and the mixture of homoeomerous parts, equated with Aristotelian matter) are not enough to explain the continuous becoming of specific entities. In his own words:

> nor is it satisfactory to say that all things were together [ὁμοῦ πάντα]; for they differ in their matter, since otherwise why did an infinity of things come to be, and not one thing? For Reason is one [νοῦς εἷς], so that if matter also is one [ὕλη μία], that must have come to be in actuality what the matter was in potentiality. The causes and the principles, then, are three [τρία δὴ τὰ αἴτια καὶ τρεῖς αἱ ἀρχαί], two being the pair of contraries [ἐναντίωσι] of which one is formula [λόγος] and form [εἶδος] and the other is privation [στέρησις], and the third being the matter [ὕλη].[35]

Here, he reinforces the idea that there must be at least three principles (form, matter and privation) as already established in *Physics*, I. Later, in Λ, 6, he further develops this idea according to his kinetic understanding of generation and corruption, stating the need for at least *two* separate movers—along with a material principle—to justify the vibrating polychrome becoming of the sublunary world, which is the realm of otherness *par excellence*:

> That actuality is prior is testified by Anaxagoras (for his thought is actuality) and by Empedocles in his doctrine of love and strife, and by those who say that there is always movement, e.g. Leucippus. Therefore chaos or night did not exist for any infinite time, but the same things have always existed (either passing through a cycle of changes or in some other way), since actuality is prior to potentiality. If, then, there is a constant cycle, something must always remain, acting in the same way. And if there is to be generation and destruction, there must be something else [in addition to the "Reason" of Anaxagoras or to the Prime Mover] which is always acting in different ways [εἰ δὲ μέλλει γένεσις καὶ φθορὰ εἶναι, ἄλλο δεῖ εἶναι ἀεὶ ἐνεργοῦν ἄλλως καὶ ἄλλως]. This must, then, act in one way in virtue of itself [αὐτό; as an Unmoved Mover], and in another [ἄλλο] in virtue of something else—either of a third agent, therefore, or of the first

34 *Physics*, VIII, 5, 256b22–28.

35 *Metaphysics*, Λ, 2, 1069b29–33; English trans. Ross in Barnes, *Complete Works of Aristotle*, vol. 2, p. 1689.

> [the Prime Mover]. But it must be in virtue of the first. For otherwise this [the Prime Mover] again causes the motion both of the third agent and of the second. Therefore it is better to say the first. For it [the Prime Mover] was the cause of eternal movement; and something else [another Unmoved Mover] is the cause of variety, and evidently both together are the cause of eternal variety.[36]

This complex passage hints at the *need* for several movers. For Aristotle, the existence of a Prime Mover is the first and necessary condition for the continuity and eternity of motion. Up to this point, he is in agreement with Anaxagoras.[37] In the sense of being *one* the Prime Mover and Anaxagoras's Reason are equal; each provides for the very first principle of change and its non-extinction. However, a single and only act, always identical to itself (τὸ αὐτό), cannot cause the diversity of the real, as claimed by Anaxagoras (and as interpreted by Aristotle); there must be some otherness (τὸ ἄλλο), also of eternal motion and pure act, whose very otherness is caused by a pre-existing logical condition of the first pure act or Prime Mover. In other words, that which is different is such by virtue of a previous referential being, the Prime Mover. This point is the basis for the unicity, ontological pre-eminence, and condition of being *one*, of the Prime Mover; it lies not in a condition of being the *only* unmoved entity. "Both [separate entities]," Aristotle states above, "are the cause of eternal variety" (Λ, 6, 1072a15–17). Therefore, an Unmoved Mover as identity giver (a giver of "the Same," in Plato's terms) cannot account for the variety of specific entities evidently present in the sublunary domain. At least one other Unmoved Mover is required to justify the fissures of the Same; in other words, the irruption of the Different (Λ, 6, 1072a4–17).

These considerations are relevant to understand the *need* for multiple movers. Once we agree on the existence of two Unmoved Movers—even if belonging to different ontological levels, one being first and the other second—it is possible to agree, without reservations, on the existence of other Unmoved Movers of the same nature as the second. That is exactly how Aristotle explains, in *Metaphysics*, Λ, 8, the phenomenal evidence of multiple celestial motions supported by previous astronomical observations.[38]

---

36 *Metaphysics*, Λ, 6, 1072a4–17; English trans. Ross in Barnes, *Complete Works of Aristotle*, vol. 2, p. 1693.

37 *Physics*, VIII, 5, 256b25–27.

38 Owens claims that there is no reason to believe that the Prime Mover is superior to the rest of the Unmoved Movers as regards their ontological structure ("The Reality of the Aristotelian Separate Movers," pp. 333–334); he considers that there is only a negative relation among the Unmoved Movers of Planetary Spheres, themselves, in that they differ

Moreover, our interpretation of passages Λ, 2, 1069b29–33 and Λ, 6, 1072a4–17 is consistent with other Aristotelian passages, in which generation and corruption—and consequently the *diversity* of entities—are explained as the result of a conjunction of the equatorial diurnal motion of the Sun (which shares its period with the motion of the first heaven) and the direct ecliptic motion of the star.[39] According to Aristotelian astronomy, the second is originated by the action of an Unmoved Mover other than that causing the daily motion of the celestial body, which imitates the motion of the fixed stars, in turn caused by the Prime Mover. This idea is echoed in *Metaphysics*, Λ, 5, 1071a15–17, where Aristotle attributes to the motion of the Sun—which results from the composition of these two motions—consequent change in the sublunary domain.

In summary, two physical conceptions hold a privileged place in the metaphysical reflections of *Metaphysics*, Λ, 7, 9 and 10: first, the need for a primary continuous, eternal motion, which, as Aristotle has demonstrated in the *Physics*, can only correspond to the circular motion of heavenly matter (ether); and secondly, the fact that such motion has a superior nature, being the kinetic

---

from one another. One could enrich this interpretation by regarding the specificity of each celestial motion as a significant factor or, in other words, by inquiring about the metaphysical specificity of the movers in view of their physical manifestations. Once again, as we noted in regard to the passage on Anaxagoras (*Metaphysics*, Λ, 6, 1072a4–17), the dynamic symphony of reality, both at a micro- and macro-cosmic level, would not be so without the *specificity* of each mover, this otherness being precisely what renders the cosmos possible (even if Aristotle does not provide a metaphysical tool to conceptualise this, namely, a principle of individuation). The differences among the movers should be considered in terms of—at a minimum—their axial tilt, speed, and position within the celestial system (more/less concentric). All of these characteristics become meaningful because they *concern* the Prime Mover, which thus becomes the first referential entity: "That the movers are substances, then, and that one of these is first and another second according to the same order as the movements of the stars, is evident" (*Metaphysics*, Λ, 8, 1073b1–1073b2; English trans. Ross in Barnes, *Complete Works of Aristotle*, vol. 2, p. 1696).

39 In *Physics*, VIII, 6, 259b34–260a11 –§ 13–, Aristotle explains that the persistence of generation and corruption is due to the composition of two different eternal motions: the motion of the fixed stars caused by the Prime Mover and—what we could very well assume to be—the annual motion of the Sun, carried by the former. In *On the Heavens*, II, 3, 286b2–3 and *On Generation and Corruption*, II, 10, 336a30–336b2 –§ 29–, he attributes the cycles of generation and corruption to the motion of the Sun. Ross also understands that the most natural interpretation of the two motions mentioned by Aristotle in Λ, 6, 1072a4–17 is that which consists in identifying—as we have done—the motion of *the Same*—being that which is most regular in the universe—with the diurnal motion of the Sun, which reproduces the motion of the fixed stars, caused by the Prime Mover; and the motion of *the Different* with the annual motion of the Sun along the ecliptic (See *Aristotle's Metaphysics*, ed. Ross, 1924, p. 371).

basis and justification of universal motion. The Aristotelian idea of the Prime Mover being a supreme singular entity and the first condition for things to move stems from this last conception, while the issue of whether there are other movers of the same nature constitutes a pending argument. However, as we have seen, *Metaphysics* Λ, 2 and 6 advance the multiplicity of moving principles as the necessary motions of the Same (τὸ αὐτό) and the Different (τὸ ἄλλο), identified with, respectively, the diurnal westward motion of the Sun (which accompanies the motion of the first heaven), and its annual eastward motion along the ecliptic.[40] Once the unicity of the Prime Mover is understood in this sense, the further consideration of a multiplicity of movers of the same nature—definitively affirmed in Λ, 8—in relation not only to the movement of the Sun but to the totality of celestial bodies, is no longer problematic, as we shall see.[41] This multiplicity of eternal motions attributed to the action of Unmoved Movers is not, indeed, a new turn in Aristotelian thought but was

---

40 Aristotle applies here (in what has been interpreted as the main motions of the sun) practically the same terms as does Plato in the *Timaeus*, 36c. Düring considers this a clear sign that Aristotle wrote the entire *Metaphysics* while at the Academy, as a way of disputing his master's ideas (Düring, *Aristoteles*, p. 210). We should bear in mind that in Plato's main astronomical texts, the *Republic*, 616d–617a and the *Timaeus*, 36b–36d, he distinguishes between the motions of the Same and the Different, giving the former both physical and metaphysical pre-eminence. For Plato, the Same and the Different in astronomy refer to the celestial equator and the ecliptic, respectively. In a physical sense, the outer equatorial circle (namely, the diurnal motion of the heavens) carries all celestial bodies, despite the opposite motion of the Different, shared by the seven wandering stars. Even though the Different moves in the direction opposite to diurnal westward motion (the motion of the Same), it does not entirely counteract this motion, which has a greater angular velocity. Consequently, the Moon, the Sun, Venus, Mercury, Mars, Jupiter, and Saturn appear to us throughout the day as if they were being carried by the motion of the Same. In a metaphysical sense, Plato attributes a higher place to what is prior and regular over what is posterior and irregular, the former being subject to the unchanging principle that governs reality.

41 Based on what we considered to be an overly restrictive interpretation of the passage characterising the Prime Mover as unmoved "either in itself or accidentally" (καθ' αὑτὸ καὶ κατὰ συμβεβηκός; *Metaphysics*, Λ, 8, 1073a24–25 –§ 15–), Wolfson ("The Plurality of Immovable Movers in Aristotle and Averroes," pp. 238–239) argues that the Unmoved Movers of Planetary Spheres move accidentally. We believe his reasoning to be problematic because—in addition to ascribing a physical nature to them—it leans towards conceptualising the Unmoved Movers of Planetary Spheres as souls of the celestial bodies. Actually, Aristotle describes the soul of higher living things as Unmoved Movers (as that which causes the motion of the body is the animal soul's desire), but not in an absolute sense, because souls accidentally move with the bodies to which they are inextricably bonded. Moreover, the idea that the Unmoved Movers of Planetary Spheres can be equated to souls is unfeasible for systematic reasons that will be explained in detail in chapter 8, at least when it comes to the mature Aristotelian development of *Metaphysics*,

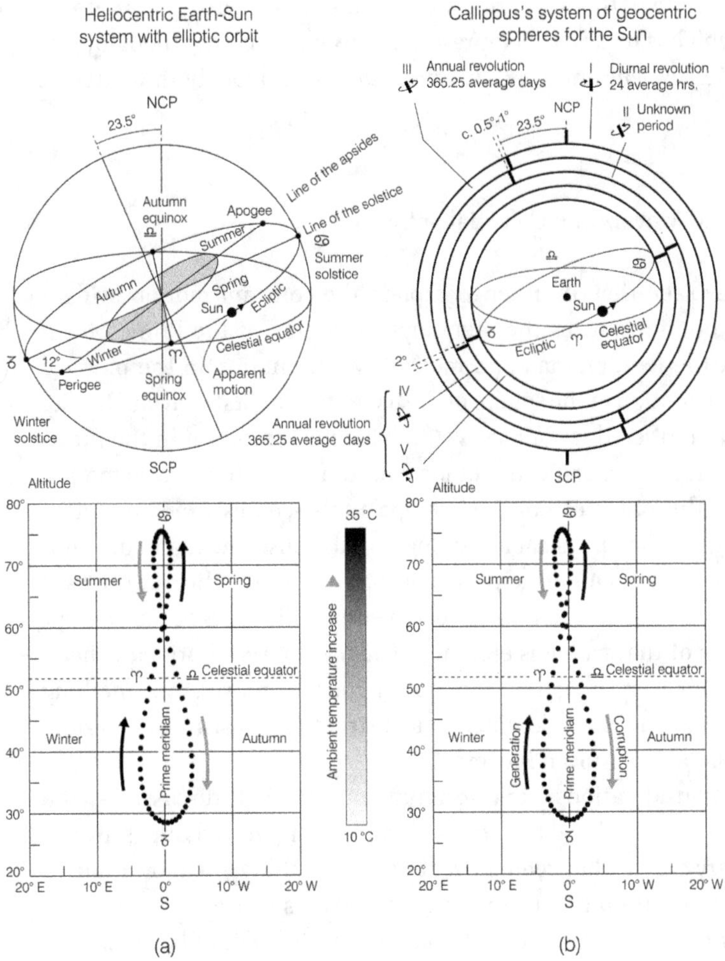

FIGURE 32 Analemma.

The apparent position of the sun for an observer located in Athens (approx. latitude 38° N) and facing south, at noon, at regular five-day intervals during a year. (a) shows the analemma estimated for the heliocentric solar system according to current astronomical knowledge. Above the figure, there is a representation of the apparent motion of the sun from a heliocentric viewpoint. (b) shows the analemma estimated according to the Callippic system of geocentric spheres, adopted by Aristotle for his own system of heavens. In the case of Callippus's system, we have disregarded the rotation of the third sphere, whose period is unknown. This simplification does not affect the analemma in a significant way since its motion was supposedly much slower than the annual revolution of the sun. The figure above the grid shows Callippus's subsystem for the sun with the assumed parameters for its calculation. The variations in solar altitude shown in both graphics are caused by the two eternal motions that Aristotle considers necessary (Λ, 2, 1069b29–33 and Λ, 6, 1072a4–17) to explain the ontic diversity in nature. The analemma illustrates how these joint motions produce the seasons responsible for life cycles or—in Aristotle's words—generation and corruption. The singular form taken by these figures is due to the ellipticity of the terrestrial orbit. This orbit is reproduced almost exactly by the Callippic model of the sun—despite its geocentricity and the assumption that the orbit of the sun is circular—through the addition to the Eudoxian model of two inner spheres, producing an extra hippopede.

already present, in hypothetical terms, both in his *Physics*, VIII, 6, 258b10–13 and 259a7–9, which is regarded as representing his more mature thought,[42] as well as in *On the Heavens* and *On Generation and Corruption*, both written at an early stage.[43]

5.2 Unmoved Movers and Celestial Spheres

We have already seen that, in order to explain the constant motion evident in nature, Aristotle introduces the metaphysical notion of the Prime Mover as the cause of the first, eternal, and most powerful motion in the physical order: the diurnal motion of the sphere of the fixed stars, or first heaven. Despite the demonstration offered in *Physics*, VII, 1, which we reviewed in the previous section, the need for a first cause of local motion—the most basic motion (κίνησις) within the order of change (μεταβολή)[44]—comes from Aristotle's metaphysical rather than physical reasoning. In this sense, the main idea supporting the need for primary principles to account for the motion of things is that everything has a cause that makes a given entity either this or that, except for the principles of that which is eternal.[45] The metaphysical principle here, according to which everything that is in motion must be moved by something acting as its cause, corresponds eminently to a structure consisting of a Prime Mover and subsidiary Unmoved Movers.

Under these considerations—and resuming our previous discussion—the experience of motion is related to a series of movers and moved things. By virtue of a necessary contact between mover and moved thing, the mover moves along with the moved thing, thus requiring a previous cause for the motion of the mover, and so on. Again, Aristotle claims that this cannot be an infinite series,[46] because in a causal series there is a beginning, middle, and end, the

Λ, 8. See also Marcelo Boeri's comment on *Physics*, VIII, 6, 259a13–20 –§ 12– in his Spanish translation *Física. Libros VII–VIII* (Buenos Aires: Biblos, 2003), pp. 231–236.

42 Düring, *Aristoteles*, pp. 50–51; Jaeger, *Aristotle*, p. 299.

43 Düring, *Aristoteles*, p. 52. In addition, Düring believes that Aristotle probably edited both texts after he returned to Athens in 334 (p. 349).

44 *Physics*, VIII, 7, 260a26–261a30 and VIII, 9, 265b17–19; *On Generation and Corruption*, II, 10, 336a19.

45 *Metaphysics*, α, 1, 993b23–31.

46 An example that the causal series cannot be infinite is the fact that humans are moved by air; air, by the Sun; and finally, the Sun is moved by Strife as part of a finite succession (*Metaphysics*, α, 2, 994a2–4). The reference to Strife, one of Empedocles's principles, involves that philosopher—clearly, without his consent—in Aristotelian considerations about causes. Such treatment by Aristotle of his predecessors is repeated—especially

first term of the series being the cause. However, if the series were infinite, there would be no beginning, hence no cause, and without a cause, no motion, as everything that exists must have a cause (a valid logical conclusion).[47] Given that motion is real and incessant, Aristotle concludes in *Physics*, VIII, 5, that there must be at least one first cause of motion in nature, and that being the very first cause, it is not previously moved by anything else. It must be an eternal Unmoved Mover,[48] moving constantly, and always in the same fashion.[49]

As we know, in *Metaphysics*, Λ, 8, 1073b14–1074a30 –§ 15–, Aristotle finally affirms that there is not only one entity of this sort, but that there are as many movers as there are homocentric celestial spheres that compose his system of the heavens; that is, fifty-five. In the previous section, we saw that a multiplicity of Unmoved Movers is necessary to explain phenomena and also to account for the diversity of entities in the sublunary domain. However, so far we have not discussed how these metaphysical entities act on celestial spheres, which constitutes one of the main issues regarding the Unmoved Movers. On account of its ontological priority, we will now try to shed light on this problem with reference mainly to the first of these entities, the Prime Mover.[50] The conclusions can be extended to the other Unmoved Movers, at least as regards their causal condition.

---

in *Metaphysics*, A—with the pre-Socratics when reviewing the doctrines of the first philosophy.

47 *Metaphysics*, α, 2, 994a1–19.

48 The argument used in *Physics*, VIII, 5, 256a13–256b24, derives from the assumption that (256a17–19) "for it is impossible that there should be an infinite series of movers, each of which is itself moved by something else, since in an infinite series there is no first term" (ἀδύνατον γὰρ εἰς ἄπειρον ἰέναι τὸ κινοῦν καὶ κινούμενον ὑπ' ἄλλου αὐτό· τῶν γὰρ ἀπείρων οὐκ ἔστιν οὐδὲν πρῶτον; English trans. R. P. Hardie and R. K. Gaye in Barnes, *Complete Works of Aristotle*, vol. 1, p. 427), in agreement with *Metaphysics*, α, 2, 994a1–19.

49 *Physics*, VIII, 6, 258b10–16.

50 *Metaphysics*, Λ, 8, 1073a24–1073b3 –§ 15–. In *Physics*, VIII, 6, 259a14–20 –§ 12–, Aristotle emphasises the need for at least one Unmoved Mover and, though he does not deny the existence of other analogous movers at this point of his theoretical development, he does not consider them to be strictly necessary. It is worth mentioning that his interest was focused on substantiating the continuous motion of φύσις, for which the necessary though not sufficient condition is at least one Prime Mover. This condition is also the source of its ontological priority over the rest of the movers, which are finally reaffirmed in *Metaphysics*, Λ, 8. In *Physics* VIII, 6, 260a3–11 –§ 13–, it is suggested that the Prime Mover is superior to other movers because it causes a simple motion, always identical to itself, while that which is moved by the first mover (we can consider here the sphere of the fixed stars, even if it moves by itself due to the action of an Unmoved Mover) does not move always in the same way. In fact, it can even come to be at rest. This is clear in the retrogradation and stations in the trajectories of Saturn, Jupiter, Mars, Mercury, and Venus, which are reproduced by each Callippic system.

Fundamentally, there have been two interpretations on the *causal status* of this eternal, uncreated, and immaterial mover that directly causes the motion of the first heaven. On the one hand, the traditional interpretation sustains that this mover is the *final cause* behind the motion of the sphere of the fixed stars. Heath, Jaeger, Ross, Düring, and Lloyd, among others, hold to this interpretation.[51] On the other hand, and taking an opposing stance, Berti, Judson, and others[52] claim that the Unmoved Mover(s) acts as an *efficient cause*, thus producing the motion of the celestial spheres and, indirectly, the motion of the stars. In the present section we attempt to 1) demonstrate, by resorting to Aristotelian texts, the sustainability of the traditional interpretation and, at the same time, 2) illustrate why, in our opinion, the mechanistic stance—which assigns to the Unmoved Movers the status of efficient causes—is incompatible with the Aristotelian doctrine of causality and, to some extent, leads to the annulment of his metaphysics by submitting it to the realm of physics.

For Aristotle, the question of "causality," implicit in the general problem of motion and change, has an explicit gnoseological scope, for he asserts that we truly know something when we know its causes.[53] From this perspective, if we intend to know any particular entity, we must be able to identify the causes of its current condition, through which we also learn how the entities taking part in the cosmological order relate to one another. Following the Aristotelian classification of the four causes (*Physics*, II, 3), "to know" means to find and specify up to the point that is appropriate for each entity: 1) the material cause (ὕλη) or "that out of which a thing comes to be" (αἴτιον λέγεται τὸ ἐξ οὗ γίγνεταί τι ἐνυπάρχοντος);[54] 2) the "efficient cause" (*causa efficiens* in the Scholastic tradition), that is, the cause in terms of the "primary source of the change or coming

51 W. D. Ross, *Aristotle* (London: Methuen, 1923), pp. 180–81; Düring, *Aristoteles*, pp. 210–211; Jaeger, *Aristotle*, p. 385; Heath, *Aristarchus of Samos*, p. 226, and Lloyd, *Aristotle*, p. 142.

52 E. Berti, "Unmoved Mover(s) as Efficient Cause(s) in Metaph. Λ 6" and "La causalità del motore inmobile secondo Aristotele," in *Nuovi Studi Aristotelici*, 2 vols. (Brescia: Morcelliana, 2005), vol. 2., pp. 440 and 465; Berti, *Ser y tiempo en Aristóteles*, Spanish trans. P. Perkins (Buenos Aires: Biblos, 2011), p. 63; L. Judson, "Heavenly Motion and the Unmoved Mover," in *Self-motion: From Aristotle to Newton*, ed. M. L. Gill and J. G. Lennox (Princeton: Princeton University Press, 1994), pp. 166–167. In this line, we could add David Bradshaw's interpretation, though for him the Prime Mover remains a final cause: "A New Look at the Prime Mover," *Journal of the History of Philosophy* 39, no. 1 (2001): 1–22, especially pp. 15–18.

53 *Physics*, I, 1, 184a13; II, 3,194b18–20; *Metaphysics*, I, 3, 983a25; *Posterior Analytics*, II, 94a20–23.

54 *Physics*, II, 3, 194b24; English trans. R. P. Hardie and R. K. Gaye in Barnes, *Complete Works of Aristotle*, vol. 1, p. 332.

to rest" (ὅθεν ἡ ἀρχὴ τῆς μεταβολῆς ἡ πρώτη ἢ τῆς ἠρεμήσεως)[55] or as what initiated the change (ἤ τί πρῶτον ἐκίνησε);[56] 3) the formal cause (τὸ εἶδος), which can be thought of as the archetype (τὸ παράδειγμα) of an entity or statement of its essence (λόγος ὁ τοῦ τί ἦν εἶναι); and lastly, 4) the final cause, or that for the sake of which it is (τὸ οὗ ἕνεκα).

This approach to a knowledge of causes applies to physical entities—physics being the field in which this classification arises[57]—as well as to metaphysical entities, including the Unmoved Movers. Aristotle does not propose a separate set of explanatory principles to address "being *qua* being". In this context, metaphysics acquires its specificity as the first and highest science, addressing the causes of being and becoming; here both the formal cause (understood as the essence of existing things), and the final cause of motion in becoming are identified by Aristotle with the Prime Mover, upon which the totality of the universe depends.[58] In contrast to physical entities, metaphysical entities are not themselves caused but are the cause of all things. Therefore, to know the Unmoved Movers is to know their causal condition.

The Prime Mover, being the object of the first philosophy, is a final and formal cause. However, it cannot be classified as a material cause because it is eternal actuality and (as pure actuality rather than potentiality) an immaterial entity.[59] The same would be true for all Unmoved Movers. The issue, then, is to establish whether it is also possible to assign to this first mover—the ultimate cause of change in φύσις—or to any of the other fifty-four Unmoved Movers, the condition of efficient cause.

Knowledge of causes—of the necessary antecedent to a particular phenomenon—is still today considered, following Aristotle, a *sine qua non* condition for scientific knowledge. Our conception of causality has, however, a much narrower sense than did that of Aristotle, this last being closely tied to a theological and metaphysical understanding of natural change. This is well beyond the reach of current scientific knowledge, whose sole aim is to address the order of phenomena. As a result, our way of dealing with the knowledge of things is supported by a notion of causality only comparable to the Aristotelian notion of efficient or motor cause. As we will see, for Aristotle, this cause has a strictly physical scope, thus it cannot be applied to the motion caused by purely metaphysical entities, such as the Unmoved Movers. Our knowledge

55 *Physics*, II, 3, 194b30; English trans. R. P. Hardie and R. K. Gaye in Barnes, *Complete Works of Aristotle*, vol. 1, p. 332.

56 *Posterior Analytics*, II, 94a22.

57 Aristotle introduces his famous four-fold classification of causes in *Physics*, I.

58 *Metaphysics*, Λ, 7, 1072b14.

59 *Metaphysics*, Λ, 6, 1071b21.

of nature is related to sensible entities, to which we attribute the condition of producers (efficient agents) and final causes of phenomena. Hence, for example, our physics assimilates the "gravitational field" (the main cause of the astronomical phenomena) to the idea of "a thing," in that it belongs to the sensible order and can be measured, and is not a mere abstraction. For Aristotle—and here lies the great difference with respect to our current understanding of phenomena—the Unmoved Movers, despite being "trans-physical" entities (i.e., able to bridge the sensible and non-sensible realms), are not mere abstractions but are existing things (even more so than sensible entities) although we cannot experience their existence directly through empirical evidence.

To understand this physical condition of efficient cause and its relation to final cause, and thereby approach the issue of which kind of causality corresponds to the Prime Mover, it is useful to analyse, in the first instance, the way in which the soul participates in motion in a human as a rational living being. (This approach, which through analogy helps us to grasp the difficult notion of the causal condition of Unmoved Movers, should not lead us to identify Unmoved Movers with a supposed soul of stars.) There are three fundamental reasons to address this preliminary question. Firstly, as we will see, Aristotle explains voluntary human behaviour by way of the action of Unmoved Movers (souls), though of a relative condition and it would be more appropriate to consider them within the order of φύσις. Secondly, he regards, like Plato, stars as living beings, rational animated beings.[60] Thirdly, he introduces the notion of the Prime Mover through an analogy with human life, assigning it self-reflecting thought:[61] this state of the very first cause, in which it remains eternally, resembles that which—he tells us—we in a contemplative life can only fleetingly experience.[62]

Let us to analyse, then, how the soul is involved in human motion, understanding the human as a relatively high-level being in the hierarchy of things, capable of rational activity, and with this make our initial approach to a

60 Aristotle distinguishes the upper (South Celestial Pole) and lower (North Celestial Pole) part of heavens as well as two directions of rotation: to the right or counter-clockwise (taking the upper pole as reference), and to the left or clockwise (*On the Heavens*, II, 2, 285b15–22), both of which movements can only be attributed to animated bodies (*On the Heavens*, II, 2, 284b15 and 284b32–34). That explains why the heavens must be animated (δ' οὐρανὸς ἔμψυχος καὶ ἔχει κινήσεως ἀρχήν; *On the Heavens*, II, 2, 285a30–33), in other words, to have a soul. Aristotle also associates the soul of stars—though less clearly—with the multiple movements comprising the visible motion of each star (*On the Heavens*, II, 292a18–27 –§ 19–).

61 *Metaphysics*, Λ, 9, 1074b34.

62 *Metaphysics*, Λ, 7, 1072b14–30.

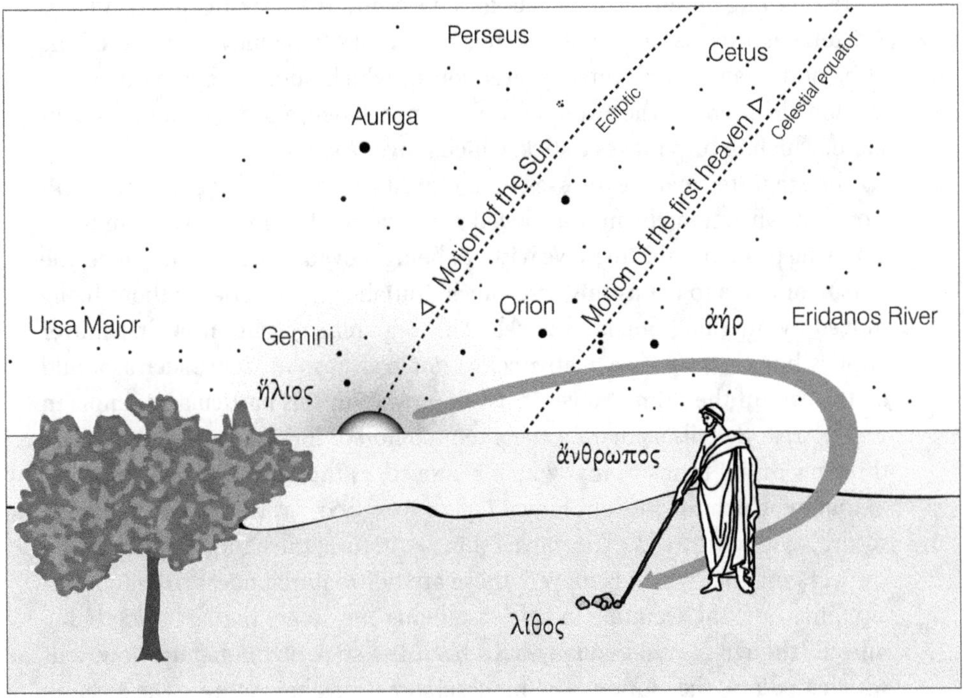

FIGURE 33 In Aristotelian physics, every motion is part of a causal chain that correlates with celestial motion and traces its origin back to the rotation of the first heaven, itself caused by the Prime Mover. Thus, the stone is moved by the person, whose soul is their essential principle of motion. Human motion is not only caused by this volitional principle, however. The body is in contact with air, for example, connecting it with every atmospheric phenomenon. In a basic sense, climatic and biological changes are prior to the beginning of the free causal series of the soul. In addition, human behaviour partly depends on both the direct eastward motion of the sun along the ecliptic throughout the year and the diurnal westward motion of the stars. Hence, interpretations that see human life as strongly influenced by the stars—though only to a certain extent—are not to be considered completely anti-Aristotelian. The thirteenth-century suggestions of celestial determination on human behaviour by heterodox peripatetic thinkers of the University of Paris based on Arabic philosophy and astronomy do, however, exceed Aristotle's ideas. These notions are inadmissible, even if Aristotle emphasises the celestial influence on the development of animal as well as social life. Man, even being soul a first entelechy, cannot stand on its own without some essential external conditions. Moreover, the same conditions are decisively influential in human behaviour. In Aristotle's example, the person moves the stone not only because they desire to do so but because the stone is there, and it lays there for a myriad of linked causes that can be traced back up to the unending actuality of the Prime Mover.

corresponding notion with respect to the Unmoved Mover. In *Physics*, VIII, 5, Aristotle assures us that "that which is moved has to be moved by something else," and analyses an everyday situation in which autonomous human self-motion is evident. In the example, already mentioned, a person is moving their hand, which in turn moves a stick, which moves a stone.

The Stagirite clearly expresses here that in this chain of events that result from human action, the mover-moved series cannot be infinite. There must be a moving principle able to move without being moved. In the first instance, the person appears to act as this first mover, initialising the series without being moved by anything outside the self. This apparent condition of first mover applies, however, only in a relative sense, not in an absolute sense such as would be the case of the Prime Mover in astronomy. With this particular example in mind, Aristotle tells us that for the phenomenon of motion to occur, there are three necessary terms: 1) that which is moved, 2) the mover or "movent," and 3) that by which the movent moves (τρία γὰρ ἀνάγκη εἶναι, τό τε κινούμενον καὶ τὸ κινοῦν καὶ τὸ ᾧ κινεῖ).[63] The moved thing is, in turn, the hand, the stick or the stone. Being that which is moved, these are not required necessarily to move anything else, the example in this case being the stone, the last thing being moved (though it could equally have been the stick or the hand). Now, that which mediates the motion—in this case, something belonging to the person's corporeality—both moves and is moved. It moves because, by definition, it is that by which the moving thing moves. It is moved because everything that moves, except the First Unvoved Mover, has a motor cause. That by which the moving thing moves (here manifested by the motion of the hand—that is, the first moved thing—understood as the first movent making physical contact with the moved thing, is expressed as follows in *Physics*, VII, 2, 243a3–5):

> That which is the first mover of a thing [the movent]—in the sense that it supplies not that for the sake of which [τὸ οὗ ἕνεκεν] but the source of the motion [ὅθεν ἡ ἀρχὴ τῆς κινήσεως]—is always together with that which is moved by it (by 'together' I mean that there is nothing between them).[64]

Aristotle tells us that the first movent, the original moving thing in a perceivable spatio-temporal series, must be interpreted as an efficient cause and not

63 *Physics*, VII, 5, 256b15.

64 English trans. R. P. Hardie and R. K. Gaye in Barnes, *Complete Works of Aristotle*, vol. 1, p. 409.

as a final cause.[65] The movent (which is physical, as it moves) has to be in contact with that which is moved. In the field of astronomical or superlunary phenomena, the notion of a first mover as an efficient cause can only correspond to the celestial spheres of the Aristotelian system of the heavens. The reasoning is that, in terms of motion as a whole, the celestial spheres are the first bodies that move and can move another, the sphere of the fixed stars being the first among them and having, itself, no prior physical mover. Celestial spheres are, in this sense, the true and first efficient causes of motion in the universe. In *Meteorology*, I, 2, 339a23–32 –§ 23–, Aristotle refers to celestial motion as follows:

> This world necessarily has a certain continuity with the upper motions; consequently all its power is derived from them. For the originating principle of all motion [ὅθεν γὰρ ἡ τῆς κινήσεως ἀρχὴ] must be deemed the first cause.... [and we] must assign causality in the sense of the originating principle of motion [αἴτιον ὅθεν ἡ τῆς κινήσεως ἀρχή] to the power of the eternally moving bodies.[66]

As will be noticed, both sections in the passage identify the first efficient causes of motion in the universe—namely, "the originating principle of all motion"—with the motion of celestial spheres, which belong to the physical order (as they are sensible bodies) rather than with the activity of the Unmoved Movers—transphysical entities that move the spheres.

Resuming the example of the stone (the moved thing), whose motion is caused by the stick and the person, Aristotle points out that, in terms of causes, we see only up to the last term of this series: the person. Now the person is a physical body that can be moved by another (namely, a movable body), however, we do not see the person being moved by something else. Rather, the person is a self-mover. So we must then infer a third term: a movent that moves without being moved. But that which moves by itself at the beginning of the series (in the example, the corporeal motion of the person) cannot be, strictly speaking, the first movent because then there would be motion with no cause. Therefore, the movent must be something else, something unmoved. If, then, in this scheme the person's body relates to the motion of the stone as "that by

65 Those who claim that the Prime Mover is an efficient cause encounter a major obstacle: Aristotle does not affirm in any passage of his *corpus* that the Prime Mover is an efficient cause. What is more, in *Physics* II, 7, 198a35–198b4, he claims that the Prime Mover is in fact a formal and final cause, rather than an efficient cause.

66 Lightly modified from the English trans. of E. W. Webster in Barnes, *Complete Works of Aristotle*, vol. 1, p. 555.

which the movent moves" (the instrument of motion), then what is the mover that, unmoved, moves the person's body while at the same time being part of it (since we say the person is a self-mover)? As regards that which moves by itself, in *Physics*, VIII, 5, 258a17–20, Aristotle simply affirms that:

> That which moves itself, therefore, must comprise something that imparts motion but is unmoved and something that is moved, but does not necessarily move anything else; and each of these two things, or at any rate one of them, must be in contact with the other.[67]

This (relative) unmoved mover must be correlated with the appetite of the rational soul, as expressed in *On the Soul*, III, 10, 433a1–433b32. There, Aristotle reaffirms that, as regards human acts and the phenomenon of motion, we have again to distinguish the three terms mentioned above: the movent; that by which the movent moves; and that which is moved. However, in *On the Soul*, Aristotle's analysis of the ultimate cause of intentional human motion goes further than in *Physics*. He explores the final cause of intentional human motion: the soul, identified with an unmoved mover but always in a relative sense. The moved thing, he says, is the animal (corporeal human), and the movent, which in the example of the person moving the stone was only identified as the unmoved mover of motion, now is correlated with the soul as the ultimate movent. To the soul as movent Aristotle attributes a double constitution. On the one hand, because that which moves in the soul is an object of appetite (τὸ ὀρεκτόν), this becomes the actual *unmoved mover* and also satisfies the condition of a final cause, which for humans is always a practical good, and as such, pleasant. On the other hand, because that which moves in the soul is also the faculty of appetite, it constitutes a movent that moves and is moved, identified with desire, which is motion in act. Aristotle says, finally, that the instrument by which desire moves is not completely corporeal but a mixture of body and soul. Thus in the background of a person's intentional motion, which is a consequence of their condition as a rational being, we have not three but four terms: 1) the unmoved mover, final cause or purpose, which is always a realisable good and object of appetite, tied to 2) the appetite that moves and is moved; 3) the instrument by which the movent moves—a corporeal one, or more precisely a mixture of body and soul; and 4) the moved thing, i.e., a person's physical body that acts to fulfil its desire. The second and third terms are interwoven and together they constitute "that which originates the

67 English trans. R. P. Hardie and R. K. Gaye in Barnes, *Complete Works of Aristotle*, vol. 1, p. 430.

movement" (δ' ᾧ κινεῖ). In addition, the (non-corporal) desire of the moving thing echoes in some sensory (corporal) organ as its instrument.[68]

That which moves—the movent—is, then, non-corporal, as it belongs to the soul. And it has a double constitution: it is the object of appetite (unmoved principle) and it is appetite itself, which could be interpreted as an efficient cause, as motion derives from it. Aristotle says: "a power in the soul [...] i.e. that called appetite [ὄρεξις], originates movement" (*On the Soul*, III, 10, 433b1).[69] Appetite, or desire, differs from the object of appetite in that it is motion towards the actualisation of that which is desired. Appetite, however, cannot exist without the object of its desire, in other words, without a yet-unfulfilled end. It is for this reason that both desire and its object are identified, in the soul, with the movent or principle of motion. This compound movent within an individual is what Aristotle calls the "desiring intellect" (ὀρεκτικὸς νοῦς), or "thinking desire" (ὄρεξις διανοητική), in his *Nicomachean Ethics*, VI, 2, 1139b4. Nevertheless, this condition of the movent in rational souls differs from that of the movent of celestial bodies: although they too have a rational soul, the end—the final cause—of celestial motions is not mere potentiality, as it is with humans. It is, on the contrary, pure actuality or what Aristotle calls a "state" (ἕξις). Thus, while in the human soul the mover conditioned by the union of desire and its object (potentiality) corresponds to the notion of efficient cause, the moving principle in celestial bodies cannot be understood as such. In *On Generation and Corruption*, I, 7, 324b13–18, Aristotle states that:

> The active power is a cause in the sense of that from which the process originates [ἔστι δὲ τὸ ποιητικὸν αἴτιον ὡς ὅθεν ἡ ἀρχὴ τῆς κινήσεως]; but the end, for the sake of which it takes place, is not active [τὸ δ' οὗ ἕνεκα οὐ ποιητικόν]. (That is why health is not active, except metaphorically.) For when the agent is there, the patient becomes something; but when [stable] states are there, the patient no longer becomes but already is—and forms (i.e. ends) are a kind of state [ἕξεις τινές].[70]

There are other reasons why the movent of human souls cannot be the same Unmoved Mover that explains motion of nature as a whole; that is, an

68 *On the Soul*, III, 10, 433b17–20. Marcelo Boeri also finds that where "that which moves and is moved" is a mixture of the corporal and non-corporal, the corporal organ must be considered the instrument of the mover, namely desire, which comes from the non-corporal soul; Aristotle, *Acerca del Alma*, Spanish trans. Marcelo Boeri (Buenos Aires: Colihue, 2010), p. 169, n. 397.

69 *On the Soul*, English trans. J. A. Smith in Barnes, *Complete Works of Aristotle*, vol. 1, p. 688.

70 English trans. Joachim in Barnes, *Complete Works of Aristotle*, vol. 1, p. 530.

uninterrupted motion. Aristotle elaborates on these reasons in *Physics*, VIII, 6, 258b10–259a7. First, things that can exist and then may no longer exist, such as human souls, cannot be absolute first causes. There must be a prior cause that accounts for this change. Also, though unmoved things that do not always exist can be the cause of some motions, they cannot be the cause of all motions, for example the continuous motion in nature, as these are infinite in the sense that as some things die others are born, and this continues indefinitely—we should remember that Aristotle rejects Plato's doctrine of παλιγγενεσία. Moreover, they are never a simultaneous totality, a necessary condition to cause eternal and continuous motion. A thing that does not always exist and does not move always in the same way cannot be the cause of that which is eternal and continuous. Neither the desire that erupts in the rational part of the soul (βούλησις) nor the purpose (προαίρεσις) arising from the process of deliberation that causes human self-motion—as we have already seen—can explain continuous motion. Clearly, the circumstances of psychic life cannot produce a continuous motion, even within the individual human.

The second reason is that a desire originating in the soul produces but one kind of human motion: a change in location. If people are, like other living beings, in continuous motion, such motion should be a natural motion that comprises, among other things, breathing, increase, and decrease. Aristotle says that these motions are caused by external factors, present in the environment (*Physics*, VIII, 6, 259b1–16). Fundamentally, that which causes a person's self-motion—that is, the object of appetite in thought—is moving, even if this mental object does not itself need to move in order to move the person—therefore being adequately characterised to some degree as an unmoved mover. However, it moves accidentally as part of that which physically moves itself, the soul being tied to the body (see *Physics*, VIII, 6, 259b16–28). A person moves and their soul moves along with them. It is precisely in the soul that we find the thought that moves a person by presenting them with a practical end to pursue. This end exists only in their thought. However, Unmoved Movers that are the final causes of continuous motion in nature must be unmoved in absolute terms: they cannot move, not even by accident.

The previous analysis shows that Aristotle introduces the notion of the Unmoved Mover as a first approach to the issue of motion caused by the rational soul, and that this first meaning is not strictly that assigned to the Prime Mover in *Metaphysics*, Λ, even if it identifies the mover with the final cause, rather than with the efficient cause. In consequence, when Aristotle assimilates the Unmoved Mover to the final cause and, thus, to the ultimate origin of motion, he is considering that the final cause is prior to the efficient cause required for

physical motion. In other words, there is no motion in any sense—whether as local change; increase or decrease; generation and corruption; or qualitative change—without an efficient cause, and far less without a final cause. The efficient cause cannot be the ultimate principle of motion, nor can it be identified at all times with the final cause, a condition that must necessarily belong to the first cause of becoming, namely the Prime Mover of *Metaphysics*, Λ. In the following passage of *Parts of Animals* (I, 639b11–16), Aristotle highlights the precedence of final cause over efficient cause:

> Furthermore, the causes concerned in natural generation are, as we see, more than one. There is the cause for the sake of which [οὗ ἕνεκα], and the cause whence the beginning of motion [ὅθεν ἡ ἀρχὴ τῆς κινήσεως] comes. Now we must decide which of these two causes comes first, which second. Plainly, however, that cause is the first which we call that for the sake of which. For this is the account of the thing, and the account forms the starting-point, alike in the works of art and in works of nature.[71]

The final cause, defined as the first principle of motion in reference not only to human products but also to the products of nature, is necessarily identified with the notion of an Unmoved Mover. This can be so in a relative sense, such as in the case of voluntary human acts; or in an absolute sense, such as in the case of the principle responsible for celestial motions, on which all the other changes in nature depend.

In Aristotle's teleological conception, all motion and rest are necessarily linked to a final or formal cause as the first principle of change. Therefore, the first causes of motion, the Unmoved Movers of Planetary Spheres, must be final causes, especially that which is first among them, the Prime Mover. Now, being eternal unmoved entities, the movers responsible for the motion of celestial spheres are fully realised immaterial οὐσίαι. Under these conditions, as we have already pointed out, following Aristotle, the final cause cannot be active, strictly speaking, as it is a particular state, the state of perfect happiness (this is the state of the Prime Mover as well as, we deduce, the rest of the Unmoved Movers). And this is what makes separate substances into beings uninvolved in and unaffected by any efficient or productive activity, because they exist above all for pure contemplation (θεωρητική).[72]

Contrary to our interpretation, Berti understands that the Prime Mover of *Metaphysics*, Λ is an efficient cause of motion, as Aristotle seems to assign to it

71 *Parts of Animals*, English trans. W. Ogle in Barnes, *Complete Works of Aristotle*, vol. 1, p. 994.

72 *Nicomachean Ethics*, X, 8, 1178b7–24.

the condition of agent (ποιοῦν), which is equivalent to that of efficient cause.[73] To support his interpretation, Berti quotes a passage from *Metaphysics*, Λ, 6, 1071b12–23, in which Aristotle affirms that the primary principles of motion and change (the Unmoved Movers), unlike the Platonic Ideas, should be able to move (κινητικόν) and produce (ποιητικόν). Such is a characteristic of Unmoved Movers, not in the sense that they are the efficient or productive cause of any change, but in the sense that the effective cause of change cannot be without end or purpose. Accordingly, the final cause is the principle that ultimately produces change, as it is necessarily prior to the efficient cause.[74]

In addition to the above, there are other obstacles to accepting the idea that the Unmoved Movers of Planetary Spheres can be efficient causes of motion. First, when Aristotle introduces the four causes (*Physics*, I), he mentions that formal, efficient, and final causes are often the same—as we have seen in the case of human souls—but this is only possible in the physical order, where that which moves is also moved. The notion of an efficient cause belongs to physics and does not apply to metaphysical entities, such as the Unmoved Movers responsible for visible celestial motions. In *Physics*, II, 7, 198a35–198b4, Aristotle explains:

> Now the principles which cause motion in a natural way are two, of which one is not natural, as it has no principle of motion in itself. Of this kind is whatever causes movement, not being itself moved, such as [1] that which is completely unchangeable, the primary reality, and [2] the essence of a thing, i.e. the form; for this is the end or that for the sake of which.[75]

That Aristotle is thinking about the Prime Mover of *Metaphysics*, Λ when he refers to "that which is completely unchangeable, the primary reality" is proven by the fact that, in the context of such statements—and immediately preceding *Physics*, II, 7, 198a26–32—he points to the difference between Physics,

73 In "Unmoved Mover(s) as Efficient Cause(s) in Metaph. Λ 6," pp. 427–45, Enrico Berti supports this idea of *Physics*, II, 3, 195a22, where Aristotle attributes the condition of agent (τὸ ποιοῦν, *efficiens*, to use the medieval expression) to the causes in the sense of "all sources whence the change or stationariness originates" (… τὸ ποιοῦν, πάντα ὅθεν ἡ ἀρχὴ τῆς μεταβολῆς ἢ στάσεως [ἢ κινήσεως], *Physics*, 195a22).

74 Following this idea, Bradshaw ("A New Look at the Prime Mover," p. 8) affirms that it is preferable to consider the Prime Mover both as an efficient and final cause rather than merely regard it as a final cause. What is undoubtedly an anti-Aristotelian idea, we add, is conceiving the causality of the Prime Mover only in terms of efficiency.

75 English trans. R. P. Hardie and R. K. Gaye in Barnes, *Complete Works of Aristotle*, vol. 1, p. 338.

Astronomy and Metaphysics as regards their object of study. The proper subjects for the latter are eternal incorruptible unmoved entities, as also stated in *Metaphysics*.[76]

The efficient cause, as mentioned before, belongs exclusively to the physical order, where things subject to becoming are susceptible to motion and change. The definition of φύσις as the "principle of motion and of stationariness,"[77] is in itself very similar to the definition of efficient cause as "primary source of the change or coming to rest." In *Metaphysics*, E, 1, 1025b18–21, Aristotle explains that Physics deals with entities that have in themselves the principles of motion and rest. By definition, where these principles exist, so does efficient cause.

Furthermore, the very definition of action (ποιεῖν) given by Aristotle clashes with the idea of the Unmoved Movers as active entities in the sense that they can produce (ποιητικόν), as does an efficient cause. In his treatise on *Categories* (11b1), Aristotle defines the categories of action and affection relative to opposites, that is to say, "action" (ποιεῖν) and "affection" (πάσχειν) only make sense if understood as correlated opposite categories, and in terms of more or less. Everything arising from the four sublunary elements (for example heaviness-lightness, humidity-dryness, cold-heat) includes this possibility of opposites.[78] However, the ether that composes the celestial spheres moved by these Unmoved Movers—supposedly efficient causes, as suggested by Berti—does not have an opposite, let alone a "more or less." Hence, in the particular case of the circular motion of celestial bodies, every place is at once the "beginning, middle and end," constituting a kind of stationary dynamic state. Bodies in such a condition are, therefore, Aristotle explains, in motion and also, in a certain sense, at rest,[79] and so can be considered to be in a final or complete state.

Berti's interpretation is that the Unmoved Movers cannot be only final causes because in Aristotelian philosophy the end of such causes must eventually be achieved, and he finds that this cannot be the case for celestial bodies given their perpetual motion. If Berti were right, the stars would suffer the fate of Ixion,[80] who, punished by Zeus, revolves eternally against his desire. Rather,

76 In his *Metaphysics*, Λ, 1, 1069a30–34, Aristotle makes a parallel distinction among three kinds of entities which belong to the disciplines of physics, astronomy, and metaphysics: corruptible (physical bodies); eternal and incorruptible (celestial bodies); and eternal, incorruptible, and unmoved (Unmoved Movers), respectively.

77 *Physics*, II, 1, 192b21; *Metaphysics*, V, 4, 1014b18.

78 *On the Heavens*, I, 3,270a6–23; II, 6, 288b19–23.

79 *Physics*, VIII, 9, 265a33–265b2.

80 *On the Heavens*, II, 1, 284a27–35.

one should consider that the stars, even if they have souls,[81] are in a desired final state. And this is the product of the regular eternal motion of celestial spheres, caused by the attraction exerted by the completeness and goodness of their respective Unmoved Movers.[82] Making a clear reference to the first heaven or sphere of the fixed stars, Aristotle tells us that its motion constitutes a perfect state that can only be understood as a final state or achieved end (*On the Heavens* II, 1, 284a2–10 –§ 3–):

> Therefore we may well feel assured that those ancient beliefs are true, which belong especially to our own native tradition, and according to which there exists something immortal and divine, in the class of things in motion, but whose motion is such that there is no limit to it. Rather it is itself the limit of other motions, for it is a property of that which embraces to be a limit, and the circular motion in question, being complete [τέλειος], embraces the incomplete and finite motions. Itself without beginning or end, continuing without ceasing for infinite time, it causes the beginning of some motions, and receives the cessation of others.[83]

In the same text, Aristotle compares the celestial bodies to gods (θεοί), as there is nothing preventing them from taking part of the noblest state, not even time, which conspires against the fulfilment of desire for human souls. In the case of

81 In an attempt to understand the Unmoved Movers from *inmanentism*, René Mugnier, in *La théorie du Premier Moteur et l'evolution de la pensée aristotélicienne* (Paris: Vrin, 1930), pp. 144–145, has suggested that they can be considered the souls of celestial spheres. Such a notion, though far from the exposition of *Metaphysics*, Λ and *Physics*, VIII, is close to some of the doctrines in *On the Heavens*, II, 1 and II, 12, according to which stars would have soul. This idea follows the general guidelines of the Academy during Plato's final years, which correspond largely with the topics discussed in the *Timaeus*. Nevertheless, the fact that Aristotle assigns intelligent life to celestial bodies does not necessarily imply—and Aristotle never affirms it to be so—that the soul is the mover of their eternal and continuous motion. From the perspective of our own human experience, it could seem curious that the motion of stars is not caused by animating souls. But it would seem even more curious if a soul were to cause an eternal continuous motion always equal to itself (the motion of the Prime Mover). For Aristotle, motion, in physical terms, is always the consequence of the action of a force. Therefore, it carries the idea of endeavour or effort by an agent, which is unworthy of divine celestial bodies (284a27–36).

82 We should remember that a celestial body is always fixed at some point of the spheres. Stars are located in their corresponding relative positions in the sphere of the fixed stars, and planets (if Jupiter is to be taken as example) are fixed at the equator of the innermost of the nine spheres that account for their motion.

83 English trans. Guthrie, pp. 131, 133.

celestial souls, desire is fully satisfied. As regards the possibility that the highest good of the Prime Mover (as a final cause) is an achievable state in which all celestial bodies take part, Aristotle says (*On the Heavens*, II, 12, 292b22–26):

> But the first heaven reaches it [the highest good (ἄριστος)] immediately by one movement, and the stars that are between the first heaven and the bodies farthest from it reach it indeed, but reach it through a number of movements.[84]

This passage can be understood to say that the stars have souls and that these souls enjoy a state of perfect happiness, having achieved the highest good, whose archetype is the simplicity and unicity of the Prime Mover as ultimate final and formal cause. It would not be appropriate to assign desire (a human-like desire) to the soul of stars, as desire relates always to a future good, and therefore involves an efficient cause, as it moves and is moved. The "thinking desire" is always desiring something one lacks, as Aristotle explains: "nothing that is past is an object of choice" (*Nicomachean Ethics*, VI, 2, 1139b6).[85] Since the stars are always moving in the same state, if they were able to desire, they would pursue a good they could never achieve, resulting in the worst possible existence, like the ever-spinning wheel of Ixion.

A possible interpretation is that if Unmoved Movers are not active, in the sense of an efficient cause, then they do not differ from Platonic forms, as Berti also suggests. But although the Prime Mover is identified with the highest good, it is not the same as the Platonic Good (*Republic*, VI, 507a–509b) and this is proven by the fact that its love (κινεῖ δὴ ὡς ἐρώμενον; *Metaphysics*, Λ, 7, 1072b4) is not the Platonic Ἔρως. For Plato, love at its highest expression is the desire to possess always the good (ὁ ἔρως τοῦ τὸ ἀγαθὸν αὐτῷ εἶναι ἀεί; *The Symposium*, 206a). Such an aspiration is never satisfied, at least within the sensible realm. In the Aristotelian worldview, the love moving the stars is act (ἐνέργεια), a state of full realisation in the superlunary world. This can be better understood by considering the specificity of supercelestial "physics," where everything is stable, the only change being the circular motion of perfect spheres. In a way, it is an unchanging physics. When Aristotle says that the Prime Mover moves as that which is loved, we should interpret it in the sense that the soul of ethereal bodies loves its own mover. However, they do not desire their movers in the way of the Platonic Ἔρως. Rather, their love is closer to the Aristotelian idea of

84 English trans. Guthrie, p. 211.

85 *Nicomachean Ethics*, trans. W. D. Ross, revised by J. O. Urmson, in Barnes, *Complete Works of Aristotle*, vol. 2, p. 1798.

self-love (φιλαυτία). This kind of love is possible because they aspire to a good that is already within the one who loves. Therefore, they can achieve its full realisation in a perfect state of happiness, where desire has meaning only as a prior state in virtual and logical terms.

Each of the planets is moved by a celestial sphere. These spheres can indeed be considered efficient causes of such states of perfection. Only by depriving the Unmoved Movers from their condition of final cause, in Aristotelian terms—a practical good that can be realised—can we consider Unmoved Movers and Platonic Ideas to be the same. The notion of final cause is the link between form and the physical world, absent in the Platonic Forms. Therefore, in some sense, the final cause is an essential principle of motion (that which things can potentially be, or actually are, derives from this condition) and, to some extent, an active principle. This is affirmed by Aristotle—and we follow Ross on this point—in *Metaphysics*, Λ.[86]

In line with our interpretation and clearly distinguishing efficient cause from final cause, Aristotle affirms that: "The origin of action—its efficient, not its final cause [ὅθεν ἡ κίνησις ἀλλ' οὐχ οὗ ἕνεκα]—is choice, and that of choice is desire and reasoning with a view to an end" (*Nicomachean Ethics*, VI, 2 1139a31–32).[87] Furthermore, in *Eudemian Ethics*, II, 6, 1222b21, he considers that:

> Such principles, which are primary sources of movements [ὅθεν πρῶτον αἱ κινήσεις], are called principles in the strict sense, and most properly such as have necessary results; God [θεός] is doubtless a principle of this kind.[88]

It is well known that the Aristotelian god is the Prime Mover from *Metaphysics*, Λ, 6–7, 9–10, and that the Prime Mover is neither a maker that intervenes to produce the physical order nor, as Berti concludes, a personal god.[89] This would imply assigning a superlative human condition to the Prime Mover, as if it were a kind of Olympian god. But such a divine condition, which could be extended to every Unmoved Mover, says Aristotle, belongs to a kind of mythical thinking that only vaguely approaches the truth.[90]

86 Ross, *Aristotle*, p. 181.
87 English trans. Ross (rev. Urmson) in Barnes, *Complete Works of Aristotle*, vol. 2, p. 1798.
88 English trans. J. Solomon in Barnes, *Complete Works of Aristotle*, vol. 2, p. 1936.
89 Berti, "La causalità del motore inmobile secondo Aristotele," *Nuovi Studi Aristotelici*, p. 465.
90 *Metaphysics*, Λ, 8, 1074b1–14 –§ 15–.

5.3 Kinematics and Dynamics

Before resuming, in the following section, "The Integration of Planetary Spheres," the issue of Aristotle's integration of Eudoxus and Callippus's subsystems, we will introduce some last considerations on how the Stagirite understands motion, to show more clearly the specificity of the integration. To do so, we will apply to Aristotelian thought the distinction between *kinematics* and *dynamics*, a notion from classical Mechanics that, regardless of the gap between Aristotelian and Newtonian physics, remains nonetheless suitable. Kinematics, in its modern sense, studies motion without considering the causes or interactions that determine that motion, while dynamics is precisely the study of these interactions or forces. Together they constitute what we call *mechanics*.

To reach a complete understanding of Aristotle's celestial system, we need to analyse thoroughly the specificities of the Unmoved Movers of Planetary Spheres, as ultimate cause of the local motion of stars. Representing a unique characteristic of the Aristotelian physics and its inextricable link to metaphysics, these movers, which move celestial spheres without effort or compulsion (ἀνάγκη),[91] take his astronomy beyond the kinematic analysis of Eudoxus and Callippus. The introduction of causes of celestial motions is a clear breakthrough towards understanding visible phenomena and is an improvement over the Eudoxian and Callippic astronomical models, which were purely kinematic. For this reason Aristotle's account of celestial motions should be considered a true *sui generis mechanics*. The superimposition of celestial motions, along with Aristotelian dynamics, whose opening statement was the introduction of the Unmoved Movers of Planetary Spheres, makes for a complex physics that can properly be described as a *mechanics*. It explains the composition of motion (celestial kinematics) along the lines of Eudoxus and Callippus's models, but also adds to that the causes of motion (celestial dynamics) providing, therefore, a complete account of the phenomena. We say a *sui generis* mechanics as such causes belong to a transphysical plane—for Aristotle just as real as the physical plane, though inconceivable and inadmissible for the Newtonian mechanics proper to the modern approach to phenomena.

In order to understand the causal implications and the effect that the introduction of transphysical movers has on celestial motions, we must trace the hints left by Aristotle in a few passages—especially in *Physics*, books VII and VIII—on the interaction between the mover and that which is moved. Equally important is his view on the superimposition of motions (celestial kinematics),

91 *On the Heavens*, II, 1, 284a27–37.

a system closely related to that designed by Eudoxus and Callippus, which, far from being rejected by Aristotle, was enriched by him.

Two levels of analysis or viewpoints can be used to broaden the previous description and help to reach a better appraisal of the Aristotelian system of the heavens: a *theoretical-predictive* level and a *theoretical-explanatory* level. This distinction between two modes of analysis (products, certainly, of our own time and worldview) does not reflect any methodological or systematic aspect of Aristotle's thought. Rather, we impose this analytic schema to help in the understanding of some elusive aspects of his celestial model, which, incidentally, was never fully systematised.

On the one hand, in order to determine whether the type of celestial "model" described by Eudoxus and Callippus, and further developed by Aristotle, provides a reasonable reproduction of visible phenomena, we have to build these systems of concentric spheres—even if just ideally—and then put them to work so as to compare the outcomes with known celestial phenomena. We call this the "*theoretical-predictive* viewpoint." The most thorough baseline study in this respect was conducted by Giovanni Schiaparelli (*Scritti sulla storia della astronomia antica*, Bologna, 1925–1927). By means of geometry and kinematics, he proved that the systems of Eudoxus and Callippus are able to describe the specific motions of the stars with a high degree of accuracy—considering the observational possibilities of Greek astronomers from the fourth century BCE.[92] We have followed a similar method (but with the advantage of software tools unavailable to Schiaparelli) to calculate the trajectory of Mars as described in Chapter 4, and the analemma shown earlier in this chapter in the section "The Prime Mover and Unmoved Movers." This approach, embraced by most later interpreters of Aristotle's system of the heavens, introduces elements that are not present in the Stagirite's description of the system but that must necessarily be added to achieve the objectives arising from this viewpoint.

One of the elements introduced by our modern perspective is the idea that spheres are fitted by "material axes of rotation" implying at first glance that the movers act *on* these axes. A mechanism of this nature is not guaranteed by the Aristotelian understanding of celestial mechanics. This is a critical point as it opens a gap between our representational models and a pure Aristotelian model. Nevertheless, it is a very productive and even valuable procedure, as long as we keep in mind its main limitation: we cannot view such a mechanism as an ultimate support to understand the full complexity of Aristotelian physics and metaphysics with respect to celestial motions. It can, however, be used as a

92 Schiaparelli, *Scritti*, vol. 2, pp. 23–42.

ladder that, like Wittgenstein's at the end of his *Tractatus logico-philosophicus*, should be discarded once the goal is reached.

We are, then, able to represent not only the Aristotelian astronomical system but also the homocentric systems of Eudoxus and Callippus by understanding their geometrical structure as the mathematical image of a mechanical system realisable in space-time. However, this *inevitable* approach is deeply *inappropriate* when applied to the Aristotelian system of the heavens.

Indeed, it is *inevitable* whenever we try to do something that neither Callippus or Eudoxus nor Aristotle seem *prima facie* to have done—that is, calculate the motions produced by their systems with great precision. It is true that our intention to show that these astronomical systems do actually describe reality was equally present in the Aristotelian perspective, but what is absent in Aristotle is the idea that it makes sense to mirror natural phenomena in an artificial way, for example, through calculus-oriented mathematical models (the images provided in this book are, just as surely, an expression of the limits of our time). Our understanding of ancient astronomical systems (*a priori* denying transphysical realities) implies the possibility of replicating their operation through a mechanism or device able to reproduce the phenomena faithfully, like a replica made to scale that does not essentially differ from its model. This requirement is, however, unfeasible in the context of Aristotelian thought, as the behaviour of the celestial bodies in the system differs from that of terrestrial bodies. As Aristotle introduces two physics, one for ethereal and one for terrestrial bodies, there is no human artifice (made with the terrestrial elements of earth, water, air, or fire) able to imitate the (ethereal) motion of the stars. Unlike our empirical, imitative, and poietic view of nature, Greek thought sees nature as if it were a work of art. In other words, the Greek aim is to understand the work by imagining all and even the slightest movements of the artist's hand, but knowing that it is not possible (nor required) to replicate it.

For Aristotle, the nature of every thing that appears before us is unique and inimitable and, thus, our knowledge of it must be direct. This notion of entity leads him to separate physics from mathematics and regard them as independent sciences in a certain sense, having distinct objects of study. Nevertheless, Aristotle uses mathematics to understand aspects of the physical world and considers astronomy to be a mathematical discipline. However, mathematics is unable to account for physical reality as a whole or even in its essential features—which is the aim of modern science—because if we consider only the mathematical aspect of an entity, it then loses its physical reality. For example, we could ask ourselves what makes an apple an apple? It is not its geometrical form or any of the numerical configurations we can impose on it,

not even the proportions of its component elements, which Aristotelian physics tends to dismiss. An apple is an apple because it appears before us as an original inimitable entity. The proportion of elements that form the matter of the apple is unique and different for each apple, but the condition of "being an apple" is what makes it knowable as such and as the holder of identifying sensible attributes. We cannot characterise the condition of being mathematically. And although the figure of the apple is somehow a necessary sensible attribute, without which we could not conceive the apple, it is not an essential attribute. For example, a stone sculpture of an apple can never be more than an image of a real apple. Informed matter is the specific entity and the subject of study of Physics. Not even the abstracted form, which constitutes its intelligible essence, can be simply identified, in Aristotelian thought, with a mathematical entity.

Clearly, this differs from our modern understanding of the relationship between mathematics and the physical world. We now identify mathematics with the intimate structure of reality and do *believe* that we can grasp the nature of things through numbers, knowing that the thing itself escapes us. We try to reach the essence of things by means of mathematical models that represent physical reality. This difference between the Greek and the current approach to objects of study may be interesting as an epochal marker, but, more importantly, our awareness of it is indispensable to achieve the self-consciousness required for a critical understanding of the Aristotelian system of the heavens. If at present we engage with categories that would have been inappropriate to the very framework in its time of development, they are, paradoxically, the best tools now available to shed light on system of Λ, 8, whose proverbial obscurity has been repeatedly pointed out throughout the Aristotelian tradition.

This *technical* approach to physical matters is characteristic of our current way of making science. It is not confined to the theories and fundamental laws of reality but is completed by a positive use of knowledge, which is consonant with our criterion to validate knowledge. For the current approach, which regards ancient astronomical systems as models of reality, the material axes linking the spheres are a technical or practical resource rather than a strictly explanatory aspect of the principles that underlie their behaviour, even if this representation never departs from a theoretical-mathematical level. However, the Aristotelian spheres do not rotate thanks to or forced by such axes. This fact compels us to shift, for a more accurate interpretation, from sublunary to superlunary physics. This is a key point, as the idea that spheres rotate on material axes—as if there were mechanical artefacts moving the axes—does not in fact come from Aristotle. That astronomical approach would, certainly, resolve a main aspect of his thought—the physical one, in a

modern sense—but it overlooks the fact that Aristotelian thought is inextricably metaphysical as well as physical, and that we can only grasp it through its intertwined nature. The first of the category duos proposed for comprehending ancient astronomical systems (the *theoretical-predictive* approach that prioritises their kinematic and geometric aspects) does not generate any difficulties when applied to systems such as those of Eudoxus or Callippus—as confirmed by Schiaparelli—but fails in the case of the Aristotelian system in so far as the introduction of transphysical Unmoved Movers concerns the highest philosophical level, the metaphysical, where the laws of (our) physics dissolve. More particularly, the notion of multiple spherical caps connected by axes of rotation is fruitful in Schiaparelli's analysis of the Eudoxian and Callippic systems, but becomes problematic when it seeks to establish a total and inclusive view able to exhaust all possibilities of the cosmological system of Aristotle. We shall see this when we look at Hanson's interpretation, which is discussed below.[93]

Faced with the limitations of the *theoretical-predictive* viewpoint, we shall turn to a complementary approach to the problem of the homocentric systems: the *theoretical-explanatory* approach. It is meant as a close study of specific aspects of the Aristotelian *sui generis* physics derived from its indissoluble link with the fundamental doctrines of the first philosophy.

Taking the *theoretical-explanatory* approach, we must bear in mind that, according to Aristotle, the Unmoved Movers that move the spherical celestial layers are the cause of circular movements—that is, their trajectories draw circumferences with a speed comparable to our lineal velocity. Current physics normally draws upon the more appropriate notion of angular velocity (unknown to Aristotle) to explain rotational motion. This distinction is significant, as it is the idea of angular velocity that induces us to think that the cause of the motion of the spheres is a physical mover, exerting its action on a material axis of rotation.

Aristotle's understanding of sphere's rotation is quite different. He believes that rectilinear and circular motions belong to different species (εἴδη) but treats both as if they were *linear* motions. Therefore, in both cases speed would depend on the time it takes the moving body to cover the length of its trajectory.[94] This consideration is comparable to the modern notion of *linear velocity* since, whether applied to rectilinear or rotational translations, it always refers to the same type of motion (or the same species if we are to use the Aristotelian term), at least from a kinematic perspective. This is despite the

93 Hanson, *Constellations and Conjectures*, pp. 33–88.

94 *Physics*, VII, 4, 248a24–248b12; 249a4–17.

fact that we use the term *tangential velocity* to denote the speed reached on a circular trajectory in the case of rotational motion. From the standpoint of classical physics, an unequivocal characterisation of all spherical caps through the notion of linear velocity is not possible, as a point on the surface of a rotating sphere will have different tangential velocities depending on its radius of gyration or radial distance from the axis of rotation in a perpendicular direction. The modern definition of *angular velocity* (the time rate of change of the angle of rotation) can be interchangeably applied to any point of a rigid rotating body.[95] Therefore, it is the most appropriate notion for the analysis of circular motion, despite the fact that it cannot be equated with the Aristotelian notion of *speed*, in any of its meanings. As a magnitude associated with the motion of the rotating whole, it is natural to locate the symbolic representation of angular velocity—the "angular velocity vector"—on the axis of rotation of the body, even if this is merely formal, given that this vector sets fixed parameters for rotational motion, such as its orientation and the invariant line around which the vector revolves. In order to fully understand the importance of the difference

95 Regarding the velocity of the planets, we should bear in mind the difference, in modern astronomy, between "angular velocity" or "angular displacement" and "linear velocity." Angular velocity measures the angular displacement per unit of time of a moving body relative to a fixed point, and it is usually the most appropriate concept to describe rotational motions. For celestial bodies with quasi-circular orbits around the sun, the natural point of reference to determine velocities, whether angular or linear, is the Sun. In turn, linear velocity equals the length covered by a celestial body along its orbit per unit of time. Because planetary orbits are actually elliptical and angular and linear velocities are not constant, a mean velocity or *mean motion* is assigned to each planet, which equals the angle swept (for angular velocity) or the arch covered (for linear velocity) per unit of time, just as if the celestial body followed a circular orbit at constant velocity. The angular velocity or, more precisely, the *mean angular motion* of the nine planets in the solar system is approximately: 4° per day for Mercury; 1.5° per day for Venus; 1° per day or 360° per year for the Earth; 0.5° per day or 180° per year for Mars; 30° per year for Jupiter; 12° per year for Saturn; 4° per year for Uranus; 2° per year for Neptune; and 1.5° per year for Pluto. In turn, the linear velocity corresponds to a more intuitive notion of velocity, thus resembling the Aristotelian *speed*, qualitatively speaking. From an astronomical point of view, the quantification of this velocity proves more difficult because it requires the measurement of distances along the real planetary orbits as regards the chosen system of reference (the Earth or the Sun) and the kind of knowledge demanded for this task was not available until the Copernican revolution. The mean orbital velocity of the planets in the solar system is approximately: 47.9 km/s for Mercury; 35.02 km/s for Venus; 30 km/s (approx. 108,000 km/h) for the Earth; 24.1 km/s for Mars; 13 km/s for Jupiter; 9.6 km/s for Saturn; 6.8 km/s for Uranus; 5.4 km/s for Neptune; 4.7 km/s for Pluto. The sun orbits around the centre of the galaxy at approximately 26,000 light-years, completing a revolution every 230 million years, approximately. It has a mean orbital velocity of approximately 220 km/s, covering an equivalent of the distance between the Earth and the Sun (149,000,000 km = astronomical unit) in just eight days.

between angular and linear velocity we should consider that in Aristotelian astronomy, unlike modern astronomy, orbital planetary motions are caused by the rotation of solid bodies, namely, the celestial spheres. Celestial bodies do not move by themselves but are carried by rigid bodies made of ether (spherical caps), each of them rotating around a symmetrical axis.

From a physical-mathematical point of view, which typically associates angular velocity with circular motions, it is not unusual to locate the *movers* on the *axes of rotations* when analysing systems of interweaved revolving spheres. Hence, a mover that generates angular velocity, or a rotational mover (such us an electric motor), would be the most appropriate choice for these systems. This idea of rotational motion, however, is impossible under Aristotle's *linear motion*: ideally, the velocity of any point on the axis of rotation of a sphere is zero (§ 25); it is, then, unnatural to imagine a mover where there is no motion at all. The notion of angular velocity is simply not appropriate for Aristotelian physics, being a mathematical notion that assigns the same velocity to all points of a sphere despite the fact that they have different linear velocities. The Aristotelian Unmoved Movers always cause motion directly and at the same speed. But if the movers of the spheres are not placed on material axes in the poles of rotation, we have to assume that, being transphysical entities, they are located nowhere. However, Aristotle states that they move on the *circumference* of the circles,[96] that is, they produce an action "exerted" on the equator of each revolving sphere (as if the equatorial circle of each sphere, driven by desire, the ὄρεξις, were to stretch itself towards the Unmoved Mover responsible for its motion, resulting in the motion of the entire sphere; see § 15; § 17). We believe that this alleged equatorial location of the Unmoved Movers confirms our previous hypothesis: from Aristotle's perspective, the Unmoved Movers cause linear displacements along circular trajectories. Therefore, each Unmoved Mover causes a single linear velocity—that we call tangential—directly moving the equator of its sphere, while the continuity between the first moved thing and the sphere is what indirectly causes the motion of the rest of the sphere.

The first conclusion one can draw from the above is that the *axes* of celestial spheres are neither adequate nor *necessary* to explain the motion of the spheres. In fact, Aristotle never claims that the spheres are fitted to one another through axes. Where we see axes of rotation, he mentions simply the existence of the unmoved poles (πόλους) of spheres.[97] One could very well imagine some kind of ring-shaped movers linking the spheres together and, thus, avoid the

96 *Physics*, VIII, 10, 267b6–9 –§ 25–.
97 *On the Heavens*, II, 2, 285b9–12.

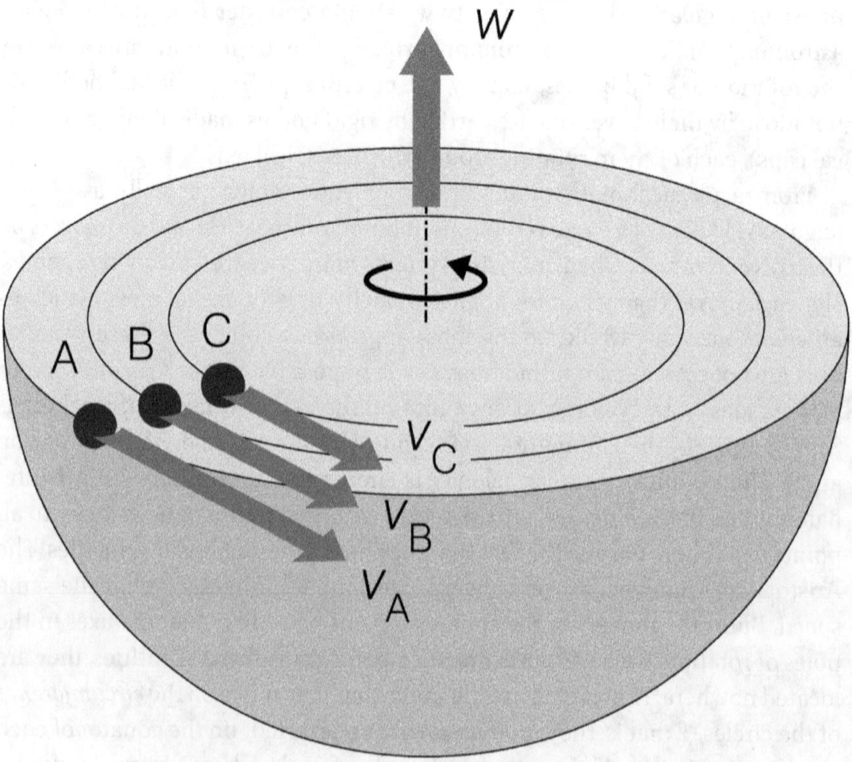

FIGURE 34 Each point of a solid rotating body located at a different distance from the axis of rotation moves at a different linear velocity. The further it is from the axis, the faster it moves (indicated by VA > VB > VC), while the angular velocity of rotation W (angular velocity per unit of time) remains the same for all.

need for axes. This would in fact constitute only a technical variation in our interpretation of the system as a *mechanism*, which remains inappropriate for the *theoretical-explanatory* approach to Aristotelian astronomy. We would be left, then, with the theoretical-predictive approach already criticised regarding its limitations. Because the Unmoved Movers are transphysical, they are incapable of materiality. Thus, there are insufficient grounds (whether textual or doctrinal) to claim that the axes (or any other physical device) occupy a primary role in the articulation of sphere motion. This is so despite some studies, based on a semi-classical mechanics (especially that of Hanson), that consider the Aristotelian system from a purely kinematic perspective, applying the same kinds and level of explanation here as to the Eudoxian and Callippic systems.

This does not mean that the poles of the spheres are inconsequential in the transmission of motion from one sphere to the other. In the majority of cases, the logical and simplest approach to the behaviour of these systems is

to assume that the outer spheres "carry," so to speak, the poles of the lower spheres. Still, this contact should not be visualised as part of a mechanical device having material axes with particular boundaries, as the Aristotelian system lacks this quality. Despite the fact that the contact between spheres at their poles may constitute the most direct way to understand the composition of motion, this contact is actually unnecessary. In the context of Aristotle's physical astronomy, one could very well assume that each sphere rotates around an axis defined by its poles while being carried by the motion of the immediately following upper sphere, without requiring any privileged point of contact between the two. In addition, nothing prevents us from alternately adopting one solution or the other, as Aristotle did not—as we shall see in the following section—consider the relation between the spheres to be always one and the same. Without these axes or other similar material means of contact, Newtonian physics cannot account for how certain motions (for instance, a pure rotational motion in a dynamic system of linked spheres, such as that of Aristotle) are caused or transmitted. Such difficulties do not arise in the geometrical analysis applied to the systems of Eudoxus and Callippus, in which one can imagine their conditions of motion without restrictions. The Aristotelian system, however, is not a mathematical but a physical one, which constitutes its most original characteristic within the Eudoxian and Callippic tradition. To seek a deeper understanding of his astronomical system necessarily means to face strictly physical issues, regardless of the viewpoint adopted.

In this sense, for Aristotle the notion of velocity, crucial for comprehending celestial motions, cannot be grasped through mathematics but is a strictly physical reality. It is easy to make the error of subjecting Aristotelian ideas to our own physical-mathematical intuition, which is always conditioned by a different approach to phenomena than that proposed by the Stagirite. The only way to comprehend the specificities of the system is to follow Aristotle's instructions, however scarce, ambiguous, or even unsatisfactory they might be according to our modern validation of physical theories. This main difference between the ancient Greek and the modern mental universe is a permanent source of difficulties in interpretation.

Here, a passage of the *corpus aristotelicum* comes to our aid. In Aristotelian physics, motions are transmitted provided that the spheres come into contact with one another.[98] The rotation of each upper sphere is then transmitted to the lower spheres, creating a superimposition of motions, and we arrive at the same result as if we had applied an axis-based model. However, nothing in Aristotelian physics leads to the conclusion that a given body will necessarily

98 *Physics*, VII, 2, 243a32–36; VII, 2, 244a14–17.

transmit its motion to a second body simply because they are in contact. As the following pages will show, Aristotle analyses motion by drawing logical consequences from his *concepts*, rather than from *empirical data*. From this perspective, and considering that the motion of a moved body is by definition accidental (κατὰ συμβεβηκός), it is possible for this body not to move but to remain at rest even when in contact with a mover.[99] Some examples of his celestial physics, such as are found in *On the Heavens*, II, 8, 289b1–290a7 –§ 27–, confirm this:

> Change is apparent in the position both of the stars and of the whole heaven [diurnal motion (motion a)], and this change must be reconciled with one of three possibilities. Either (1) both are at rest, or (2) both are in motion [solution S1], or (3) one is at rest and the other in motion [solution S2]. (1) For both to be at rest is impossible, if the earth is at rest, for that would not produce the phenomena; and the immobility of the earth shall be our hypothesis. There remain the alternatives that both move [solution S1] or that one moves and the other is at rest [solution S2].
>
> [*Description of solution S1: the celestial body and the sphere move independently*] (2) If both move, we have the improbable result that the speeds of the stars and the circles are the same, for each star would then have the same speed as the circle in which it moves, seeing that they may be observed to return to the same spot simultaneously with the circles [motion a]. This means that at the same moment the star has traversed the circle and the circle has completed its own revolution, having traversed its own circumference. But it is not reasonable to suppose that the speeds of the stars are related to one another as the size of their circles. That the circles should have their speeds proportional to their magnitudes is no absurdity, indeed it is a necessity, but that each of the stars in them should show the same proportion is not reasonable. If it is by necessity that that which moves in the path of the larger circle is the swifter, then it is clear that even if the stars were transposed into each other's circles, still that in the larger circle would be swifter, and the other slower; but in that case they would possess no motion of their own, but be carried by the circles [S2]. If on the other hand it has happened by chance, yet it is equally unlikely that chance should act so that in every case the larger circle is accompanied by a swifter movement of the star in it. That one or two should show this correspondence is conceivable, but that it

99 *Physics*, VIII, 5, 256b4–11.

should be universal seems a fiction. In any case, chance is excluded from natural events, and whatever applies everywhere and to all cases is not to be ascribed to chance.

[*Specific case within S1 that we will call S1C*] (3a) But again, if the circles are at rest and the stars move by themselves, the same absurdity arises and in the same way: for the effect will be that the stars which are far out will move faster, and the speeds will correspond to the size of the circles.

[*Description of solution S2: the moving sphere carries the celestial body*] (3b) Since then neither the motion of both [S1] nor the motion of the star alone can be defended [CS1], we are left with the conclusion that the circles move and that the stars stay still and are carried along because fixed in the circles. This is the only hypothesis that does not lead to an absurdity. That the larger circle should have the higher speed is reasonable, seeing that the stars are dotted around one and the same centre. Among the other elements, the larger a body the more swiftly it performed its proper motion, and the same is true of the bodies whose motion is circular. If arcs are cut off by lines radiating from the centre, that of the larger circle will be larger, and it is natural therefore that the larger circle should take the same time as the others to revolve. This too is one reason why the heaven does not spring apart, and another is that the whole has been demonstrated to be continuous [συνεχὲς ὂν τὸ ὅλον].[100]

In an attempt to demonstrate that celestial bodies are not capable of self-motion but are moved by the spheres that contain them, Aristotle explains the same phenomenon under two seemingly incompatible conditions—at least from our point of view. They are *solution S1*: the celestial body and the sphere that contains it move at the same speed and in the same direction; and *solution S2*: only the sphere moves by itself and the sphere carries the celestial body. And he considers a third solution (*CS1*) that can be ignored as it is a specific case within *solution S1*, that is, only the body moves while the sphere remains at rest. Aristotle considers that both solutions (*S1* and *S2*) are equally capable of explaining how the celestial body and the sphere move at the same speed (motion a). For current physics, solutions *S1* and *S2* contradict each other. That is, if *S1* is correct, the motion of the sphere would not affect the motion of the body and, if the body does not move by itself as per *S2*, it must be at rest while the sphere rotates. Therefore, only solution S2—wherein the body rotates with

100 *On the Heavens*, II, 8, 289b1–290a7; English trans. Guthrie, pp. 183, 185, 187.

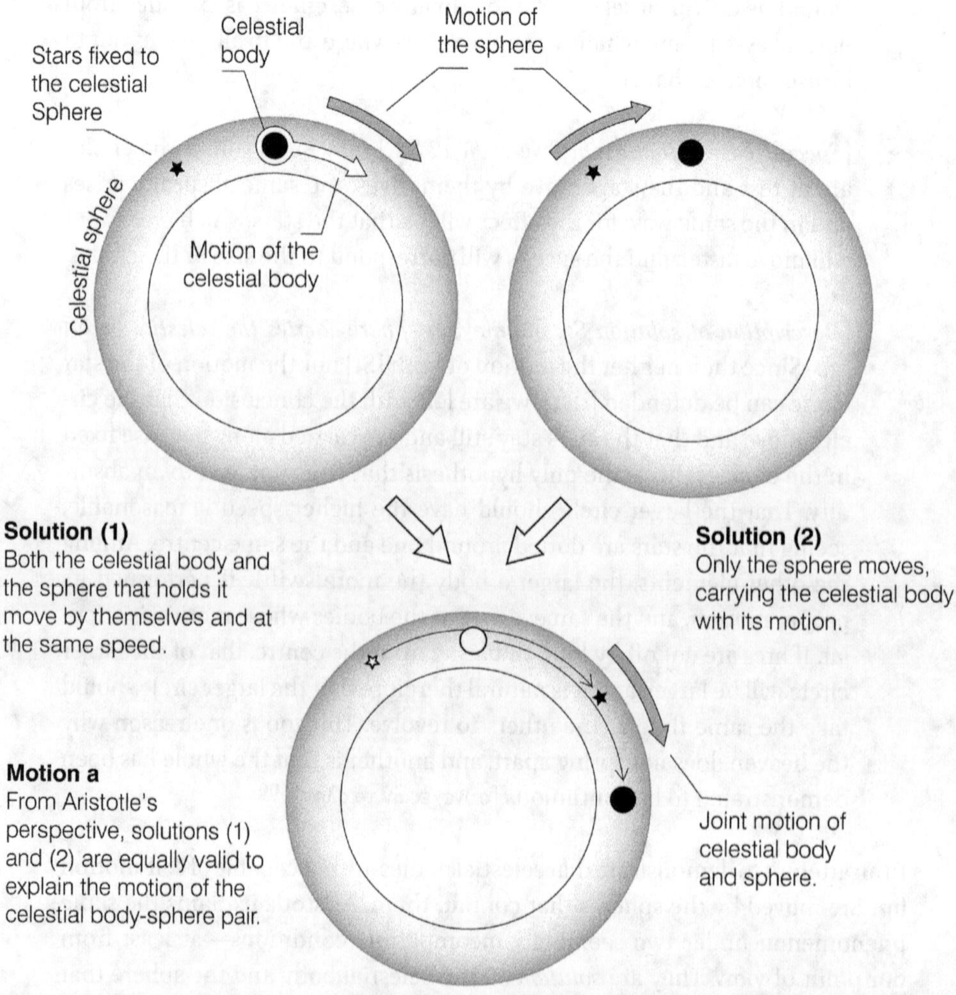

FIGURE 35 A representation of the two solutions presented by Aristotle in *On the Heavens*, II, 8, 289b1–290a7 –§ 27– to explain the motion of the celestial spheres and the bodies within them. That fact that solutions 1 and 2 are considered equally valid is closely linked to the Aristotelian notion of accidental motion.

the motion of the sphere—can explain motion a. If one accepts solution S2 and if the speed of the body equals the speed of the sphere, as suggested in *S1*, then the body would move twice as fast as motion a. Therefore, only *S2* can account for motion a. In conclusion, though both solutions would be possible under our notion of motion, they cannot be simultaneously valid nor can they cause the same phenomenon.

The essential difference between today's way of understanding and analysing phenomena and Aristotle's is that the former is purely physical, therefore the choice between solutions *S1* and *S2* is based on our knowledge about the nature of the contact between bodies. Referring always to the same sphere and celestial body, either the ether that constitutes the sphere flows around the body (also made of ether)[101] without friction, the body moving itself independently (*S1*); or the body attaches itself to the substance of the sphere and is carried by the sphere (*S2*). Either way, both possibilities cannot be true at the same time. Aristotle's reasoning follows a less empirical path and concludes that the correct solution is *S2*, given the "accidental" nature of the motion, as if the definition of accidental was an axiom of deductive reasoning.

From Aristotle's perspective, if we regard the sphere as the first mover (meaning that closest to the moved thing) and the celestial body as the moved thing, then any motion caused to the body would be accidental, admitting two possibilities: 1) the body moves or 2) the body is at rest relative to the mover. If the self-moving sphere does not cause the motion of the body, then it is possible that the body moves by itself: solution *S1*. However, if the sphere acting as mover moves a body incapable of self-motion, we would be left with solution *S2*. Thus it is possible to arrive at the same result (the visible motion of celestial bodies) from two completely different but equally valid approaches, in so far as both conform to the notion of "accidental motion." If both solutions (*S1* and *S2*) are equally able to account for motion a, and as they cannot apply simultaneously according to the principle of non-contradiction,[102] we need another argument to ultimately decide which of them is the most reasonable. Aristotle ends this passage dismissing—quite reasonably—solution *S2* as he deems it illogical that celestial bodies would move by themselves at proportional speeds according to their distance to the centre of rotation. He finds no valid reason for independent isolated bodies to behave in such an orderly fashion. On the contrary, it is reasonable that the spheres move at higher speeds as their radius decreases, and since the daily motion of celestial bodies manifest this behaviour, it must result from being carried by the spheres that contain them (solution *S2*). Ultimately, the Aristotelian solution has analytical and semi-empirical logical-deductive grounds, even if it always maintains its reference to sensible

101 *On the Heavens*, II, 7, 289a13–19.

102 As Aristotle says, in *Metaphysics*, Γ, 3, 1005b18–20: "which principle this is, we proceed to say. It is, that the same attribute cannot at the same time belong and not belong to the same subject in the same respect"; English trans. Ross in Barnes, *Complete Works of Aristotle*, vol. 2, p. 1587.

phenomenon given that it aims to explain what is perceived by the senses. However anachronistic it is to use modern categories for the classification of ancient thought, one could say that, in this respect, Aristotle was an empiricist forced towards rationalism by virtue of the limitations imposed on perception by the distance between himself and the phenomena studied.

That the Aristotelian solution to the problem of motion is less empirical than ours does not mean that his physics is less attached to phenomena. In fact, our own understanding of physical phenomena is also theoretical and speculative, though not in the metaphysical plane but in that of formal mathematics. Thanks to mathematical abstraction, which we assume as the matrix of the real, we approach phenomena that we cannot access through direct experience. In this sense, we could compare Aristotle's research on the heavens to our research on the origin of the universe or black hole physics. He is just as aware of the limitations he faces in his research as we are with respect to problems not yet solved by theoretical physics. Regarding sphere motion, he states in *On the Heavens*, II, 3, 286a3–7 –§ 16–:

> Now since there exists no circular motion which is the opposite of another, the question must be asked why there are several different revolutions, although we are far removed from the objects of our attempted inquiry, not in the obvious sense of distance in space, but rather because very few of their attributes are perceptible to our senses. Yet we must say what we can.[103]

The idea that a mover in contact with a body can either move it or not—implicit in the notion of accidental motion—is not surprising if we consider that Aristotelian physics ignores the *principle of action and reaction* as well as the concept of *inertial mass*. Let us consider, for instance, two consecutive spheres of any planetary subsystem taken from the Eudoxian and Callippic systems interpreted by Aristotle: spheres III and IV, which form the *hippopede*. While both spheres are very similar (they are made of ether and are of similar size), the outer sphere moves the inner sphere without affecting its own motion, thus violating the principle of action and reaction. In addition, the two spheres in contact move due to their corresponding Unmoved Movers but Aristotle considers that just the outer sphere III moves the sphere IV and not the other way around. In other words, the inner sphere IV, despite having its own motion, which earns it the title of mover (it moves the celestial body in its equator),

103 English trans. Guthrie, p. 149.

does not affect the motion of the sphere III in any way. Once again, this is a violation of the Newtonian Laws of mechanics, which demand certain symmetry in the interaction of bodies.

In Aristotelian physics, in order to produce motion, the mover must lean on something at rest, different from that which is moved (*Movement of Animals*, 2, 698b6–18). We will call that something "the point of support." The state of rest of the support is due to a power (δύναμις) or force (ἰσχύς) of its own. Thus, if the force of the mover is less than the force keeping the support at rest, then the support will remain at rest relative to the mover, allowing the mover to move by itself or move another body (699a33–699b6). In some cases, Aristotle views this force that maintains the state of rest of a given body against external force as being proportional or even equivalent to the idea of weight (βάρος), understood as an inherent property of the body (*Physics*, VII, 5, 250a8–20). He frequently uses this notion to refer to the tendency or gravity that drives terrestrial bodies towards their natural resting place. Even though this quality or property (gravity) is absent in the element of fire, it can be attributed to all the sensible sublunary bodies as they are composed of different proportions of all four elements, and the others (air, water and earth) do possess gravity to varying degrees (*On the Heavens*, IV, 5, 312b14–19). In this sense, the idea of βάρος and the modern idea of *weight* are quite similar but they differ significantly—as we shall see—regarding the force responsible for immobility. Weight is for us the measure of the gravitational force acting on a body due to the gravitational force of the Earth, and offers resistance only against upward motion along the line aiming towards the centre of the Earth. Resistance to motion in the Aristotelian element may, however, also be caused by lightness; that is, a body in its natural resting place resists both upward motion (through heaviness) and downward motion (through lightness). Moreover, when a body is at rest and—we may assume—in its natural place, the resistance that the mover has to overcome to put the body in motion does not seem to privilege any particular direction according to Aristotle's analysis. In this sense, resistance to motion would be similar—though just in appearance—to the modern idea of *inertial mass* rather than to our notion of weight. Aristotle's affirmation of a force that resists the change of state from rest to motion is clearly based on phenomenal evidence, but lacks solid explanatory grounds.

This understanding of motion leads, for instance, to his statement that it is not possible to move a ship if the point of support is within the ship itself (*Movement of Animals*, 2, 699a6–11). In such a scenario, if the force that tries to move the vessel is strong enough to do so, it would then also move the support, and in the opposite direction since, as Aristotle affirms: "For as the pusher

pushes [the point of support] so is the pushed pushed, and with equal force" (ὡς γὰρ τὸ ὠθοῦν ὠθεῖ, οὕτω τὸ ὠθούμενον ὠθεῖται –699b5, § 14–).[104] Clearly, such motion would not be possible as two equal and opposite forces applied to the same body cancel each other (699a37). This is not an Aristotelian explanation, strictly speaking, but can be deduced from his principles and is quite similar to that which we could provide using Newtonian physics. Aristotle, however, does not argue the impossibility of motion on the basis of a balance of forces—which is implicitly there regardless—but on the basis of a simpler and more intuitive reason: the point of support of the mover is not valid because it is within that which the mover is trying to move.

This conclusion is valid at a cosmological level as well. The motion of the universe cannot be caused, as some suggest, by an Atlas who leans on the resting Earth to move the Whole because, as Aristotle points out, it is evident that the force needed to move the heavens must be greater than the force maintaining the Earth at rest. Although the Stagirite does not justify this statement, he probably considers that the size, weight, or heaviness of the body of the heavens is much superior than that of the Earth, thus offering greater resistance to the planet's motion (699b1–12 –§ 14–). Under such conditions, if Atlas were to move the heavens, he would first—and to a larger extent—move the Earth. For the Greeks this is not possible, for they believed that the Earth was at rest at the centre of the universe. This might seem contradictory, as it puts into relation a certain resistance of a superlunary mass (which does not apply to ether) with the resistance to motion of a sublunary mass (which does makes proper sense). In fact, any kind of relation between these orders will prove problematic as, for Aristotle, they are subject to two different physics.[105] This apparent contradiction may be resolved by the idea that in that passage Aristotle is hypothetically agreeing with borrowed ideas, which do not reflect his views on this matter, in order to analyse and refute them.

---

104 *Movement of Animals*, trans. A. S. L. Farquharson in Barnes, *Complete Works of Aristotle*, vol. 1, p. 1088.

105 The resistance of Earth to motion is congruent with the Aristotelian notion of gravity, as the Stagirite believes that the natural resting place of Earth, towards which all bodies move by virtue of their gravity, is the centre of the universe and not some other place at which, in time, a majority of the element earth might be located. It is as if gravity responded to the famous formulation: "like moves to like" (τὸ ὅμοιον φέροιτο πρὸς τὸ ὅμοιον). If, for example, the Earth were to occupy the place of the moon, bodies made of earth would not move towards this new position but towards its previous place at the centre of the universe (*On the Heavens*, IV, 3, 310b3–5). It is only in this sense that one could agree with the statement "like moves to like" since, according to Aristotle: "to move towards its own place is to move towards its like" (*On the Heavens*, IV, 3, 310b11; English trans. Guthrie, p. 347).

Although the Aristotelian description of the way that bodies are set into motion conforms—to some extent—with the empirical evidence, it is an incorrect account of motion. The statement "as the pusher pushes so is the pushed pushed, and with equal force" (remarkably similar to Newton's *principle of action and reaction*: "to any action there is always an opposite and equal reaction") lacks, however, a clear formulation of the notion of force, or action, and it is different from the notion of *inertial mass*, drawn by Newton in the principles—later called "laws" by the didactic tradition that constitute the cornerstones of classical Mechanics.[106] The Aristotelian statement involves three different notions: the mover, the moved thing, and the point of support that the mover uses to move and which has to be outside the mover. Aristotle affirms that the mover exerts the same force on the moved thing and the point of support, which is not the same as claiming that the mover receives from the moved thing a reaction force equal in magnitude and opposite in direction than that exerted by the mover itself, as affirmed by the *principle of action and reaction*. The reaction force opposite to the motion and transmitted to the point of support is a consequence of the inertial mass of the body, not an obscure and diffuse "force of immobility," as Aristotle seems to suggest. According to the *principle of action and reaction*, a mover does not need a stationary point of support to move another body. For example, when a person walks on a the deck of a ship, their motion is produced by the reaction force of the ship as their feet exert a force against the floor. The ship, pushed, exerts—with friction—a force equal in magnitude and opposite in direction to that of the foot, moving the person forward. While the reaction force moves the person, the force exerted on the floor by the person moves the ship in the opposite direction without the need of a third body serving as an external point of support. All other things being equal, the motion of the ship will be proportional to the ratio between the inertial mass of the person and the inertial mass of the ship, and will thus be virtually imperceptible for a large ship, though never zero. The same would

106 The interrelated notions of *inertial mass*, *force* and *action-reaction*, which support our whole explanation of body interaction and motion, are defined by Newton in three laws included in his emblematic *Philosophia naturalis principia mathematica*, first published in 1687: "Law I: Every body perseveres in its state of being at rest or of moving uniformly straight forward except insofar as it is compelled to change its state by forces impressed"; "Law II: A change in motion is proportional to the motive force impressed and takes place along the straight line in which that force is impressed"; "Law III: To any action there is always an opposite and equal reaction; in other words, the actions of two bodies upon each other are always equal and always opposite in direction"; this English version from Newton, *The Principia: Mathematical Principles of Natural Philosophy*, trans. I. B. Cohen and A. Whitman, assisted by J. Budenz (Berkeley/Los Angeles: University of California Press, 1999), pp. 416–417.

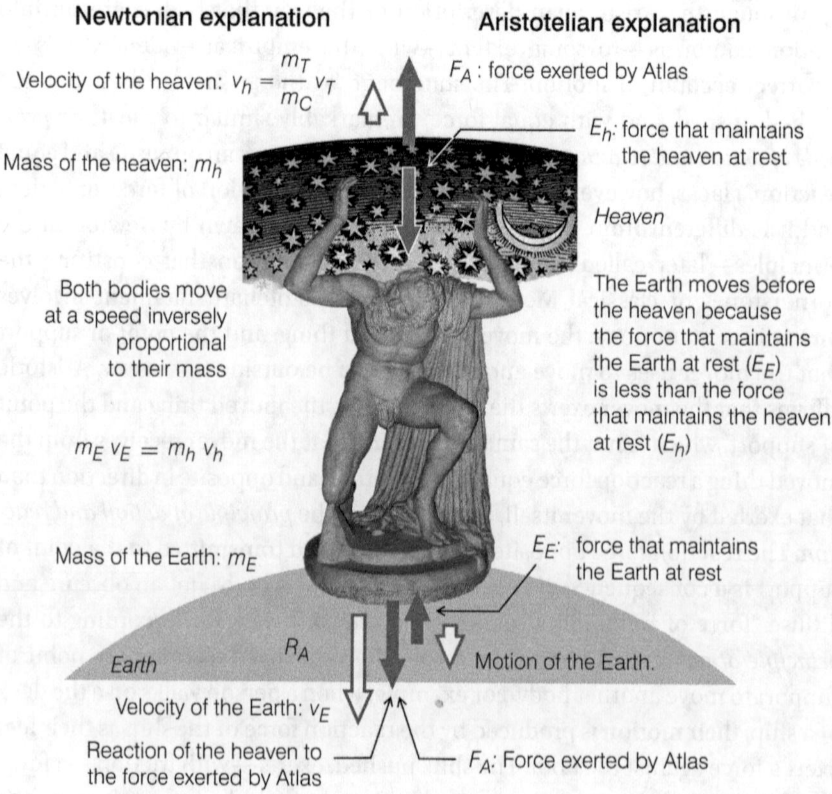

FIGURE 36 A physical explanation of the myth of Atlas, condemned by Zeus to carry the weight of the heavens on his shoulders (Hesiod, *Theogony*, 517). Aristotle mentions this myth in *Movement of Animals*, 2, 699b1–17 but rejects the interpretation according to which the giant leans on the surface of the Earth to exert a force causing the rotation of the heavens. On the right side is a synthesis of the Aristotelian explanation and on the left, an interpretation of the same phenomenon under Newtonian physics. According to Aristotelian physics, the resistance of a body against motion is proportional to its weight (βάρος). Therefore, since the heavens are heavier than the Earth, only the latter would move as a result of the effort exerted by Atlas simultaneously and equally on both the heavens and the Earth. According to classical physics, under the law of action and reaction and the law of conservation of linear momentum, both bodies would move, though at speeds inversely proportional to their masses.

be true in a scenario in which weight is absent. For instance, if an astronaut, without a point of support, were to push a spaceship in outer space where the gravitational interaction can be disregarded, both the astronaut and ship would move, though at velocities inversely proportional to their masses.[107]

This understanding of phenomena demands not just a vague intuitive perception of the *principle of action and reaction*, as that exhibited by Aristotle, but also a comprehension of the nature of resistance offered by the bodies acting as points of support in Aristotle's formulation. This involves the notion of inertial mass, which was unknown to the Stagirite. For the case under consideration, inertial mass is a kind of resistance that maintains bodies at rest, provided that they be at relative rest with respect to the mover acting upon them. Resistance is not, however, a force but a property of the state of aggregated matter that we call mass. Aristotle also believes that bodies at rest have an intrinsic quality that allows them to maintain their state, but he wrongly attributes this to a force or potentiality of the same nature as that which attempts to move the body. Under this assumption, it is reasonable to believe, as Aristotle does, that the moving force must exceed this "intrinsic stationary force" if there is to be motion.[108] The inertial mass of a body does not prevent massive bodies from moving, regardless of the magnitude of the moving force. The tendency to rest is not caused by a force opposite to the moving action,

---

107 Two or more bodies that form an isolated system (in the sense that no force acts on them) interact by pulling and pushing each other due to the forces mutually exerted under Newton's Third Law, and will move in a way that validates the *law of conservation of linear momentum*, which estates that: "If the sum of the external forces on a system remains zero, the total momentum of the system remains constant." The linear momentum is an important physical magnitude used to establish the dynamic state of a body. It is expressed as the product of the mass of an object (m) and its velocity ($\mathbf{v}$) at each moment in time: $\mathbf{p} = m\mathbf{v}$ (the bold type indicates that these are vector magnitudes). In the example provided, if a person whose mass is initially at rest ($\mathbf{v}_o = 0$) starts walking on a ship (or pushing a spaceship in outer space), whose mass is M and which is also initially at rest relative to the person ($\mathbf{V}_o = 0$), the momentum of both will be $\mathbf{p} = m\,\mathbf{v}_o + \mathrm{M}\,\mathbf{V}_o = 0$. Under the law of conservation of linear momentum, given the interaction between the person and the ship (or spaceship), as soon as the person starts moving, the ship (or spaceship), with velocity $\mathbf{v}$, will star moving with velocity $\mathbf{V}$ so that the total momentum remains zero ($\mathbf{p} = m\,\mathbf{v} + \mathrm{M}\,\mathbf{V} = 0$). As a result, they will move at velocities inversely proportional to their masses and in opposite directions, that is, $\mathbf{v} = -\mathrm{M}\,\mathbf{V}/m$ and $\mathbf{V} = -m\,\mathbf{v}/M$. All the velocities correspond to an inertial reference system in which Newtonian laws apply.

108 In classical physics, the idea of an actual force offering resistance to motion in conditions similar to those considered by Aristotle can only be related to the phenomenon of friction. If, for example, one tries to move a table, the force exerted has to be greater than the friction between the table and the floor if there is to be motion. Friction is not intrinsic to the bodies but a force originated by two surfaces coming into contact with each other, as well as a consequence of the principle of action and reaction.

even if it seems so intuitively (hence the name *fictitious force* or *pseudo force*), but by a certain resistance that conditions, rather than prevents, the motion caused by the actual force or action.

From a Newtonian understanding of motion, if two isolated bodies (such as the celestial spheres) are in contact with each other and one of them pushes the other, causing it to move, then, under the *principle of action and reaction* both bodies must move in opposite directions and their motions must be proportional to their corresponding masses, without the need to overcome any intrinsic force of immobility. This will be true also in Aristotelian physics, in which ethereal bodies have *inertial mass* but no *gravitational mass*,[109] the latter seeming to explain the Aristotelian notion of βάρος meaning "heaviness" or "weight," a property that ether clearly lacks. However, it is not possible to attribute *inertial mass* to celestial bodies even if the passage quoted above (*Movement of Animals*, 2, 699b1–12 –§ 14–) suggests that there would be a kind of potentiality of force to overcome in order to move the heavens. It would be closer to Aristotelian thought to assume that the heavens move without any mediating effort.

If we introduce the notion of inertia into the Eudoxian-based planetary systems, which are much like gyroscopes superimposed one within the other, they would cease to function the way Eudoxus, Callippus, or Aristotle imagined. The precession of the axes of rotation of the lower spheres caused by the upper spheres would generate inertial forces (or fictitious forces, in Newtonian terms) that would alter their motion and cause the breakdown of the entire system. To apply the concepts of classical physics to the *sui generis celestial mechanics* of Aristotle is, then, inappropriate. Firstly, being *transcendent*, the transphysical movers of celestial spheres (even if regarded as physical movers, such as a group of vigorous Atlases) would not be supported by anything from this world in order to move. In other words, it would be wrong to assume, for example, that the mover of the second sphere of Saturn, which is in contact with the first sphere of the planet, moves this first sphere by leaning on it. And this is true even when the first sphere pushes the second causing the daily motion of the planet. If the mover of the second sphere does not lean

---

109 For classical physics, the notions of inertial and gravitational mass are inseparable. One can deduce from the equivalence principle of Einstein's General Theory of Relativity that the inertial and gravitation mass are the same. These two notions are introduced to physics concerning different phenomena. The inertial mass determines the dynamic response of a body to the action of a force, while the gravitational mass is a property of material bodies that draws them to each other. It is well known that Newton laid down the relation between the mass of a body and the force of mutual attraction in his Law of Universal Gravitation.

on the first sphere, it should only move its own sphere. Secondly, though the modern notion of the *inertial mass* of material bodies (for us inseparable from the notion of *gravitational mass*) is unknown to Aristotle, it is not fair to apply the notion to his system because the matter of superlunary bodies, ether, being neither light nor heavy, is free of inertia and alien to our notion of "mass" (despite being a material entity). If ether existed, it would not be subject to Newton's laws of mechanics. This disagreement between Aristotelian and Newtonian mechanics is yet another proof that in order to comprehend Aristotelian physics, we have to study it on its own terms and dispense with the Newtonian conceptual framework. We must acknowledge and accept that Aristotelian astronomy, despite being a faulty or insufficient explanation of physical phenomena, is not without internal consistency.

We will return now to the initial fragment analysed (*On the Heavens*, II, 8, 289b2–290a7 –§ 27–). It is worth noting that Aristotle's reasoning to demonstrate the *accidental* nature of celestial motion is only valid if we regard celestial spheres as being part of a "continuous medium" that rotates with the period of the fixed stars. Only under these conditions can Aristotle justify that celestial bodies on spheres with a larger radius move at a higher speed than those on spheres with a smaller radius. He states expressly: "that the larger circle should have the higher speed is reasonable, seeing that the stars are dotted around one and the same centre" (289b35).[110] In view of the fact that in the system of *Metaphysics*, Λ, 8, these motions, decreasing as we move inwards, are produced by the first spheres of Callippus's subsystems, the analysis provided in *On the Heavens*, II, 8 supports the idea that the totality of these spheres must be regarded as a single system linked to the motion of the first heaven. (We shall examine this in more detail in the section "Two Celestial Systems.") We believe that this hypothesis is deduced from Aristotle's characterisation, at the end of the fragment, of the heavens as a *continuum* (290a6).

This is not strictly the Aristotelian notion of continuity (συνεχής)[111] but must be understood as a dynamic continuity underlying the system of fifty-five celestial spheres. There, the coordinated motion of the first spheres would be in a sense penetrated by other spheres responsible for the specific motions of each celestial body, whose motions differ from that of the first spheres. However, if we dismiss the idea that the celestial spheres are interconnected by material axes, as the *theoretical-predictive* approach suggests, the first spheres of Eudoxus and Callippus can be interpreted, as a group, as the continuous medium responsible for the daily motion of every celestial body (motion a),

110 English trans. Guthrie, p. 187.

111 *Physics*, V, 3, 227a10–17.

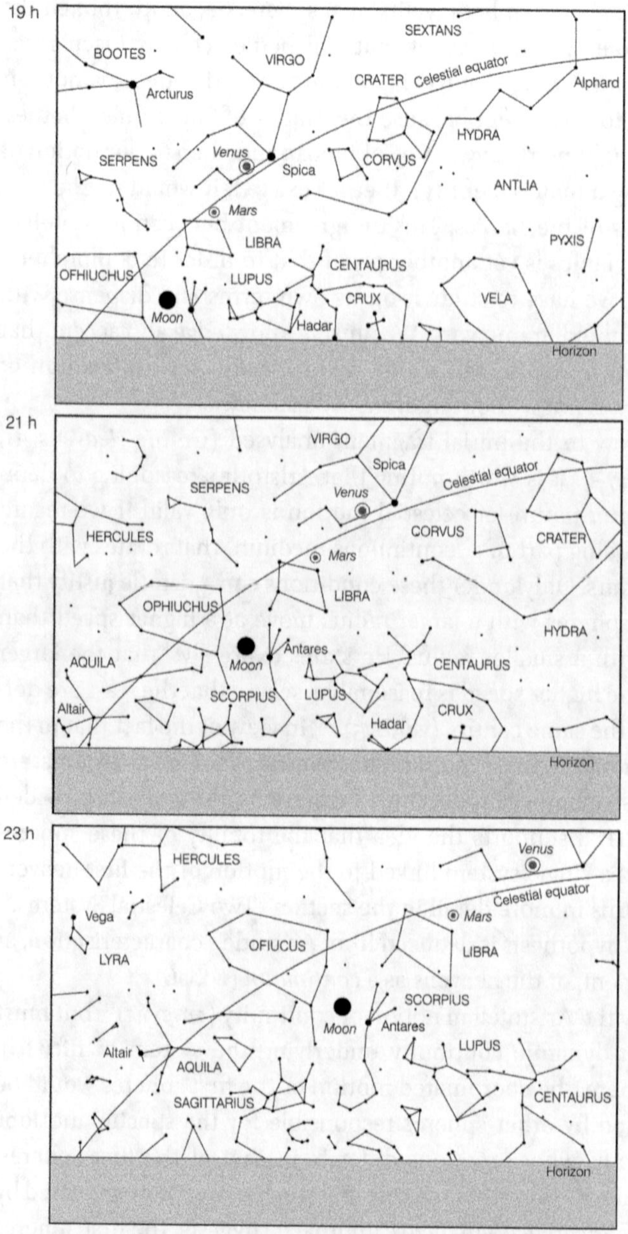

FIGURE 37 The night sky over the Aegean Sea as seen from the Eastern coast of the Attic peninsula during March, 343 BCE, estimated with the Stellarium simulation software. The images, taken with a few hours difference, show that the relative positions of the Moon, Mars, and Mercury are almost unchanged with respect to the background of the fixed stars, which gives the impression that all the stars move at once.

which, rather than being penetrated, penetrates the spheres responsible for specific planetary motions, thus transmitting the first movement of the heavens to each of these motions.

Our interpretation is supported by the fact that in the reviewed passage Aristotle considers only one celestial motion, completely ignoring the others, as part of the phenomenal evidence that celestial bodies move at speeds directly proportional to their distance from the centre of rotation. This is true if we consider solely the daily westwards motion that affects all celestial bodies (motion a) and gives the impression of the entire heavens as a continuum, a compact body that contains the celestial bodies within it and rotates without end in continuity with the fixed stars.

This way of understanding the relation among first spheres should not appear so strange to us. Anyone watching the night sky prior to the Copernican revolution would have felt the powerful first impression of a great force carrying the totality of celestial bodies in a single, regular, circular motion. In fact, Copernicus (following Aristarchus of Samos) was the first to question this, and most especially the *origin* of such a force. He believed that this show of the heavens—the daily revolution of stars—was better explained by the rotation of the Earth on its own axis. It is easy to imagine Aristotle, the most celebrated of theoretical minds, devoting time to the patient contemplation of the celestial night parade over the Aegean.[112] He notes in a few simple words: "but we ourselves see the heaven revolving in a circle" (οὐρανὸν ὁρῶμεν κύκλῳ στρεφόμενον; *On the Heavens*, I, 5, 272a5 –§ 27–).[113]

It remains now to clarify how the notion of a continuous first heaven can be brought together with the visible variation of the spheres that cut across it, rotating at different speeds and in different directions. Once we surrender to the specificity of Aristotle's supercelestial physics, such a problem, unconceivable for classical mechanics, can lead to a solution for the enigmatic astronomical scheme of *Metaphysics*, Λ, 8.

112 This way of approaching phenomena is similar to certain thought experiments in classical physics and akin to the mathematisation of nature that characterises our theoretical contemplation. It would be valid to assume that in *On the Heavens*, II, 8 Aristotle disregards the remaining planetary motions as belonging to a second order of things that can overlap with the central phenomenon: the daily motion. Similarly, when Galileo considers falling bodies, he disregards the air friction, which allows him to comprehend the fundamental laws of kinematics.

113 English trans. Guthrie, p. 37.

5.4 The Integration of Planetary Spheres

In order to understand the specificity of the Aristotelian proposal for the integration of planetary spheres, with its increase of the number of the Callippic spheres from thirty-three to fifty-five, we will now look at the composition of the overall system taking into consideration ideas so far expressed in isolation. The physics and metaphysics pertinent to Aristotle's approach to integrating his predecessors' planetary subsystems were expounded in the previous sections. The precise characteristics of the integration have, however, aroused controversy, especially as regards the number of rewinding spheres introduced by Aristotle in order to couple the Callippic subsystems to one another. If we adhere strictly to *Metaphysics*, Λ, 8—the only direct source of significance regarding the constitution of his astronomical system—we can accept the following:

1. There are as many Unmoved Movers as spheres necessary to describe celestial motions (1074a11–22 –§ 15–); Aristotle counts fifty-five movable spheres, thus the Unmoved Movers are fifty-five (putting aside the remark that reduces the number of spheres to forty-seven);[114]

114 Serious difficulties have arisen regarding the interpretation of this reduction in the number of spheres. Here we summarise two reasonable interpretations. The first is held by Ross, based on Pseudo-Alexander and described by Tricot in note Λ, 8, 1074a of his translation of *La Métaphysique*. It proposes that in saying "... if one were not to add to the Moon and to the Sun the movements we mentioned" (trans. Ross in Barnes, vol. 2, p. 1697), Aristotle considers returning to the Eudoxian system (three spheres) in absolute terms for these two bodies, excluding his own proposal of rewinding spheres in the case of the Sun (for the Moon, the rewinding spheres had already been deemed unnecessary in 1074a13 –§ 15–, as there are no stars below it). According to this, six solar spheres would be removed (the two added by Callippus + their two corresponding rewinding spheres [4 and 5 of Callippus] + the two rewinding spheres corresponding to the Eudoxian-Callipic spheres [2 and 3 of the Sun]). The second, a simpler interpretation that consistently conforms to the *Aristoteles astronomus*, proposes that only six spheres should be removed, instead of eight, which is why Aristotle must have counted forty-nine spheres instead of forty-seven. We agree that the reduction made by Aristotle is from fifty-five to forty-nine spheres. Therefore, the number forty-seven at the end of the passage can only be an addition error of either Aristotle or, more probably, his editors. This becomes clear if we focus on the astronomical description in the passage. Aristotle considers removing from the scheme the two spheres of both the Sun and the Moon added by Callippus plus the two rewinding spheres of the Sun, adding up to a total of six removed spheres. Most interpreters, however, choose to trust the Aristotelian text, even if the allusion to forty-seven spheres lacks astronomical meaning: the first proposal becomes a conceptual admixture to "save the textual phenomenon" (*Philosophus dixit*); the second proposal, in turn, is an attempt to "save the celestial phenomena," which certainly seems more reasonable.

2. The Unmoved Movers are hierarchically arranged according to the order set by the motions of the celestial bodies, the first being the Prime Mover, responsible for the motion of the first heaven or the outermost sphere of the subsystem of Saturn, which also contains the fixed stars (1073a34–1073b2 –§ 15–);
3. There are N-1 rewinding spheres for each Callippic subsystem of N revolving spheres (excluding the lunar subsystem) and they are arranged in such a manner as to counteract every motion of the revolving spheres, except for the motion of the first sphere (1073b38–1074a4 –§ 15–);
4. The first spheres of each Callippic subsystem integrated in the Aristotelian model of fifty-five spheres move but are not moved by other spheres, while the remaining spheres both move the lower spheres and are moved by the upper spheres.[115]

The most important and original feature of the Aristotelian system of fifty-five spheres is how it interposes the rewinding spheres to connect the Callippic subsystems of the seven planets. Through these spheres, Aristotle constructs a dynamic unit that reproduces the main planetary motions completely and simultaneously, rather than in the isolated way of the models of Eudoxus and Callippus. The paucity of first-hand information has led to several interpretations regarding the number of spheres and the way in which they act, and even the viability of the system has been challenged. In this respect, the mechanistic interpretation of Norwood Russell Hanson stands out, even when the matter had previously been addressed and commented—without further development—by renowned Aristotelians, such as David Ross. We will see now how Aristotle describes the integration of the spheres through the interposition of rewinding spheres in order to outline the difficulties brought about by this integration when adopting a mechanistic perspective (in a modern sense) such as Hanson's, which subscribes to what we have called the

115 If the whole system is to reproduce phenomena, this becomes a necessary condition on account of the number of rewinding spheres considered by Aristotle. As the following lines will show, this statement may constitute the corollary of our solution for the integration of motions in the Aristotelian system. In addition, the fact that the first spheres of each planetary subsystem move but are not moved by any of the prior spheres shows their ontological pre-eminence over the remaining planetary spheres as well as the pre-eminence of their movers, as this characteristic is shared with the Prime Mover. Moreover, the Prime Mover does not lose its superior hierarchical position among the spheres, as all the other motions are subordinated to that first principle in some way, as stated by Aristotle (*On Generation and Corruption*, II, 10, 337a17–22 –§ 18–). The *aesthetic* origin of such pre-eminence can, we suppose, be attributed to the strong impression caused by the diurnal motion of the heavens on those philosophers and, for that matter, on any terrestrial observer.

theoretical-predictive approach. Then, we will introduce our own interpretation, which attempts to follow the Aristotelian text in a strict sense while keeping in view the fundamental principles of his physics and metaphysics.

The only reference made by Aristotle to the number and arrangement of the rewinding spheres can be found in *Metaphysics*, Λ, 8, 1074a1–1074a5 –§ 15–:

> But it is necessary, if all the spheres combined [the spheres of Callippus's planetary subsystems] are to explain the phenomena, that for each of the planets there should be other spheres (one fewer than those hitherto assigned) which counteract those already mentioned and bring back to the same position the first sphere of the star which in each case is situated below the star in question; for only thus can all the forces at work produce the motion of the planets.[116]

All interpretations of this passage agree that the arrangement and action of the spheres must be as described below, taking the transition from Saturn's to Jupiter's planetary subsystem as the archetypal example. Given that the Callippic planetary subsystem for Saturn has four spheres, including the sphere of the fixed stars that gives the planet its diurnal motion, Aristotle states that there should be three interposed spheres between Saturn and Jupiter. We will call them rewinding spheres II', III', and IV', corresponding to the spheres II, III, and IV of Saturn. Sphere IV', placed immediately following sphere IV (the two having collinear axes and rotating with the same period but in opposite directions), cancels the latter's motion so that sphere III', the sphere that follows inwards in the system, receives only the motion of sphere III. In turn, the rotation of sphere III' cancels the rotation of sphere III. Sphere II', in immediate contact with III', does the same with sphere II. Consequently, any point on this last sphere (II'), the innermost sphere of the subsystem of Saturn (4 revolving + 3 rewinding spheres), will remain at relative rest with respect to sphere I.

In other words, if all the rewinding spheres are laid out in this manner, the last one (II') will be carried by the set of all the upper spheres, mirroring the motion of sphere I, which in this case is the sphere of the fixed stars.

Up to this point, all agree that this must be the arrangement and the action of Aristotle's rewinding spheres, and that this is the resulting motion of the ensemble of spheres. But now problems arise with respect to how the motion of the last rewinding sphere affects the first sphere of the next connected planet, in this case the first sphere of Jupiter. Hanson reasonably understands

116 English trans. Ross in Barnes, *Complete Works of Aristotle*, vol. 2, p. 1697.

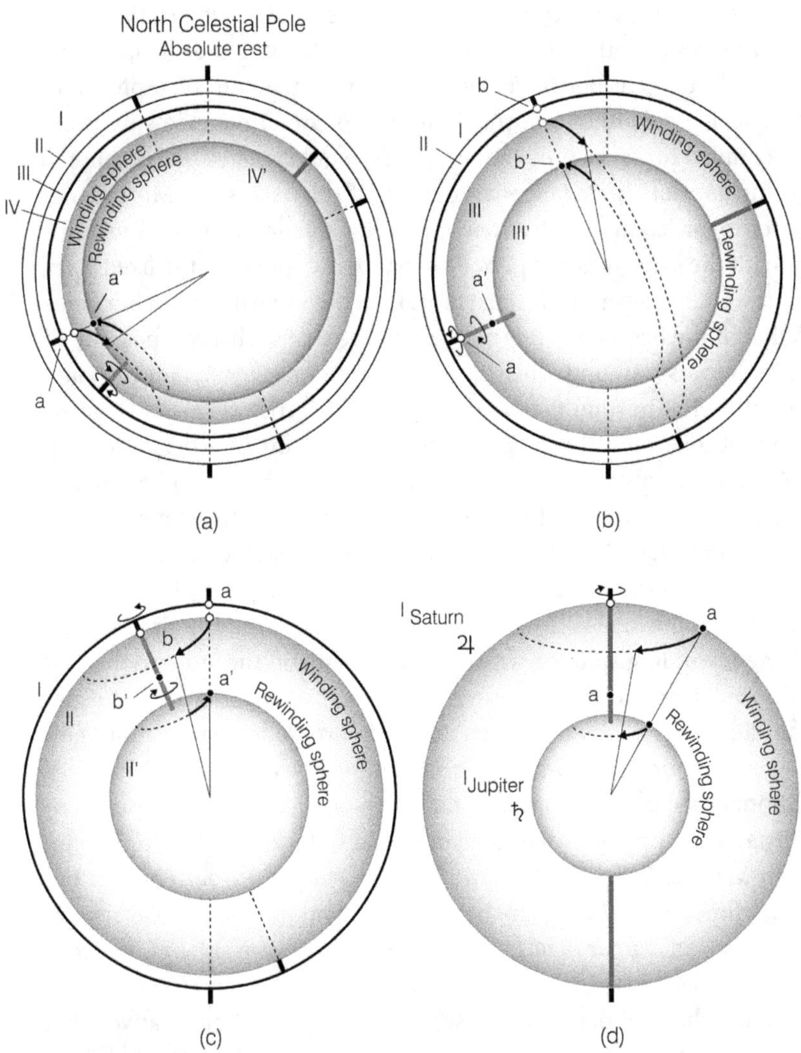

FIGURE 38 Examples (a), (b), (c), and (d) show the behaviour of the rewinding spheres introduced by Aristotle between Saturn and Jupiter.

In (a) the rotation of the first rewinding sphere IV' counteracts the rotation of planetary sphere IV, so that all the points on rewinding sphere IV' remain at rest relative to the revolving planetary sphere III, particularly at the point where the axis of inner rewinding sphere III' is fixed (indicated by letter a'). Sphere IV' cancels the revolution of sphere IV causing the axis of sphere III' to behave as if it was directly fixed to sphere III, as shown in (b). The same reasoning applies to the remaining pairs of revolving and rewinding spheres, as shown in (c) and (d). The interposing of rewinding (or counteracting) spheres between Jupiter and Saturn (one less than the revolving spheres) establishes a link between the planetary subsystems such that the first sphere of Jupiter is linked to the sphere of the fixed stars as if there were no mediation between them and they had coaxial rotation. As we move inwards, the subsequent planetary subsystems are in like manner linked to the system of Jupiter and to one another, and the first spheres of each subsystem behave as if connected to one another by their poles and sharing a coaxial rotation with the fixed stars.

that if, in the previous interactions between spheres, the motion of the upper spheres carries along with it the lower spheres, whatever their disposition, the same would happen between the last rewinding sphere of Jupiter (11') and first revolving sphere of Saturn. Under these conditions, the subsystem of Jupiter, coupled to the last rewinding sphere, would receive—according to Hanson—the *uncounteracted* diurnal rotation of the fixed stars. Moreover, if (as is proper to all Callippic subsystems) the first sphere of Jupiter moves by itself also with the same rotation period as that of the sphere of the fixed stars, then Jupiter would receive a rotational motion in the same direction but twice as fast as that of the fixed stars. It is not hard to see that such a composition of motions would cause, among other inconsistencies, an extension of the planetary orbit of Jupiter beyond the band of the zodiac. Aristotle was, of course, well aware that this is not the case.[117] Therefore, according to this interpretation, the Aristotelian system would not correspond to observable phenomena.

If the integration of the remaining planetary spheres into the system proved to be as it is interpreted by Hanson—and as we have described for Saturn and Jupiter—then the diurnal rotation of the stars, all sharing a 24-hour period, would increase progressively. Thus, according to Hanson's interpretation of the action of the rewinding spheres, as we move inwards in the system, the first sphere of Mars, linked to the subsystem of Jupiter, would move three times faster than the sphere of the fixed stars. Instead of reproducing the motion of the fixed stars, the first spheres of each planetary subsystem would increase their rotational speeds as we approach the Earth, reducing the supposed periods of diurnal revolution of the celestial bodies as follows: 24/2 hrs. for Jupiter; 24/3 hrs. for Mars; 24/4 hrs. for Mercury; 24/5 hrs. for Venus; 24/6 hrs. for the Sun; and 24/7 hrs. for the Moon. Such a system would produce a solar revolution every 4 hrs. and a lunar revolution approximately every 3.25 hrs., which is clearly not the case.

To overcome these difficulties, Hanson proposes several alternatives that invariably deviate from the Aristotelian text and can only be admitted if we accept that the source text (*Metaphysics*, Λ, 8) contains some errors of description, that its intention was to convey something different from what is actually expressed.

The figure below summarises—for clarity and to better understand our main criticism of Hanson's interpretation—different interpretations of the systems of Eudoxus, Callippus, and Aristotle.

The table following this provides is a brief account of the three solutions proposed by Hanson to the problem he detects in Aristotle's integration of the

117 *Meteorology*, I, 8, 345a19–22.

Callippic systems. In the table, these solutions appear under the labels Aristotle C and Aristotle D.[118] The first solution (Aristotle C) is to add one rewinding sphere to each planetary subsystem in order to counteract the motion of the first sphere of the subsystem. This prevents the rotational motion of the first spheres from being transmitted from the upper subsystems to the lower celestial bodies as—Hanson assumes—occurs in the system of fifty-five spheres.

If we agree that these spheres have collinear axes with the first planetary spheres—which supposedly rewind—and with the first heaven, we can easily see that when joined with the remaining upper spheres, they would come to be at absolute rest. Thus, the spheres connected to these (the first spheres of each planetary subsystem, as Hanson proposes) could rotate only by their own motion; that is, the motion caused by their corresponding Unmoved Movers. The result would be a system of sixty-one spheres.

Although this solution restores the proper functioning of the system within Hanson's interpretation, it presents two obstacles. Firstly, it is mere speculation, for nothing indicates that Aristotle considered things to be so. Secondly, and strangely enough, this solution adds to the Aristotelian scheme of the heavens several bodies at rest (those corresponding to the last rewinding spheres added by Hanson), a condition which is exclusive to the sublunary realm. About this, Aristotle refers to the divine nature of the heaven and its rotational motion (*On the Heavens*, II, 3, 286a12–15 –§ 16–):

> Because when a body revolves in a circle some part of it must remain still, namely that which is at the centre, but of the body which we have described no part can remain still, whether it be at the centre or wherever it be.[119]

The second solution suggested—and preferred—by Hanson is to remove the first revolving sphere of each subsystem, as he considers them redundant. With this change (Aristotle D), the system works as expected, with forty-nine integrated spheres. Again, there is no reason to believe that Aristotle actually thought about this possibility, and indeed in Λ, 8, which provides the basis for these models, he expresses that if eventually some spheres should be removed, it would not be these same as are suggested by Hanson. This in itself

118 The Aristotle E scheme (65 spheres) is based on a hypothesis by T. H. Martin, in "Mémoire sur les hypothèses astronomiques d'Eudoxe, de Callippe, d'Aristote et de leur école," *Mémoires de l' Académie des Inscriptions et Belles-Lettres* 30, 1, 1881, pp. 263–264, mentioned by Hanson in *Constellations and Conjectures*, pp. 78–79.

119 English trans. Guthrie, pp. 149, 151.

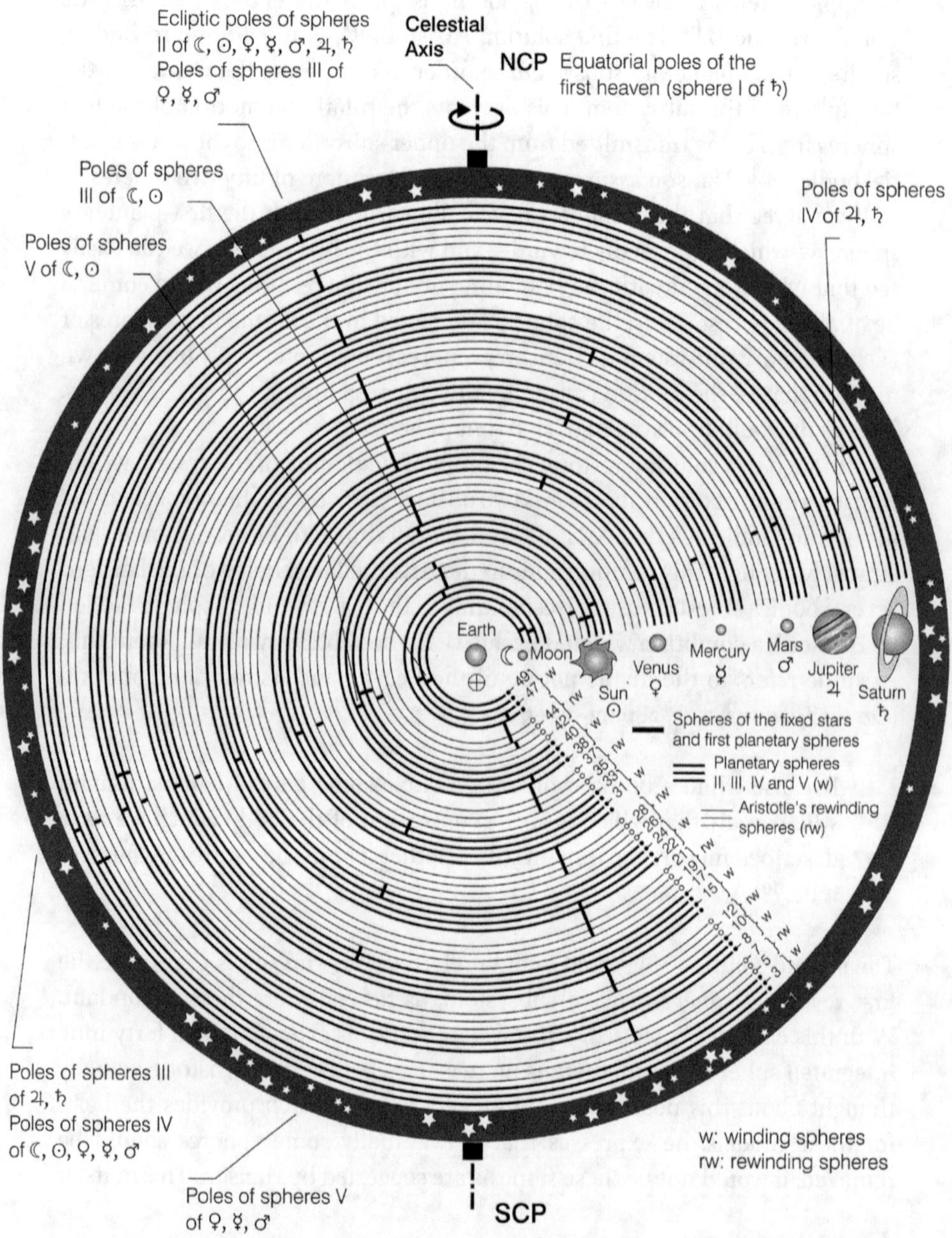

FIGURE 39 System of forty-nine spheres modified by Hanson. If the first planetary spheres were carried by the rotation of the fixed stars, they would lack the necessary motion for the system to work properly, rendering them redundant and reducing the system to forty-nine spheres. Such a system can be objected to for two reasons: it deviates from the text of *Metaphysics*, Λ, 8 and disregards the *continuity* between the sphere of the fixed stars and the first planetary spheres (*On the Heavens*, II, 8, 290a6–7).

Celestial bodies	Eudoxus Aristotle	Eudoxus Heath	Callippus Aristotle	Callippus Heath	Aristotle A	Aristotle B	Aristotle C Hanson	Aristotle D Hanson	Aristotle E Martin-Heath-Hanson
	GEOMETRIC SYSTEMS				INTEGRATING SYSTEMS				
Fixed stars	1	1	1	1	$1_w +$	$1_w +$	$1_w +$	$1_w +$	$1_w +$
Saturn	3	4	3	4	$3_w + 3_{rw} = 7$	$3_w + 3_{rw} = 7$	$3_w + 3_{rw} = 7$	$3_w + 4_{rw} = 8$	$3_w + 4_{rw} = 8$
Jupiter	4	4	4	4	$4_w + 3_{rw} = 7$	$4_w + 3_{rw} = 7$	$3_w + 3_{rw} = 6$	$4_w + 4_{rw} = 8$	$4_w + 4_{rw} = 8$
Mars	4	4	$4 + 1 = 5$	$4 + 1 = 5$	$5_w + 4_{rw} = 9$	$5_w + 4_{rw} = 9$	$4_w + 4_{rw} = 8$	$5_w + 5_{rw} = 10$	$5_w + 5_{rw} = 10$
Mercury	4	4	$4 + 1 = 5$	$4 + 1 = 5$	$5_w + 4_{rw} = 9$	$5_w + 4_{rw} = 9$	$4_w + 4_{rw} = 8$	$5_w + 5_{rw} = 10$	$5_w + 5_{rw} = 10$
Venus	4	4	$4 + 1 = 5$	$4 + 1 = 5$	$5_w + 4_{rw} = 9$	$5_w + 4_{rw} = 9$	$4_w + 4_{rw} = 8$	$5_w + 5_{rw} = 10$	$5_w + 5_{rw} = 10$
Sun	3	3	$3 + 2 = 5$	$3 + 2 = 5$	$5_w + 4_{rw} = 9$	$(5 - 2)_w + (4 - 2)_{rw} = 5$	$4_w + 4_{rw} = 8$	$5_w + 5_{rw} = 10$	$5_w + 5_{rw} = 10$
Moon	3	3	$3 + 2 = 5$	$3 + 2 = 5$	5_w	$(5 - 2)_w = 3$	4_w	5_w	5_w
comets, shooting stars									
Total	**26**	**27**	**33**	**34**	**55**	**(¿47?) 49**	**49**	**61**	**65**

It accounts for motions a, b and c and proposes independent kinematic geometric systems for the wandering stars, as it does not take into consideration the causes of the motions. The sphere of the fixed stars is the first sphere of Saturn.

It accounts for motions a, b and c, as long as we regard the systems as independent from one another. The sphere of the fixed stars is different and separated from the first sphere of Saturn.

It accounts for motions a, b and c, as long as we regard the systems as independent from one another. The sphere of the fixed stars is different and separated from the first sphere of Saturn.

It accounts for motions a, b and c and improves the Eudoxian systems by introducing the uneven duration of seasons, correcting the irregularities in longitude of the lunar orbit and improving the apparent motions of Venus, Mercury, and specially Mars. They are kinematic independent systems, like those of Eudoxus.

It accounts for motions a, b and c. In order to understand this model, we first have to agree that the first planetary spheres reduplicate the motion of the fixed stars, thus constituting a dynamic continuum. It is an improved overall system intended as a *sui generis* mechanical integration of the Callippic systems, which become subsystems in the Aristotelian interpretation.

It is essentially similar to **Aristotle A**, except for the unexplained and confusing reduction in the number of solar and lunar spheres. There may also be an error in *Metaphysics Λ*, 8 (made by either Aristotle or his tradition): where it says 47 it should say 49. If we follow this interpretation, the resulting system (49 spheres) would function in a similar way to **Aristotle A**, though not entirely accounting for motions c of the sun and the moon.

It accounts for motions a, b and c. Hanson interprets that the first sphere of each planetary subsystem moves by itself. If that was the case, the Aristotelian system of 55 spheres (**Aristotle A**) would be inadmissible. Under this interpretation, the system only works if the first sphere of each subsystem is removed, except for that of Saturn, resulting in a system of 49 spheres (**Aristotle C**, similar to **Aristotle B**, though only in number); or if another rewinding sphere is added to each counteracting subsystem, resulting in a system of 61 spheres (**Aristotle D**). Apart from seriously deviating from *Metaphysics Λ*, 8, these interpretations only consider the kinematic aspect of the Aristotelian system disregarding some fundamental philosophical points of his doctrine.

If we adopt the remaining characteristics of Hanson's interpretation in **Aristotle D**, this model would account for motions a, b and c as well as for some sublunary phenomena, such as comets and shooting stars. It adds some of further meteorological aspects raised by Aristotle to the system developed in *Λ*, 8.

w: winding spheres, rw: rewinding spheres

FIGURE 40 Interpretations of the homocentric systems of Eudoxus, Callippus, and Aristotle. The celestial motions explained by these systems are: a) a uniform rotational motion that carries all visible bodies westward, completing a revolution approximately every 24 hrs. and maintaining the relative position of the fixed stars; b) a circular motion of the planets in the opposite direction, within independent orbits, all of which are contained within the band of the zodiac, and with particular periods for each celestial body; and c) the irregularities in the observed motions of the Sun and the Moon as well as the wandering motions of Venus, Mercury, Mars, Jupiter and Saturn.

renders this interpretation completely inadmissible. Finally, Hanson's third solution proposes that, excepting the first sphere of the overall system (the sphere of the fixed stars), the first sphere of each planetary subsystem would not be self-moving. Under these conditions, the first sphere of each planetary subsystem would be carried by the motion of the first heaven, completing a revolution every 24 h, as expected. Again, there is no reason to concede this possibility, for it is inconsistent with Aristotle's celestial dynamics. Even if this solution maintains the Aristotelian number of fifty-five spheres, it contradicts the statement that there are as many Unmoved Movers as celestial spheres.

In order to understand how it is possible for the system of spheres to work as expected under the exact conditions described by Aristotle, we must consider the nature and the dynamic characteristics of the systems in light of Aristotelian principles. First, we should note that the Callippic subsystems integrated by Aristotle are purely geometrical and are, through his act of integration, forced into a sensible world, with material spheres made of ether. The model, however, does not need to be physical to work. It is enough to imagine each sphere as an abstract geometric body, rotating freely with its poles linked to and carried by the surface of the upper sphere. Ether is the substance that translates the perfect properties of geometric bodies into the physical world, as it lacks the impure properties of sublunary bodies, which are closer to our notion of material bodies. Although Aristotle attributes matter to celestial bodies, he does not at all modify the operating conditions of the inherited systems. As each upper sphere revolves on its axis, it carries the poles of the next lower sphere, causing it to move along a circular trajectory. Within these geometrical systems, motion is transmitted from one sphere to the next provided that the axes of rotation of both spheres form an angle different from 0°, and such is the case in the models of Eudoxus and Callippus. Two rotating coaxial spheres connected by their poles—if we regard them as mathematical points—cannot interfere with each other in any way, as the rotational pole represents a point at rest relative to the sphere itself and, as such, cannot cause any motion, either in Aristotelian physics or our own. The following passage attests to Aristotle's understanding of Eudoxus and Callippus's celestial spheres in a purely geometrical sense. In answer to objections regarding the circular motion of the heavens as a whole, he refers to the poles of celestial spheres as points at relative rest with respect to the sphere itself (*Movement of Animals*, 2, 699a13–24 –§ 14–):

> Here we may ask the question whether if something moves the whole heavens this mover must be immovable, and moreover be no part of the heavens, nor in the heavens. For either it is moved itself and moves the

> heavens, in which case it must touch something immovable in order to cause movement, and then this is no part of that which cause movement; or if the mover is from the first immovable it will equally be no part of that which is moved. In this point at least they argue correctly who say that as the sphere is carried round in a circle no single part remains still; for then either the whole would necessarily stand still or its continuity be torn asunder; but they argue less well in supposing that the poles have a certain power, though they have no magnitude, but are merely termini or points. For besides the fact that no such things have any substantial existence it is impossible for a single movement to be initiated by what is twofold; and yet they make the poles two.[120]

This passage clearly shows that Aristotle conceives the celestial spheres as geometric bodies. He affirms that the force of rotational motion cannot come from the poles, as they are geometrical points. In other words, they do not exist as bodies and are always at rest relative to their own sphere. This, however, does not prevent him from stating that no part of the sphere is at rest when the sphere rotates around such poles. The passage is consistent with the Aristotelian statement that the principle of physical motion of celestial spheres must originate in their equator and not in another place, particularly not in the poles.[121]

At this point, we need to shed some light on the Callippic subsystems for Mars, Mercury, and Venus. As regards their arrangement, we have assumed that the second and third spheres have a coaxial articulation, thus the upper sphere transmits its motion to the lower sphere, which opposes Aristotle's statement that the poles cannot produce motion by themselves. When judging this seeming contradiction, we should bear in mind that, as Heath points out, quoting Simplicius, there are no textual grounds to affirm that this was the exact disposition considered by Callippus when he added one more sphere to the Eudoxian subsystems. Following what we have called the *theoretical-predictive* approach, Schiaparelli was the one who—adopting a criterion of his

120 English trans. Farquharson in Barnes, *Complete Works of Aristotle* vol. 1, p. 1088.

121 The Aristotelian claim that motion must originate in the equator is not enough to assert that the Prime Mover is an efficient cause. It seems that Aristotle approaches the first cause of motion first from empirical evidence and then, upon reaching a point at which physics can offer no further explanation, looks to metaphysics. This also implies a shift from efficient cause to final cause. In the analysis of sphere motion, his physics demands that the motion originates in the equator. Of course, this is an explanation arising from the physical order; in the metaphysical realm it makes no sense to "situate" an Unmoved Mover and much less to think of its contact with a moved thing.

own—placed the sphere added by Callippus between the second and third spheres of the Eudoxian model for these planets, as indicated and graphically represented earlier in the section, "The Prime Mover and Unmoved Movers." Schiaparelli assigns the revolution periods and orientations to the three inner spheres of these planets with the intention of reproducing, as truly as possible, the motions of the planets according to what we assume was Callippus's knowledge of their trajectory parameters, knowledge which apparently bettered that of Eudoxus.

In summary, the arrangement that has three inner spheres replace the two that together form the hippopede in the Eudoxian subsystems—normally adopted for the systems of Mars, Venus and Mercury—remains merely conjectural, even if it does reproduce phenomena quite accurately and fully satisfies the intellectual approach that conceived it. Again, this arrangement answers only our requirement that it reproduce as fully as possible the phenomena. Therefore, considering the scope intended for Schiaparelli's interpretation, objections that may be valid for Aristotle's astronomical physics do not apply to these systems. Moreover, we must take into consideration that, whatever the original structure thought by Callippus, these systems are not Aristotelian. They were thought by geometers and there is no evidence that Aristotle studied them in detail or tried to strictly adapt them to his own physics. It is interesting, however, that Callippus added—as we suppose—an additional coaxial sphere to the second sphere of Eudoxus's subsystems for Mars, Venus and Mercury, since the effect of this addition is exactly equivalent to that of a single sphere with a period equal to the difference between the periods of the first and second spheres, given that they move in opposite directions. These considerations raise questions about the actual order of spheres formulated by Callippus and allow us to think—contrary to reasonable assumptions—that he may have imagined some kind of angle between the second and third spheres of these subsystems to justify its presence.

We must be more rigorous when interpreting how Aristotle understood the behaviour of rewinding spheres, as they constitute an original contribution to the astronomy of his predecessors. To begin with, each rewinding sphere in the Aristotelian system is aligned, pole to pole, with the upper spheres that they compensate. This kind of coaxial connection is most evident between the revolving spheres in the inner part of Callippus's planetary subsystems and the first rewinding spheres introduced by Aristotle, for example between the spheres IV and IV' of Jupiter. But in such a case, if the poles are understood as mathematical points, two spheres connected only by the poles cannot interfere with each other's motion in any way. For the rewinding sphere IV' of Saturn to counteract the rotation of sphere IV—as supposedly intended by

Aristotle—we have to admit another kind of contact between them. Such contact must be related to the physical nature of the spheres, not to the kind of contact considered so far between the abstract revolving spheres *per se*—it cannot be explained from a purely geometrical perspective. In that, it differs from what animates the planetary models of Eudoxus or Callippus. If the sphere IV' of Saturn is to counteract the rotation of sphere IV, we have to assume that sphere IV' is somehow dragged or carried by the rotation of sphere IV in such a way that, as it rotates in the opposite direction, it counteracts the motion received. As we have already explained, in Aristotle's celestial geometrical physics, this kind of effect on the motion of a sphere cannot occur merely by the mutual contact of poles, as no motional force can originate there. It must be caused by some kind of contact between sphere surfaces, as if there is no dissolution of continuity between them, such that the motion of the upper sphere carries the lower sphere while at the same time allowing the lower sphere to move with an equal and opposite motion to that above. This last condition would allow the rewinding sphere to serve its role as a compensator of motion.

It is true that an approach based on Newtonian physics—such as our own—cannot easily grasp this phenomenon. We would have to accept the idea of a lower rewinding sphere that "clings" to the upper sphere in order to be carried by it while, at the same time, "detaching" from it to move freely in the opposite direction. A similar contact takes place when we walk on an escalator at the same velocity as the machine but in the opposite direction. To the outside observer it would seem that we are at rest. More similar to the behaviour of the spheres would be the motion of a person walking in circles on a rotating platform, moving at the same velocity as the platform but in the opposite direction.

Now, this difficulty, which belongs to our way of understanding physics, is not an obstacle to accepting the Aristotelian system. Our physics is not that of Aristotle, and even his terrestrial physics—which may bear some intuitive resemblance to ours—is not that of his astronomy. His incomplete knowledge of the mechanics of celestial bodies—of which he was well aware, on account of the distance of observation—allow us some concessions for the sake of understanding the phenomenon as appropriate within Aristotelian thought. Centrally, as superlunary entities are composed of ether, a substance that cannot be fully experienced from the distance of Earth, knowledge about them is inevitably vague, restricted to perception through observation. Aristotle is aware that his knowledge is limited with respect to celestial bodies, as our own knowledge remains limited today.[122] It is—we believe—in an attempt to

122 *On the Heavens*, II, 3, 286a3–7 –§ 16–.

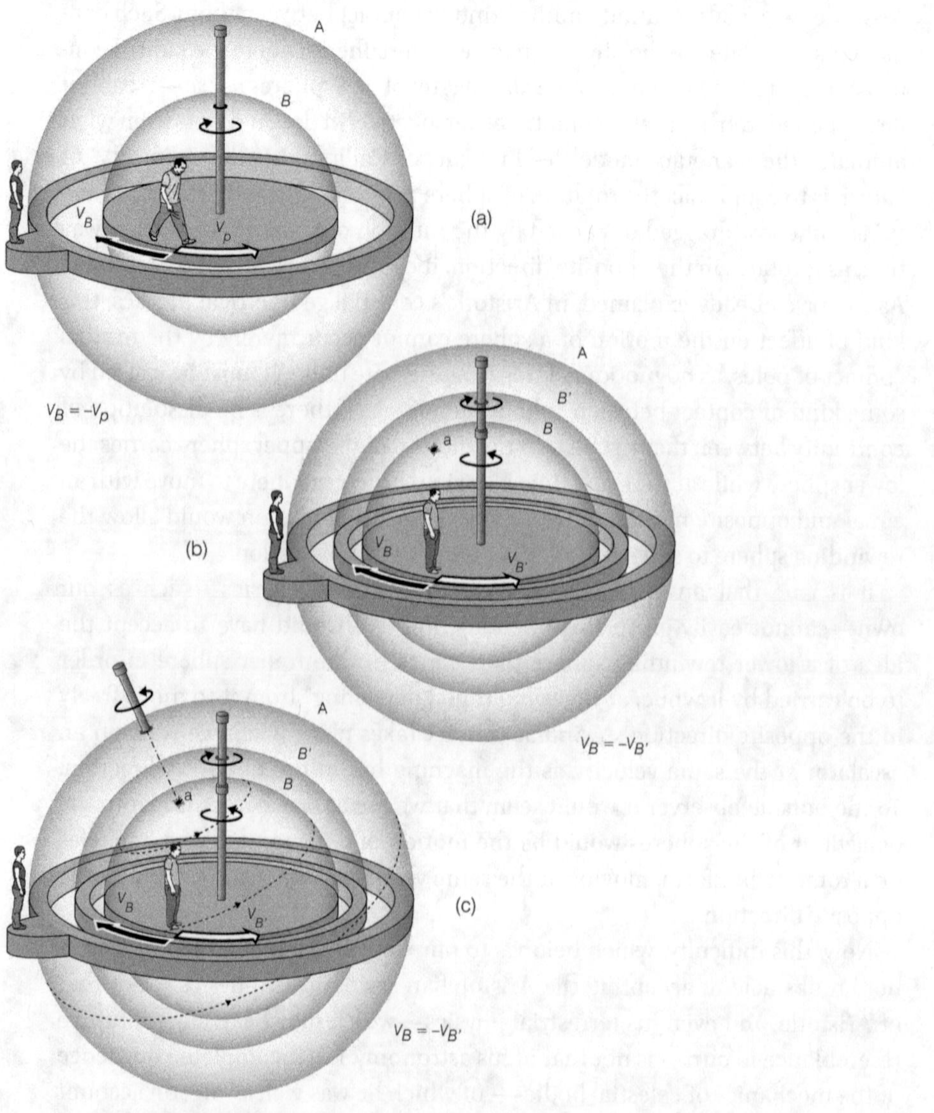

FIGURE 41 The rewinding spheres introduced by Aristotle in his astronomical system produce an effect similar to that shown in these figures.

(a) A person walking in circles on a rotating platform mounted on the equator of sphere B. If the person moves at the same speed as the platform but in the opposite direction, they will remain in the same position for an outside observer at rest (sphere A). (b) An equivalent situation—closer to the composition of rewinding spheres—would show sphere B if rotated by sphere B', while sphere B "walks"—like the person on the platform—against the motion caused by sphere B' at the same speed but in the opposite direction. The superimposition of motions determines that not only the person but any fixed point on the inner sphere B—like the spider at point a—remains at relative rest for the outside observer (semisphere A). (c) The composition of the rotational motions of B and B' allows any rotation of sphere A to be transmitted to any other body connected to sphere B—for example, another sphere with its axis fixed to point a. This motion is neither modified nor influenced by the rotation of B and B', which cancel each other.

overcome these limitations that, as we have seen (in the section "Kinematics and Dynamics") Aristotle considers the motion caused by a mover, and especially by a celestial sphere, from an analytical as well as an empirical perspective. We have analysed *On the Heavens*, II, 8, 289b1–290a7 –§ 27–, in which he states that the pushing and carrying of the celestial bodies by their spheres is accidental, therefore accepting, *prima facie*, that these movements may or may not occur, and that the phenomenon itself is the only means to decide the result in each particular case. We can accept, then, concerning the rewinding spheres, a relationship such as that analysed between the mover and the moved thing, and yet agree that this does not necessarily happen in the case of the forward-revolving spheres.

All rewinding spheres across the planetary subsystems exhibit a similar interaction to that described between spheres IV and IV' of Saturn. Looking particularly at the last rewinding sphere of Saturn (sphere II') and the first sphere of Jupiter, which is connected to it, the reversion of motion caused by the rewinding spheres of Saturn leaves at rest the exact point on sphere II' where the pole of the first sphere of Jupiter connects. This sphere and the sphere of the fixed stars are coaxial, and if we maintain the geometrical condition of Callippus's revolving spheres, this sphere cannot receive any motion from its only point of contact with the upper revolving sphere simply because the point is at absolute rest. Therefore, the first sphere of Jupiter needs to replicate by itself the diurnal motion that it transmits to the planet. According to Aristotle, the sphere is moved by its own Unmoved Mover, just like the remaining spheres of the overall system. Certainly, from this point of view, and contrary to Hanson's assumptions, it is not necessary for the spheres supporting the first planetary spheres of the Callippic subsystems to be at absolute rest. Only the points where the poles of the first planetary spheres are connected need to be at rest.

In summary, considering that only translational motions can be transmitted through the poles of the spheres but never pure rotational motions, as demanded by the phenomena, we need to assign a different nature to the contact between spheres, namely: 1) the revolving spheres of Callippus and Eudoxus's subsystems only connect and transmit their motion to the lower revolving spheres through their poles, and 2) the rewinding spheres are carried by their surface but do not exert any motion on the revolving spheres by themselves. It should not be difficult for us to admit this difference in the nature of revolving and rewinding spheres if we consider the secondary role that Aristotle attributes to the rewinding spheres. Revolving spheres reproduce the visible motions of celestial bodies and are the cause of their participation in the divine. Rewinding spheres, on the other hand, are unable to produce

positive self-motion. They merely have, in this forced Aristotelian construct, the subsidiary function of preventing the tearing apart of the heavens by the complex motions of the celestial bodies.

Hanson's account is not convincing, for there are no compelling reasons to consider that the total number of spheres would be different from the fifty-five proposed by Aristotle, or that the first spheres of the Callippic subsystems are unnecessary or lack self-motion. On the contrary, there are reasons to accept that the fifty-five spheres given in *Metaphysics*, Λ, 8, along with an identical number of Unmoved Movers, are actually necessary within Aristotelian physics and metaphysics. We will turn to this in the following section, as we try to show the importance and singularity of the first spheres of planetary conglomerates and their essentiality to the integrated dynamics of the heavens from the perspective of Aristotle.

5.5 The First Heaven and Wandering Stars

The Aristotelian understanding of sphere motion is more complex than it initially appears. He attributes to the diurnal motion of heavens (motion a) the capacity to affect the speed of zodiacal revolutions, in consequence of his teleological understanding of universal dynamics. In order to understand this, we need to focus on the first sphere of the Aristotelian cosmos, the sphere of the fixed stars or first heaven (πρῶτος οὐρανός). This sphere governs the entire celestial system, as it is the first visible body directly moved by the "Prime Mover," which Aristotle describes as "the best" (τὸ ἄριστον) and highest good towards which all things aim.[123] The sphere of the fixed stars is also the first sphere of the subsystem of Saturn in the Callippic model. It produces the diurnal motion of the planet by directly moving the second sphere and indirectly moving spheres III and IV of the subsystem. This motion is also transmitted to the three rewinding spheres introduced by Aristotle between the subsystems of Saturn and Jupiter.

The relation between the exemplary motion of the sphere of the fixed stars and the motions of the second planetary spheres (responsible for the zodiacal periods of the stars) transcends the much simpler *superimposition of velocities* that characterised the models of Eudoxus and Callippus. In the words of Aristotle (§§ 27–28), the sphere of the fixed stars not only carries every lower sphere in the subsystem of Saturn, along with their corresponding rewinding spheres, but also determines the speed of the direct motion of the six planets

123 *On the Heavens*, II, 12, 292b17–25 –§ 19–.

that follow and it does so in a progressive manner: its influence is strongest over the nearest celestial bodies and becomes weaker with distance.[124] This kind of interaction exceeds kinematic analysis and involves consideration of the *causes* of motion, whereby we enter into the domain of Aristotle's *sui generis mechanics*. As he states in *On the Heavens* (II, 10, 291a32–291b10 –§ 28–):

> One characteristic is that their movements [of the stars] are faster or slower according to their distances. That is, once it has been admitted that the outer most revolution of the heaven is simple and is the swiftest of all, whereas that of the inner spheres is slower and composite (for each in performing its own revolution is going against the motion of the heaven [along the ecliptic]), then it becomes natural for the star nearest to the simple and primary revolution [Saturn] to complete its own circle in the longest time, and the one farthest away in the shortest, and so with the others—the nearer in a longer time, the farther in a shorter. This is because the nearest one [Saturn] is most strongly counteracted (μάλιστα κρατεῖται) by the primary motion, and the farthest [the Moon] least, owing to its distance. The others are influenced in proportion to their distances, and how this works out is demonstrated by the mathematicians.[125]

Here Aristotle is not strictly comparing the speed of celestial bodies but the time it takes to complete their circles on the ecliptic plane. In other words, he is comparing their periods of zodiacal revolution. To the terrestrial observer, the different lengths of the zodiacal periods of celestial bodies appear as a faster or slower direct speed relative to the fixed stars. Therefore, following Aristotle's reasoning, which is clearly based on observation, we should assume that a shorter period of direct revolution, which equals a higher relative speed in relation to the fixed stars, indicates less retardation or drag in the direct motion of the body from the diurnal motion of the heavens. This retardation

124 *On the Heavens*, II, 12, 293a4–11 –§ 19–: "Here is a second reason why the other motions carry one body: the motions before the last one, which carries the one star, move many bodies, for the last sphere moves round embedded in a number of spheres, and each sphere is corporeal. The work of the last one, therefore, will be shared by the others. Each one has its own proper and natural motion, and this one is, as it were, added. But every limited body has limited powers"; English trans. Guthrie, p. 213. In this passage, Aristotle states that the moving action of the last sphere over the planet is shared by the other spheres as they are superimposed onto the motion of the last sphere. Thus, he seems to equate the notion of "speed" with the notion of "work": the speed that the last sphere transmits to the planet is a compounded speed, shared from the motion of the penultimate sphere, and so on through the succession of spheres.

125 English trans. Guthrie, p. 199.

produced by the first heaven can also be explained in *teleological* terms: the fixed stars exhibit the perfect motion that all remaining celestial motions wish to imitate. This explains why, as the direct motion of Saturn (in the opposite direction to the fixed stars) is retarded, it comes closer to the perfect motion of the first heaven. However, Aristotle also applies this idea to the other celestial bodies. He states that their direct motions (motion b) are retarded in inverse proportion to their distance from the sphere of the fixed stars. Thus, the action exerted by the sphere of the fixed stars—that we will call an *attracting* action—explains not just the speed of the direct motion of Saturn (being the body directly moved by this sphere) but also the regular increase in the direct speed of all celestial bodies as we move away from the outer rotational limit of the universe. According to Aristotle, the *attracting* force exerted by the motion of the first heaven is weaker on the stars farther from it. That is why their direct speeds relative to the fixed stars increase or, in other words, why their periods of zodiacal revolution do not decrease as much as that of Saturn. That is to say, the farther a body is located from the fixed stars, the faster will be its direct motion relative to the fixed stars and, consequently, the more dissimilar will be the motion of the two.

The fact that the first heaven retards the direct motion of Saturn does not present a problem if we understand it as another feature of Aristotle's *sui generis* celestial mechanics. The Stagirite simply affirms that, in addition to carrying the second sphere of Saturn along its same revolution period, the sphere of the fixed stars also exerts a retarding action or "force" opposite to the independent motion of that sphere. To understand how the first heaven exerts its retarding action on the totality of celestial bodies, we need to conduct a more insightful analysis that provides answers to the question of how is it possible for the sphere of the fixed stars to act on remote spheres not in immediate contact with it. These spheres are responsible for the direct motions of stars whose revolution periods are "slowed," according to Aristotle, by the motion of the first heaven. The action cannot, clearly, be direct, as it was with Saturn. Even though the text seems to indicate that the regular decreasing retardation of direct planetary motions is caused by the first heaven exerting its influence remotely, there are reasons—in Aristotle's celestial mechanics—to believe that, strictly speaking, the first sphere of each Callippic planetary subsystem is responsible for the retardation of the corresponding star. But even this action is ultimately determined by the first heaven, given that all these spheres rotate in coordination and with the revolution period of the first heaven. Aristotle uses the metaphor of a strategist who governs their army to represent how

the first rotation of the firmament imposes its rhythm on the totality of the universe.[126]

We should bear in mind that Aristotelian physics, as well as classical physics, does not admit remote interactions. If a body acts on the motion of another that is not in direct contact with it, this is thanks to a third body placed between and in contact with both. So if the first sphere of the heavens retards the direct motion of the spheres responsible for the zodiac motions of the stars, it should also retard the motion of all the spheres interposed between them.

Clearly, in the Callippic subsystem for Saturn, the sphere that produces the direct motion of the planet is that in immediate contact with the sphere of the fixed stars. Therefore, the retardation of the direct motion of Saturn is due to the direct slowing action exerted by the first sphere of heavens. Aristotle does not confirm—nor can we deduce from his description of celestial motions—if this effect (directly originating from the sphere of the fixed stars) extends to the remaining spheres of Saturn or of other planetary subsystems. Everything seems to indicate, however, that this is not the case. If it were, every sphere, not only those producing the direct planetary motions, would exhibit speeds relative to that of the first heaven (the farther from the fixed stars, the faster their motion), but this does not happen. Given that in the astronomical theories of Eudoxus and Callippus, the synodic and zodiacal planetary revolutions determine the motion of each celestial body,[127] it is reasonable to think that Aristotle may have made those previous affirmations having in mind the values set by Eudoxus for these periods, as he did not have any other means to quantify speed. At the end of the quoted passage, he points out that the regularity of direct planetary motions had been proven by mathematicians (291b9–10), probably a reference to Eudoxus, who he knew well. Following this approach, the faster the motion of the body relative to the fixed stars, the shorter its period of zodiacal revolution.[128] If we consider Schiaparelli's

126 *Metaphysics*, Λ, 10, 1075a11–16.

127 The zodiacal revolution is the direct motion of a star along the middle line of the zodiac, as illustrated in the figures concerning the behaviour of the planetary subsystems of Eudoxus in the section within this chapter, "The Prime Mover and Unmoved Movers." While this motion is caused by the revolution of the second sphere, the period of synodic revolution is caused by the joint revolution of the inner spheres that form the *hippopede*.

128 Aristotle lacked an accurate mathematical definition of velocity that would allow him to quantify what he termed the "speed of motion," let alone the notion of "angular velocity." Despite this limitation, his comparative analysis of velocities can be regarded as a comparison of angular velocities. The uniform circular motion of celestial spheres presents angular velocities that are inversely proportional to their revolution periods given

	Planet	Synodic Revolution	Zodiac Revolution
♄	Saturn	390 days	29.5 years
♃	Jupiter	390 days	11.9 years
♂	Mars	260 (780) days	2 years
☿	Mercury	110 days	1 year
♀	Venus	570 days	1 year
☾	Moon	29.53 days	27.32 days

FIGURE 42 The table values are very similar to modern calculations for the same periods, except for Mars, whose synodic revolution is 780 and not 260 days, as Eudoxus might have thought. Given that Eudoxus so accurately determined the other periods, Schiaparelli (*Scritti*, vol. 2, pp. 70–71) hints at the possibility that this incorrect value is the result of an error introduced by the written tradition. If we adopt the correct value for the synodic period of Mars, the irregularities in the succession of synodic periods would be even more evident.

table of the Eudoxian values, we notice that indeed the periods of zodiacal revolutions of the celestial bodies regularly decrease as we move away from the sphere of the fixed stars. Consequently, we can agree with Aristotle that there is an increase in the direct speeds of the bodies relative to the fixed stars.[129]

The preceding table shows that the periods of synodic revolution do not decrease in a regular manner as we move away from the first heaven. They correspond to the innermost spheres of each Callippic subsystem, that is, the spheres interposed between each planetary conglomerate and the first heaven. From Mars to the Moon, one of these spheres is placed between the first heaven and the sphere responsible for the direct motions of the planets. If the first heaven is responsible for the retardation of the direct motions of every celestial body, this action should propagate to these intermediate spheres as well, which is not the case as we can see from the table. Therefore, one can argue that the retarding effect exerted by the sphere of the fixed stars on the

that $w = 2\pi/P$ (w being the angular velocity and P, the revolution period). Therefore, a decrease of the period equals an increase in velocity, as understood by the Stagirite. In addition, considering his understanding of phenomena, a comparison of periods equals, in observational terms, a comparison of lineal velocities, as a slower rotation of the body (related to a longer period of revolution) translates into a circular trajectory with a slower lineal velocity relative to any star, thus closer to the motion of the fixed stars. Even small reductions in the direct linear velocity of a body with respect to the fixed stars are thus evidenced by an increase in the corresponding period of revolution, even if it remains unclear just how this retarding action operates.

129 The pseudo-Aristotelian treatise *On the Cosmos* (399a8–11) mentions the same periods of zodiacal revolution detailed by Eudoxus and indicated in the table.

second sphere of Saturn—that responsible for the direct motion of the planet and in immediate contact with the first heaven—does not affect the remaining revolving and rewinding inner spheres of Saturn. That being so, this mode of action—which is also an attribute of the motion of the sphere of the fixed stars—cannot be transmitted in the same way that the simple superimposition of velocities is transmitted to the inner spheres of each subsystem, that is, in a mediated way. Consequently, the retardation of the direct motion of each celestial body associated with the diurnal motion of the whole of the heavens should be caused by a sphere in immediate contact with the sphere responsible for the direct motion of each celestial body, that is, by the first spheres of Callippus's planetary subsystems, which, to the observer, move the bodies with same the diurnal motion as the fixed stars.

This explanation seems reasonable if we take into account the Aristotelian idea of *speed*. According to this notion, if body A travels a greater distance than body B during the same period of time, then body A moves faster than body B (*Physics*, VII, 4, 248a22–248b2). Such is the case of the first spheres of Callippus's planetary subsystems. Because all spheres move with the same revolution period (24h), those with a greater radius will cover a longer trajectory in the same period than those with a smaller radius. One can conclude that the velocity or speed of the spheres decreases as their radii decrease, the sphere of the fixed stars being the fastest among them. The Aristotelian text seems to point out that the slowing effect is proportional to the speed responsible for it. The farther one gets from the first heaven, then, the slower the spheres responsible for diurnal planetary motion will move and, consequently, the direct motion of the corresponding celestial bodies will have less retardation.

To reach a better understanding of the retarding action exerted by the motion of one body on that of another with which it is in contact, we should bear in mind that Aristotle establishes a proportionality between certain ideas of force or power and his notion of speed. Namely, that the speed of motion is proportional to the moving force.[130] Certainly, the retardation attributed by the Stagirite to the sphere of the fixed stars depends on this force, since this is the fastest sphere, the most powerful, and the one that exerts the greatest influence on the bodies moved by it. Finally, this retarding action needs to be considered in terms of speed of motion rather than period of revolution.

130 For Aristotle, the action of a force is necessary for any motion, even at a constant speed (velocity), in which case the (constant) speed is proportional to the magnitude of the force. For classical physics, and under Newton's principles, only accelerated motion is caused by the action of a force. That is, a motion with constantly varying speed. Newtonian mechanics states that it is the rate of change of velocity, rather than the velocity itself, that is proportional to the force.

It is interesting that classical physics also acknowledges some retarding effects on relative motions, though not under the conditions of uniform motion (constant speed) but regarding accelerated motions, in which there are forces acting on the direction of the motion. The figure below, for example, shows a person running on a surface that is accelerating. If the person and the surface move in the same direction, the acceleration of the person relative to the surface would decrease to match the acceleration of the surface. Unlike the phenomenon described by Aristotle, this effect is due to the inertial mass of the person. Therefore, the greater the acceleration applied to the surface, the greater the retardation of the accelerated motion of the runner. An equivalent effect—though not caused by relative motion—is achieved by the gravitational mass that acts on a person climbing an inclined plane, as illustrated in figure c. An increase in the plane gradient implies an increase in the gravitational force in the directional line of motion and, consequently, a greater retardation of the person's motion.

The previous analysis shows clearly the pre-eminent role of the first planetary spheres of Callippus's subsystems in the Aristotelian system of the heavens, despite the fact that some interpreters (e.g., Hanson) have deemed them redundant. According to our interpretation of the Aristotelian text, these spheres cannot be dismissed, as they are responsible for the specific periods of synodic revolution of the celestial bodies.

The exclusive influence of the first planetary spheres on the direct motion of each body distinguishes them from other celestial spheres, whether forward-revolving or rewinding. This, in addition to the fact that Aristotle seems to attribute to the first spheres the continuity of heaven and the continuity of the motion of the fixed stars (motion a), allows us to regard them altogether as a kind of autonomous system. Moreover, all of these spheres have the same axis and direction of rotation as the sphere of the fixed stars. Thus, from a teleological perspective, they are closer to the perfect first motion. They constitute a kind of community with the sphere of the fixed stars, altogether giving the impression of the heavens as a single moving continuum. This places the first six inner spheres in greater ontological proximity to the sphere of the fixed stars, as they are jointly responsible for the central phenomenon of Aristotelian cosmology: diurnal motion. In addition, these seven governing spheres share a unique mode of acting on the planets, distinct from the other revolving and rewinding spheres that form Aristotle's system of fifty-five spheres.

Aristotle applies an interesting reasoning to justify the supremacy of the motion of the first heaven, which can also be used to justify the primacy of the first spheres of the Callippic planetary subsystems. He believes that its superiority is demonstrated by proportion, and affirms that nature arranged for the best (unique) motion to move many bodies, while the other (multiple) motions,

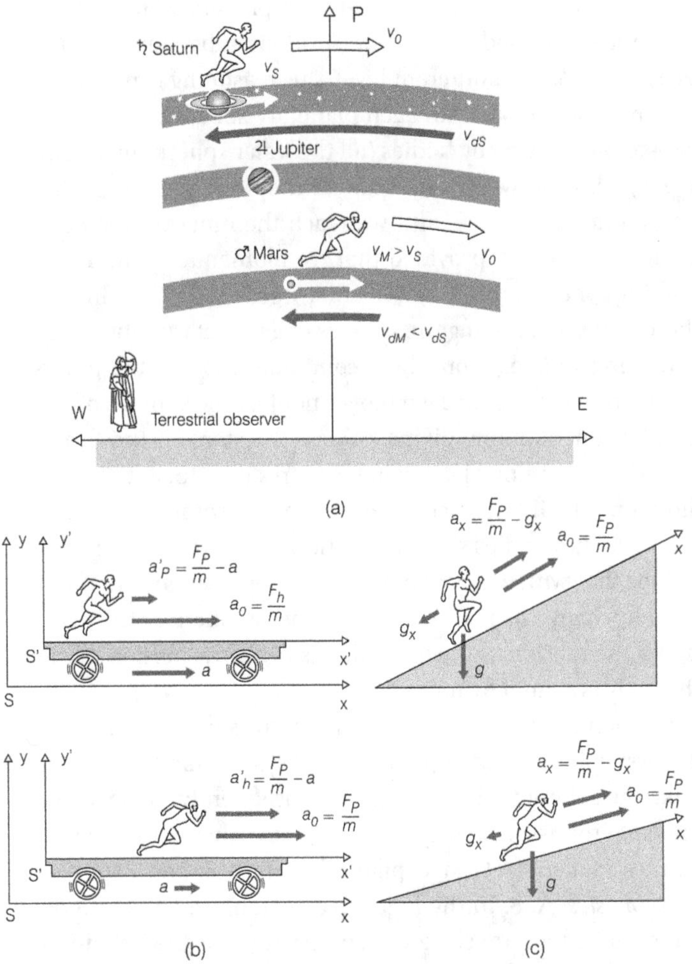

FIGURE 43 The retardation caused by the diurnal motion of heavens on direct planetary motions.

(a) A schematic representation of the retardation caused by the diurnal motion of heavens on direct planetary motions, taking the cases of Saturn and Mars for example. The indicated velocities represent the movements as follows: V_o being the presumed original speed of the second planetary spheres of Callippus's subsystems, responsible for the direct motion of the celestial bodies; V_{dS} and V_{dM} being the speed of the diurnal motion of Saturn and Mars respectively, caused by the rotation of the first planetary spheres of the corresponding systems; V_S y V_M being the speed of the direct motion of Saturn and Mars relative to the fixed stars respectively, which results from the retardation of V_o caused by V_{dS} and V_{dM}. Newtonian physics includes similar slowing phenomena but these affect acceleration rather than velocity, as shown in (b) and (c). Image (b) shows the slowing or retardation of an accelerated motion caused by an inertial mass m in a system accelerated in the same direction as the human motion, where a is the acceleration of the system S' relative to S, a_o would be the acceleration of the person if the system S' was at rest with respect to S, and a'_h is the acceleration of the person relative to the system S' when in motion with respect to S, assuming the same effort F_h of the runner in all cases. Image (c) shows the slowing or retardation of accelerated ascent on an inclined plane, caused by the gravitational mass m, where a_o is the presumed acceleration of the walker on a horizontal surface, a_x is their acceleration as they ascend the inclined plane, and g is the local gravitational acceleration, assuming that the person makes an equal effort F_h in all cases.

corresponding to the spheres that cause the specific planetary motions in each subsystem, move just one body, the planet in their subsystem (*On the Heavens*, II, 11, 292b27–295a3). On a different level, this reasoning can be used to establish a hierarchy of spheres within each planetary subsystem: the first sphere of each subsystem moves many bodies (all the inner spheres including the rewinding spheres) while, as we move inwards the system, the remaining spheres move fewer and fewer bodies, until we reach the innermost sphere, which moves one body, the planet, provided that we momentarily minimise the role of the rewinding spheres (295a4–12). In other words, following this criterion, the first sphere ranks first among the spheres of a planetary subsystem, followed by the remaining spheres, from the second outermost to the innermost sphere.[131] The fact that the first heaven moves not just the spheres of the subsystem of Saturn but uncountable divine bodies (the stars on the firmament) and influences the remaining first spheres, ensures its primacy.

This primacy should not be interpreted in the sense that something is better just because many (divine) bodies share its motion but, conversely, in that nature has arranged for that which is best in itself (the diurnal motion of the heavens) to be shared by many, the result being that every sphere of the planetary subsystems share the motion of the first spheres while the motion of the spheres carrying the body is shared by none.

If the previous interpretation is correct, it is not surprising that Aristotle considered that the first sphere of each planetary subsystem moves without being moved, just like the sphere of the fixed stars. Consequently, he need not introduce rewinding spheres associated to the motions of the first planetary spheres. This constitutes a salient fact that explains the total number of spheres in the system of *Metaphysics*, Λ, 8. In the following section, we will build on this idea to offer an enriched approach to the composition of the system of fifty-five spheres described so far. The goal will be to prove that the specificity of the seven first spheres can be imparted to the remaining spheres without altering the exact description provided by Aristotle in *Metaphysics*, Λ, 8.

131 A systematic problem that Aristotle has left unsolved is the hierarchical order of spheres and movers. In principle, there are reasons to assume that the hierarchical order matches the outside-in geometrical order. However, our interpretation suggests another order for the pure forms. As we move inwards in the system, the first spheres in order of appearance would be the noblest movers, followed by the inner spheres from Saturn to the Moon. The movers of rewinding spheres would rank lower and be ordered according to their proximity to the first heaven. These are only conjectural given the lack of any indication by Aristotle on the matter, yet it seems reasonable to attach less importance to the rewinding spheres and their movers as they serve a secondary function—namely, to prevent the specific motions of each planet from propagating beyond their subsystem—and do not produce any observable motion.

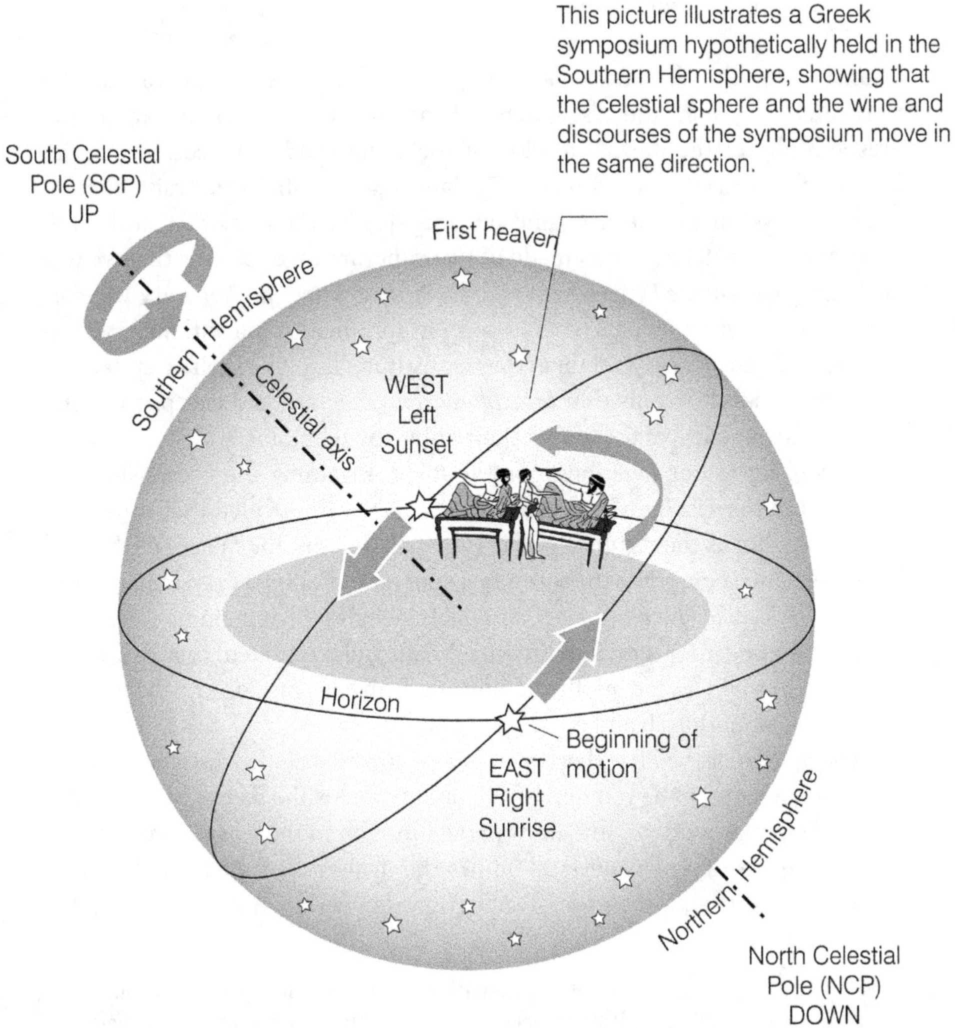

FIGURE 44 Left and right, up and down the universe. Aristotle says: "Of the poles, the one which we see above us [NCP] is the lowest part, and the one which is invisible to us the uppermost [SCP]. For we give the name of righthand to that side of a thing whence its motion through space starts. Now the beginning of the heaven's revolution is the side from which the stars rise, so that that must be its right, and where they set must be its left. If this is true, that it begins from the right and moves round to the right again, its upper pole must be the invisible one [SCP], since if it were the visible, the motion would be leftward, which we deny" (*On the Heavens*, II, 2, 285b14–23; trans. Guthrie, pp. 145, 147). The circulation direction indicated as the preferred one is equally the direction of circulation of drinks and discourses at Greek feasts, that is, clockwise (ἐπὶ τὰ δεξιά; see the note of Miguel Candel to the quoted passage in *Aristóteles, Del cielo—Meteorológicos* [Madrid: Gredos, 1996], p. 113). Aristotle notices that for an observer situated in the Northern Hemisphere the motion of the celestial sphere is opposite to the axiologically superior clockwise motion, from which he deduces that the superior and absolute pole of the universe is the South Celestial Pole.

5.6 Two Celestial Systems

Considering the specific characteristics of the celestial spheres in the Aristotelian system (the first spheres of the planetary subsystems are jointly responsible for the diurnal motion of the Whole and produce the retardation of the zodiacal revolution of the planets), our main hypothesis concerning the physical, geometrical, and metaphysical astronomy of Aristotle is to regard his celestial model as made of two superimposed systems that we will call the *System of the First Heaven* and the *System of Specific Planetary Motions*. This interpretation will allow us to account for phenomena without encountering contradictions. Together, they constitute the Aristotelian system of fifty-five spheres described in *Metaphysics*, Λ, which reproduces phenomena as expected and leaves the foundations of his thought intact.

According to our hypothesis, the problem regarding the composition of motions in the system of fifty-five spheres is solved provided that we visualise the system as the result of two juxtaposed systems: the *System of the First Heaven*, which comprises the seven first spheres of Callippus's system, and the *System of Specific Planetary Motions*, which includes the remaining forty-eight spheres adding up to a total of fifty-five. While the first system easily accounts for the diurnal motion of the totality of heavens (motion a), the second accounts for motions b and c.

The System of the First Heaven, which of course includes the sphere of the fixed stars, governs the rotation of the totality and is the *means* by which the diurnal circular motion—the most perfect motion in the Aristotelian physics and metaphysics—regulates and unifies the dynamics of the overall system, and[132] exerts its action on every wandering star. It should be noted that this

132 The unity principle related to the celestial spheres and revealed by our twofold interpretation was outlined by Plato in *Laws*, x, 893c–d, where the various celestial spheres are portrayed as if they were subject to the same harmonically distributed rotation, which maintains the angular velocity of the assemblage and preserves the speed of each sphere (tangential velocity) according to their distance from the centre: "Athenian: 'O stranger, do all things stand still then, and is there nothing in motion? Or is the case entirely the opposite to this? Or are some of the things in motion, and some remaining still?'—'Some things are presumably in motion,' I will declare, 'and some things remain still.'—'And then isn't it in a certain place that the standing things stand and the things in motion move?'—'How could it be otherwise?'—'And some, at least, would presumably do this in one location, but some in many.'—'Are you speaking of those that obtain the power of things that stand still in the middle,' we will declare, 'and thus move in one place, just as the circumference turns in circles that are said to stand still?'—'Yes.'—'And we learn, at any rate, that in this rotation such motion carries the largest and smallest circles around

characteristic is somehow similar to the modern notion of solar gravitational field—that is, the force that determines the conditions of motion of the planets as a set. In the same way, the seven spheres that belong to the System of the First Heaven move with the same period of revolution causing the diurnal motion of the stars, just as if they were immersed in a *continuous medium*. And this is despite the fact that, when incorporated to the system of fifty-five sphere, the first spheres of Saturn (sphere of the fixed stars), Jupiter, Mars, Mercury, Venus, the Sun and the Moon give the appearance of a discontinuous structure made of seven independent bodies.

We believe that the spheres comprising the System of the First Heaven reflect the Aristotelian idea of a continuous dynamic heaven, which can function without counteracting spheres, as he states in *Metaphysics*, Λ, 8, 1074a1–3 –§ 16–, because all its spheres move in harmony with the same period of revolution and together they constitute the very first motion of the heavens. We can imagine this system as a continuous mass of ether eminently driven by the attraction of the Prime Mover but also requiring the presence of six other Unmoved Movers (those driving the first spheres from Jupiter to the Moon), placed right at the points where the spheres responsible for specific planetary motions are coupled. This is the reason why—in Aristotle's words—"the heaven does not spring apart" (*On the Heavens*, II, 8, 290a7 –§ 10–) as the spheres of the System of Specific Planetary Motions penetrate it. In *On the Heavens*, II, 8, 289b34–290b7 –§ 27–, Aristotle states that the continuity of the heavens is ensured because the circles, whose radii coincide with the first planetary spheres of each planetary subsystem, move together as if they were part of one continuous body. The synchronised motion of the Unmoved Movers of these spheres maintains the dynamic continuity of the Whole when the real continuity of heavens is interrupted by the interposition of the spheres responsible for specific planetary motions.

The System of Specific Planetary Motions, which comprises the rest of the spheres in the Callippic systems in addition to Aristotle's rewinding spheres (a total of forty-eight) and is superimposed to the first system, completes the model of fifty-five spheres. This new perspective does not introduce changes to the structural distribution of the spheres and offers the advantage of stressing

together, distributing itself proportionally to the small and the large, being less and more according to proportion. That is how it has become a source of all wonders, conveying the large and small circle at the same time, at slow and fast speeds that are in agreement, an effect that someone would expect to be impossible'"; English trans. T. L. Pangle (Chicago: University of Chicago Press, 1980), pp. 290–291.

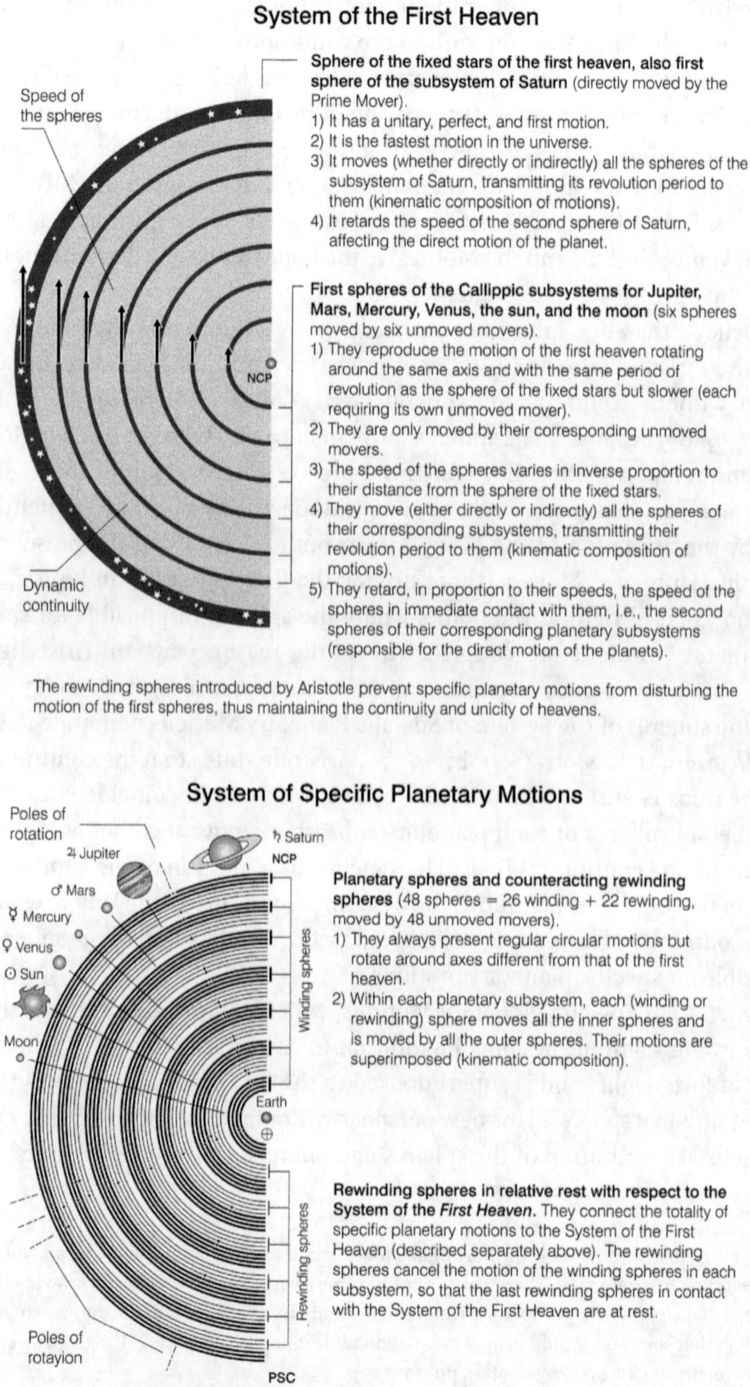

FIGURE 45 A scheme of Aristotle's astronomical system and its composition of motions

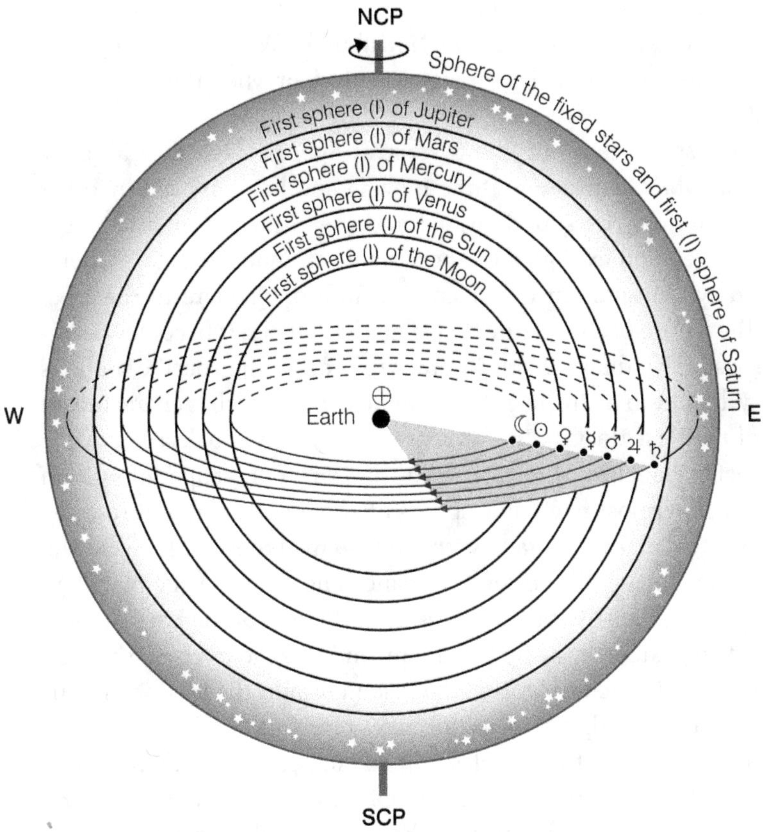

FIGURE 46 System of the First Heaven.

Aristotle's rewinding spheres cancel every motion of the revolving planetary spheres except for the motion of the sphere of the fixed stars (the first sphere of Saturn), the motion of the first sphere of each Callippic subsystem (first spheres of Jupiter, Mars, Mercury, Venus, and the Sun) and, finally, the motion of all lunar spheres (as they are the innermost spheres of the system and there are no subsystems below them that could be affected by their motions, they do not require counteracting). Thanks to the interposition of rewinding spheres, the first spheres of each planetary subsystem behave as if they were in continuity with the spheres of the fixed stars. The result, as shown in the figure, is the motion of all first celestial spheres as if they were part of a continuous medium, that is, moving with the same period of revolution (that of the fixed stars) but at different speeds, given that they are at different distances from the common centre. In addition, Aristotle addresses this kind of motion in *On the Heavens* to prove that celestial bodies do not move by themselves but are moved by the spheres that hold them. By considering the diurnal motion shared by all celestial bodies abstracted from other motions, he is able to affirm that: "Since then neither the motion of both [stars and spheres or circles] nor the motion of the star alone can be defended, we are left with the conclusion that the circles move and that the stars stay still and are carried along because fixed in the circles. This is the only hypothesis that does not lead to an absurdity. That the larger circle should have the higher speed is reasonable, seeing that the stars are dotted around one and the same centre" (*On the Heavens*, II, 8, 289b33–36 –§ 27–; trans. Guthrie, pp. 185, 187). Later, he concludes that: "This too is one reason why the heaven does not spring apart, and another is that the whole has been demonstrated to be continuous" (*On the Heavens*, II, 8, 290a6–7 –§ 27–; trans. Guthrie, p. 187).

the idea that the group of first planetary spheres carries the remaining spheres, thus producing the diurnal revolution (motion a) of the Whole. By analogy, the same happens when a ship carries someone on deck or when the sea current carries the ship.

In turn, each sphere forming the System of Specific Planetary Motions moves by itself, while at the same time moving and being moved by other spheres. Their motions are superimposed and the entire set maintains the same kinematic relation exhibited by the systems of Eudoxus and Callippus. In the second system, motion is transmitted from one sphere to the next. Then, given the nature of the relation between the spheres, it becomes necessary to interpose rewinding spheres in order to prevent the specific motions of each subsystem from disturbing the dynamic continuity that characterises the System of the First Heaven. In this sense, an isolated visualisation of the System of Specific Planetary Motions reveals that, due to this superimposition of motions, the innermost rewinding sphere of each subsystem is at rest just at the point where it makes "contact" with the System of the First Heaven. The rewinding spheres thus fulfil their role by cancelling each specific planetary motion. It is natural, then, that the spheres in the System of the First Heaven, which connect and extend away from these rewinding spheres at relative rest, move by themselves in such a way that the system regains the motion of diurnal revolution (motion a) that governs the entire orb.

If the specific motions of each celestial body are cancelled, then we can conclude that nothing from these moves the spheres of the System of the First Heaven, but that they move by virtue of their associated Unmoved Movers. This is a consequence of their particular place in the overall model rather than a natural condition of motion.

As shown in the previous section, "The First Heaven and Wandering Stars," the retardation of the direct motion of Saturn that Aristotle attributes to the sphere of the fixed stars is not a property that can be transmitted from one body to the next. Consequently, Hanson's suggestion that the motion of the first planetary spheres is produced by the carrying action of the first sphere of heaven (and not by their corresponding movers), would involve no decrease in retardation on the direct motion of the remaining celestial bodies, which disagrees with the Aristotelian text. In conclusion, the first planetary spheres must have their own motion, caused by their associated Unmoved Movers. These last share with the Prime Mover the ability to locally retard—through the motion they transmit—the direct motion of each body (motion b), but differ from the Prime Mover in that the diurnal motion they produce in the

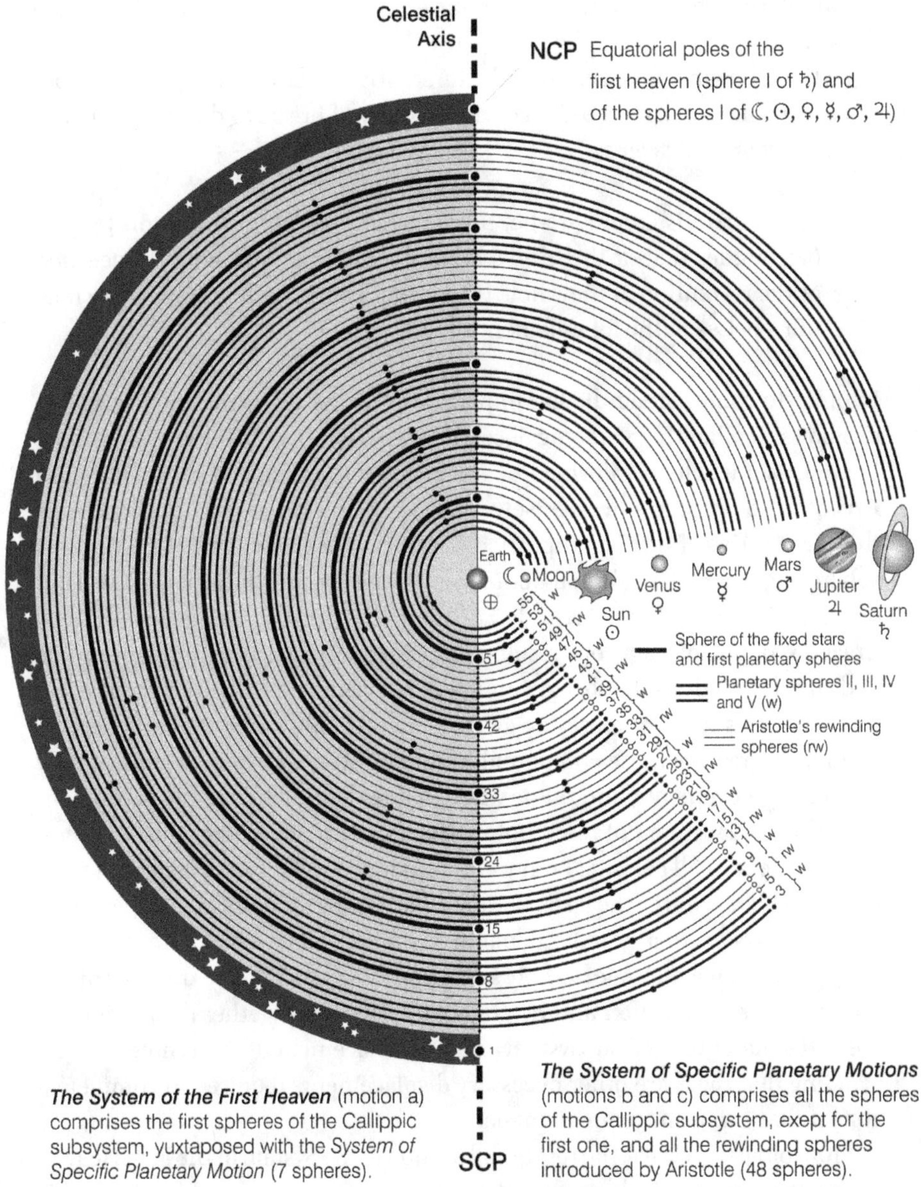

FIGURE 47 Aristotle's astronomical system understood as a superimposition of the two partial systems that we have called the *System of the First Heaven* and the *System of Specific Planetary Motions*.

six wandering stars becomes, as we move inwards in the system, slower and slower.

In the following passage, Aristotle differentiates between at least two types of celestial motions in accordance with our way of breaking down and analysing his system of heavens:

> But if there is coming-to-be, there must be another revolution [φοράν] (sc. besides that of the fixed stars), or more than one; for the operation of the revolution of the whole [ὅλου] could only result in leaving the relations between the four elements unchanged.[133]

In order to comprehend the full implications of this and similar passages, we should remember that what the observer understands to be the motions of the heaven, as a whole, is informed by the motion of celestial bodies. These, for Aristotle, do not move by themselves but are transported—more specifically they are carried (ἕλκειν) and pushed (ὠθεῖν) at the same time (§ 26)[134]—by spheres of ether (the celestial bodies are themselves a kind of ethereal "condensation" –§ 10–). From Aristotle's phenomenal perspective, the continuous irregular motion of each body results from the superposition of different simple and circular motions that occur simultaneously and are caused by the motion of the spheres that compose the given planetary subsystem. Nonetheless, he often disaggregates these compound motions and treats them independently as isolated movements, and this is what he does here. When the passage mentions the displacement of the Whole, it does so with reference to the common displacement of all celestial bodies (diurnal motion) as if other motions were absent, which is never the case. In other words, Aristotle sees the superimposition of motions in a systematic manner, similar to our own approach to the passage: it is not a matter of simply superimposing the fifty-five motions of the fifty-five spheres one after another, but of considering together the motions of the first spheres in each subsystem as the "displacement of the Whole," understanding that there are other necessary displacements required to complete a system that conforms to phenomena.

From our perspective, the rotational motion of the System of the First Heaven equals the diurnal displacement of the Whole. Motions that differ from this motion of the Whole—analysed by Aristotle in other passages as well[135]—and

133 *On the Heavens*, II, 3, 286b2–5 –§ 16–; English trans. Guthrie, p. 153.

134 *Physics*, VII, 2, 244a5–7 –§ 26–.

135 *Metaphysics*, Λ, 6, 1072a4–17; *Metaphysics*, Λ, 8, 1073b9–14 –§ 15–.

that are necessary for generation and corruption, belong unequivocally to the specific motions of the planets, which we consider to constitute a juxtaposed second partial system: the System of Specific Planetary Motions.

Also significant is that Aristotle considers the displacement of the Whole as something equivalent to the simple rotational motion of a continuous body, which he characterises under the notion of "period of revolution" rather than "speed." As indicated in the section "Kinematics and Dynamics," Aristotle regards local motion as a displacement in length (*Physics*, VII, 8), even though he differentiates between movement in a straight line and movement along a curve, as in rotation, as separate kinds of movement (*Physics*, VII, 8). In Aristotelian terms, one motion is faster than another when the same length is covered in a shorter time or when a greater length is covered in the same time. This is how, in *On the Heavens*, he analyses the different speeds of celestial bodies in relation to the diurnal motion of the Whole, understanding this last as if it were carried by the force of a continuous medium. He says, in this respect:

> That the larger circle should have the higher speed [τὸ τάχος] is reasonable, seeing that the stars are dotted around one and the same centre.[136]

In the surrounding section of *On the Heavens*, II, 8, 289b1–290a7 (§ 27), Aristotle analyses the motion of the heavens as a continuous whole and we see that celestial bodies do not move by themselves but are transported by celestial spheres. In the fashion of classical physicists, he considers diurnal rotational motion in abstraction from other planetary motions and he agrees that the speed of all celestial bodies is directly proportional to their distance from the centre of rotation. The specific motion of individual bodies is due to the fact that they remain fixed to a monoaxial continuum in rotation, such that bodies at a greater distance from the centre (those closer to the sphere of the fixed stars) move more quickly than those located nearer the centre of the cosmos. He is referring exclusively to the diurnal motion of the heavens as a whole, as can easily be seen through simple observation: apart from the dominant motion of the heavens (motion a), the celestial bodies exhibit various direct eastward motions along their orbits (motion b) that, when brought all together, do not make up—unlike the diurnal motion—a single circular motion. This is because the motion-producing spheres of the System of Specific Planetary Motions rotate with different periods of revolution, and around axes with different orientations.

136 *On the Heavens*, II, 8, 289b34–35 –§ 27–; English trans. Guthrie, p. 187.

Both in *Physics* and *On the Heavens*, the only type of speed considered in relation to rotational motion is similar (all differences considered) to our canonical notion of *tangential velocity*. The assertion, however, that points at different distances from the centre of a solid rotating sphere move at one and the same speed, requires the modern notion of *angular velocity*, which is not found in Aristotelian physics. Angular velocity defines a change of location in terms of the *angles* rather than *length*. Therefore, in a continuous rotating sphere, the points located at different radii from the centre and covering different distances in the same time will sweep out the same angle in the same time, thus presenting the same angular velocity. By contrast, if we were to understand local motion in terms of displacements of length, as Aristotelian physics does, it would be impossible to assert that all the first planetary spheres move at the same speed, regardless of the fact that they complete a revolution in the same time.

Even if the first spheres of the planetary subsystems forming the System of the First Heaven provide a single overall motion to the entire system (motion a, the diurnal motion of the heavens), this does not mean that they introduce just one single speed. Given the real discontinuity introduced into this kind of rotating platform by the interpenetration of the specific motions of each planetary subsystem, each sphere in the System of the First Heaven must be moved by its corresponding Unmoved Mover. As each Unmoved Mover generates the appropriate speed for the sphere according to its position in the complete system, all spheres achieve a full revolution at the same time, ensuring the dynamic continuity of the Whole. The Unmoved Movers of the first planetary spheres differ from one another in terms of the speeds at which their spheres are made to move but have in common that these speeds synchronise with the period of revolution of the first heaven caused by the Prime Mover.

We have now resolved the *prima facie* inconsistency that inspired this digression; that is, the apparent incompatibility of the following two affirmations from *Metaphysics*, Λ : 1) that the Prime Mover moves all thing in a single motion, a motion that can be only understood in reference to the diurnal period of revolution related to this first Unmoved Mover, and 2) that each first planetary sphere has its own Unmoved Mover and that all these movers are different from one another as well as from the first Unmoved Mover because they generate different lineal speeds or velocities in their spheres. These two aspects of the system—the unicity and the multiplicity of the diurnal motion of the heavens—are then rendered compatible inasmuch as each first planetary sphere must have its own Unmoved Mover by virtue of the fact that

its lineal velocity or speed is exclusive to itself, and differs from that of the first heaven. And this is unchanged by the fact that the angular velocity or, in Aristotelian terms, the period of revolution of the overall system is one and the same for every celestial body, and that this single motion ultimately refers back to its upper limit, the first heaven moved by the Prime Mover.

CHAPTER 6

Metaphysics, Λ, 8 and the Genetic Interpretation

Non è una esagerazione l'affermare, come qualcuno ha fatto, che tutta la letteratura sullo Stagirita posteriore al 1923 è, in qualche modo, una presa di posizione pro o contro le conclusioni dell' Aristotele di Jaeger.

It is not an exaggeration to state, as some do, that the whole literature on the Stagirite after 1923 is, in some way, a stand in favour of or against the conclusions of this work.

GIOVANNI REALE, *Il concetto di filosofia prima e l'unità della Metafisica di Aristotele*, Milan: Società Editrice Vita e Pensiero, 1967, p. 386

To tackle the problems discussed so far, we have approached the Aristotelian system as if it is timeless, unique, and conclusive. And, to be sure, the subheading of this work specifies it to be an interpretation. This presentation must, however, be qualified if our study is to take into account the evolution of Aristotle's work—an essential aspect for any interpretation after Werner Jaeger's (1888–1961) influential *Aristotle: Fundamentals on the History of His Development*.[1] We will now therefore address certain aspects of the chronological evolution of Aristotle's thought to highlight related aspects pertinent to his astronomical system.

First, we shall briefly review Aristotle's intellectual development (384–322 BCE).[2] At the age of seventeen or eighteen Aristotle entered the Athenian Academy, where he remained for nineteen years, up to the year 348 or 347, the time of Plato's death. Speusippus took over the Academy, promoting a Pythagorean-like intellectual programme that tended to reduce philosophy to mathematics, which Aristotle found compromising. Given this context,

1 English trans. by R. Robinson (1934) of Jaeger's *Aristoteles: Grundlegung einer Geschichte seiner Entwicklung* (Berlin: Weidmann, 1923).

2 For a succinct and enjoyable reading of Aristotle's biography see (currently available in Spanish only) A. Vigo, *Aristóteles. Una introducción* (Santiago de Chile: Instituto de Estudios de la Sociedad, 2007).

© GERARDO BOTTERI AND ROBERTO CASAZZA, 2023 | DOI:10.1163/9789004525535_008

and probably also compelled by the anti-Macedonian atmosphere in Athens after Philip's incursion in Olynthus in 348 BCE,[3] Aristotle accepted an invitation from Hermias, a Platonic tyrant from Atarneus and Assos, in Mysia. The Stagirite moved to Assos in 347, where he spent three years and married Hermias's granddaughter, Pythia, with whom he had a daughter. He then moved to Mytilene, island of Lesbos, perhaps invited by Theophrastus, where he continued his investigations and writing. His years in Assos and Mytilene were mainly devoted to studies on mineralogy and zoology, as most biographers agree.

Just a few years later, in 343 or 342, at the age of forty, he was summoned by Philip of Macedon to educate his son, Alexander, who was thirteen at the time. Aristotle instructed the young successor on the works of Homer—especially *The Iliad*—and other tragic poets such as Aeschylus, Sophocles, and Euripides, as well as on political theory and government affairs, as suited to the instruction of a future monarch.

Returning to Athens about ten years later, Aristotle then passed the most fruitful thirteen years of his life. There he rented a few houses at the foot of Mount Lycabettus, in the outskirts of Athens northeast of the city, and established his school.[4] Mornings he would walk with students (hence the name "peripatetics"; the verb περιπατεῖν means "to walk," "to wander," "to stroll"), and in the afternoons would deliver courses for a broader audience. At this school, Aristotle is thought to have gathered (probably partly financed by Alexander the Great) a significant library, as well as maps and natural history pieces. By the end of this period, it is presumed that he centred on more scientific interests, focusing on specific investigations on the natural world. Most Aristotelian works available today belong to this period, his second stay in Athens between 335 and 323, known as the "Lyceum Period," at the end of which he was sixty-two years old. During this time, Aristotle himself penned almost all the works that make up the current *corpus aristotelicum*, only a few being compilations from students' notes.[5]

3 Düring, *Aristoteles*, p. 9.

4 There is no certainty that the reference to the "Lyceum" in the life of Aristotle means anything other than the small forest where he would discuss with his disciples and other professors. Some authors, including Düring (*Aristoteles*, pp. 13–14) hold that it is unlikely that Aristotle would have been able to establish a school resembling the Academy; considering only material and legal questions, as a foreigner he would have been denied property rights. Lloyd (*Aristotle*, p. 24) believes that it was only with Theophrastus, after Aristotle's death, that the Lyceum bought properties and gained the legal status of θίασος or religious fraternity.

5 It is worth noting that the material that has reached us consists mainly of his class notes made during this period, while his actual works have been lost. It would not, then, be appropriate to describe Aristotle's frugal prose as monolithic, for we have access to a very particular

During the Lyceum period, he adopts the two teaching methods noted above: the peripatetic and the lecture. Esoteric or acroamatic teaching was for the morning, delivered while walking and intended for an intimate audience of just a few disciples. The more exoteric teaching of the afternoon, intended for a broader audience, was delivered to a seated audience. With the death of Alexander in 323, Aristotle's stay in Athens became more difficult, and entrusting his school to his disciple and friend Theophrastus, he left for Chalcis where he passed away in 322.

Throughout his productive life Aristotle's thought never ceased to evolve, to the extent that Jaeger affirms that any systematic study that neglects this intellectual evolution falsifies his thought. Indeed, much of what precedes Jaeger is subject to such criticism. Knowledge that the papers left by Aristotle reflected a personal evolution was soon lost, however, after the master died. The Aristotelian tradition increasingly privileged systematic over genetic interpretations, when, in fact, the second approach is more suitable if the aim is to understand Aristotle's thought at each stage of his work. Instead, monumental, systematic, and monolithic constructions prevailed, such as that forged by medieval Christian thought—following in the steps of Arab thinkers—to justify faith by means of philosophy (and we by no means wish to deny with such a general characterisation the richness of nuance in the *Aristoteles latinus*).

Although the perspective of Jaeger is *prima facie* irrefutable, it is not always well substantiated:[6] the premise of Aristotelian evolution often puts too much at stake to allow any solid reading at all. For example, in the case of theology, are we to conceive of Aristotle as the monotheist of Λ, 6–7 and 9–10? Or as the polytheist of Λ, 8? Failing that, should we be arguing that Aristotle was first a monotheist and then became polytheist? These issues—which arise also in other topics—add to the difficulty posed by text insertions made to the canonical texts by Aristotle's disciples, some closer than others, which can just barely be detected through the admittedly rather conjectural method of stylistic comparison. Jaeger's analysis is enabled by his philological mastery. He bases most of his conclusions on intuition about which particle or verb Aristotle would

---

sample. It would be fair to be cautious and admit that we know very little of his more exoteric literary style.

6 Jaeger's ideas had been anticipated by Hermann Bonitz, as Jaeger points out in his *Aristotle* (p. 194) and were validated by twentieth century interpreters such as Ross, Mugnier and Owens, who praised the significant thesis of Jaeger, but were also qualified by Guthrie and John Burnet, as Guthrie indicates. Hermann Bonitz was one of the most prominent nineteenth-century Aristotelians, and author of several critical reviews of Aristotle's works, such as his *Observationes criticae in Aristotelis libros metaphysicos* (1842) and his influential *Aristotelis Metaphysica* (1848–1849).

have used if a text were to belong to one given period rather than another, and in the same manner suggests passages or insertions that should be considered as spurious, having been modified or inserted by later editors. Some high profile Aristotelians, such as Burnet, Guthrie, Reale, Judson, Vigo, etc., have noted that an overuse of this method undermines some of his conclusions.

As regards Λ, 8, Jaeger's idea has been widely accepted: it is thought to be a fragment written by Aristotle after the rest of book Λ, probably two decades later, shortly before his death and after been briefly acquainted with Callippus. The chapter was likely inserted in the final edition of his work on first philosophy, probably by Aristotle himself, though it is also possible that his students decided its position within book Λ, in the midst of the theological discourse.[7] According to Jaeger, during his last years Aristotle experienced a sort of *Kehre* (a return to the most empiricist principles of his intellectual development), turning away from a more metaphysical approach to problems and looking instead for the empirical basis of his objects of investigation. In this context, devoted to the study of nature, and of animals in particular, Aristotle spent the end of his life reformulating some of his earlier conceptions. The text of Λ, 8 is thought to be a product of this period.

Jaeger's core hypothesis is acceptable for several reasons. The most important is the fact that it would be difficult to reconcile the *prima facie* inconsistent exaltation of the Prime Mover as the unique and ultimate cause of all motion in Λ, 6–7 and 9–10 with the unequivocal statement in Λ, 8 on the plurality of Unmoved Movers. The passage, inserted between those two other texts, is clearly related to his account of the wandering motion of planets, as we have repeatedly pointed out. According to Jaeger, Λ, 8 was written between 330 and 324 BCE, belonging to his last Athenian period, his most prolific years at the Lyceum, which ended shortly before his death. Among the reasons that lead Jaeger to this conclusion is the use of some imperfect verb tenses[8] in reference to Callippus. This would indicate, as suggested by Simplicius,[9] that Aristotle had met Callippus in person (supposedly, by 330 the astronomer from Cyzicus

7 Guthrie ("The Development of Aristotle's Theology—II," p. 97) assumes that this section of text was inserted later by disciples who may or may not have been close to their master, and adds that the insertion must have been the work of "a not too intelligent editor," as he believes—and here we disagree—that the astronomical text was written for a context other than that in which it was finally inserted.

8 Jaeger, *Aristotle*, p. 343.

9 Simplicius, *In Aristotelis de Caelo commentaria*, 493, 6–8.

was in Athens working on the reform of the calendar),[10] and/or that Callippus had died by the time that Aristotle wrote his astronomical excursus.

As regards theology, Jaeger is visibly uncomfortable with the content of Λ, 8. Inserted into a metaphysical speculation about the supposedly first and unique divine substance, whose characteristics are enumerated and substantiated, it introduces strange astronomical considerations derived from the theses that: 1) there exist many separate substances, and 2) these substances are almost indistinguishable, in metaphysical terms, from the Prime Mover. Jaeger states, bewildered:

> Chapter 8 interrupts this continuous train of thought and breaks it into two parts. Remove it, and chapters 7 and 9 fit smoothly together. After reading chapter 8, on the other hand, it is impossible to take up again the

10 Guthrie, "The Development of Aristotle's Theology—II," p. 93. Düring (*Aristoteles*, p. 192) on the other hand, notes that though the reform in the calendar—establishing the beginning of a new cycle of 76 years by 330/229 BCE—made by Callippus is certain, there is no sure indication of his stay in Athens during those years, and that he most likely left Athens along with Eudoxus, his master, for Cyzicus in 360 BCE, after which he never returned. This reform in the calendar is important for Aristotle's understanding of astronomy: the Callippic period of 76 years that equals 940 lunar years sought to replace the Metonic cycle of 19 years. These cycles tried to reconcile the solar calendar determined by the duration of the tropical year, and the lunar calendar related to the cycles of the moon or lunar months. The Callippic cycle is better suited to the phenomena as it has one day less (27,759 days) than the equivalent of four Metonic periods (4 × 6,940 days = 27,760) resulting in a tropical year of 365.25 days, which is closer to the now accepted value of 365.242199 average days, in contrast to the corresponding value derived from the Metonic cycle (365.263158 days = 6,940 days/19). The hypothesis of Düring, which questions whether a meeting between Callippus and Aristotle would actually have taken place during the latter's second stage in Athens, requires the assumption—not unreasonable—that either the young Aristotle was aware of both astronomical theories represented by these calendars, and that these theories were already challenging one another—which is difficult to prove—or that Aristotle became acquainted with Callippus's reforms to Eudoxus's planetary systems indirectly after his stay at the Academy. This last possibility is in particular favoured by the fact that not until *Metaphysics*, Λ, 8 did Aristotle demonstrate his preference for the planetary systems of Callippus. In other works, such as *On the Heavens*, II, 12, 291b35 –§ 19–, the Eudoxian systems still seem to underlie his reasoning: in this treatise, probably corresponding to his last years in the Academy, Aristotle suggests that the Sun and Moon exhibit fewer motions than the other wandering stars, something consistent with the Eudoxian but not the Callippic systems. Indeed, according to Eudoxus's geometrical astronomy, the motions of the Sun and Moon are composed by the rotation of three spheres while those of the remaining planets are composed of four spheres, whereas in the Callippic subsystems, the Sun and Moon have five spheres, equalling or outnumbering those of the other planets (four for Saturn and Jupiter, and five for Mars, Mercury, and Venus).

> speculative meditation broken off with chapter 7. From soaring flights, from Platonic religious speculation, we plunge headlong down to the monotonous plain of intricate computation and specialised intelligence. Simplicius was right when he said that such a discussion belonged rather to physics and astronomy than to theology for it loses itself entirely in subsidiary matters, and shows far more interest in ascertaining the exact number of the spheres than it does in understanding the fact that this grotesque multiplication of the Prime Mover, this army of 47 or 55 movents, inevitably damages the divine position of the prime mover and makes the whole theology a matter of mere celestial mechanics. Hence Simplicius transferred his explanation of this astronomical passage to his commentary on the *De Caelo*, and it has been a favourite subject for astronomers from Sosigenes to Ideler.[11]

Yet Jaeger's own genetic interpretation solves this discomfort: book Λ must be read without chapter 8, which in turn has to be considered as a later reflection made in a different intellectual context. Thus, Aristotle would have gone through a Platonic stage (that includes his first travelling years after leaving the Academy), and later became interested in the empirical realm, losing interest, during his last years, in transcendental and metaphysical issues. Therefore, in the same way that the theological meditation of *On Philosophy* reflected the thinking of his youth, the chapters of Λ, 6–7 and 9–10, written shortly before he left Athens (according to Jaeger), would reflect his more mature philosophical thought, and Λ, 8 would belong to his last and more scientific period (in the current sense of the term), during his last years at the head of the Lyceum.[12]

Nevertheless, it is worth considering that there is an alternative to the interpretative line that regards Λ, 8 as a sloppy interpolation—made either by Aristotle or by his editors—lacking proper systematic integration with the rest of the book. Contrasting with this view, another line of reasoning, represented for example by Düring, views book Λ as a coherent unit, written all together at one time, probably when Aristotle was still at the Academy, confronting the Platonic cosmology of the *Timaeus*.[13] As far as we are concerned, though we

11 Jaeger, *Aristotle*, pp. 346–47.

12 Ross also considers that book Λ, is one of the oldest parts of the *Metaphysics*; see Ross, *Aristotle*, p. 14.

13 Düring, *Aristoteles* on the chronology of *Metaphysics*, Λ, see pp. 189–194 and p. 226; on the confrontation with Plato's *Timaeus* see pp. 212–213. Philip Merlan ("Aristotle's Unmoved Movers," pp. 1–30) considers that *Metaphysics*, Λ, 8 does not represent any internal contradiction in so far as chapters 6 and 7 harmonise without difficulty with the discussion opened in Λ, 8 regarding the plurality of the Unmoved Movers. Merlan considers that the

believe that Λ, 8 is a late text, we accept that Düring is right to emphasise that there is no incoherence in Aristotle's theology. This opinion is also shared by other critics, such as Merlan or Guthrie. In other words, there is no incompatibility between the metaphysical chapters and geometrical-astronomical chapter of Aristotle's theological discourse, regardless if the two sections belong to different periods of his intellectual evolution.

However, apart from the question of any theological evolution, it is necessary to clarify the evolution, if there is one, of his thought on the motion of the heavens. In this regard, Jaeger identifies a first stage, purely Platonic and mainly expressed in the dialogue *On Philosophy*, in which the desiring souls of stars would be responsible for celestial motions.[14] In a second stage, Aristotle's thought shifts to consider that it is not the stars but rather the spheres of the system that are animated, since their souls are moved accidentally by the influence of the first heaven. This last view is based in two ideas, namely 1) that there are first movers that move accidentally (*Physics*, VIII, 6, 259b20–32 –§ 13–)[15] and 2) that the motion of the first heaven drags the other spheres (*On Generation and Corruption*, II, 10, 337a17–22 –§ 18–; *On the Heavens*, II, 10, 291a32–291b10 –§ 28–). But Aristotle was not satisfied with the idea that the spheres move accidentally on account of the first heaven. In a third stage he introduces a transcendental mover for each sphere, reaching the final astronomical and metaphysical scheme of his intellectual trajectory.[16] According to

---

doctrine presented in *Metaphysics*, Λ, is consistently polykinetic (rather than polytheist) and not monokinetic (or monotheist), regardless of the monoarchetic Homeric quote (εἷς κοίρανος ἔστω, *Iliad*, II, 204) that concludes book Λ (§ 15).

14 See also Mugnier's, *La théorie du Premier Moteur*; another representative of the genetic model, Mugnier distinguishes three periods in the metaphysical and theological evolution of Aristotelian thought. The first period, Platonic, correlates with *On Philosophy*; the second, immanentist, conceives the Prime Mover as the soul of the first heaven (*Physics*), which admires a certain pure separate form that, by tradition, is usually called "God." The third and last period is evidenced by the introduction of several Unmoved Movers in Λ, 8, following the fundamental lines of Jaeger's interpretation. Mugnier also believes that celestial matter is a *sui generis* matter; perhaps a suitable characterisation given that its physical behaviour does not resemble that of sublunary matter (the divide separating sublunary from superlunary physics should always be kept in mind).

15 Ross admits that a section from this passage, *Physics*, VIII, 6, 259b28–31 (§ 13), could be a late insertion, as suggested by Jaeger; the section reads: "We must distinguish, however, between accidental motion of a thing by itself and such motion by something else, the former being confined to perishable things, whereas the latter belongs also to certain principles of heavenly bodies, of all those, that is to say, that experience more than one locomotion"; English trans. Hardie and Gaye in Barnes, *Complete Works of Aristotle*, vol. 1, p. 433. The doctrine expressed in this passage agrees with that of Λ, 8, which, Ross also agrees, is a late addition to the theological section of *Metaphysics*.

16 See Jaeger's *Aristotle*, pp. 360–361.

this scheme, the souls of celestial spheres would be something of a transitional step between the Platonic-like original doctrine of celestial souls, and the transcendental movers of Λ, 8.[17]

If we take into account both the general evolution of his work and the particular evolution of the relationship between spheres and their principles of motion, the heart of Jaeger's hypothesis would seem to be accurate: we suggest that Λ, 8 was not written at the same time as the other theological chapters (though we believe this does not lead to incompatibility between Λ, 6–7–9–10 and Λ, 8).[18] Accepting this hypothesis, with a partial reformulation, we distinguish at least four moments in Aristotelian evolution as regards the metaphysical grounds of celestial motions (as the stages are not cleanly enough delineated to date them with precision, we have given them qualitative names—note also that the second and third moments are to be considered as simultaneous, as consequences of parallel speculative developments that are only resolved by the fourth moment):

1. *Psychic Moment*: A first moment, expressed in the dialogue *On Philosophy*, coincides with the first evolution on Aristotle's thought, and proposes, in a Platonic style, that stars are animated. This moment has its roots in the discussions of the Academy and, especially, in Plato's *Laws*, book x. There he declares the soul to be the unique principle of autonomous motion (896b), and attributes souls to the sphere of the fixed stars, the Sun, and the planets (899a–b), being the soul of the first heaven, the supreme god. Guthrie's analysis on the few Ciceronian fragments of Aristotle's incomplete work, *On Philosophy*—our most reliable source of information according to Jaeger—allows a reasonable reconstruction of Aristotle's

17 For a complete and careful review of the bibliography about Aristotle's Unmoved Movers prior to 1950, see Joseph Owens's "The Reality of the Aristotelian Separate Movers," pp. 319–337.

18 According to Guthrie ("The Development of Aristotle's Theology—I") there would be three stages in Aristotelian thought: an early stage, represented by *On Philosophy*, in which stars have souls; an intermediate stage, represented by *On the Heavens*, in which the characteristic φύσις of ether would suffice to explain the eternal motion of heavens; and a final stage, unfolded in *Metaphysics* and *Physics*, in which the transphysical Unmoved Movers make their appearance. Miguel Candel, Spanish translator of *On the Heavens* (*Acerca del cielo*, Gredos), makes reference in his introduction to the three evolutionary stages in Aristotelian thought as considered by Guthrie but, in contrast, Candel believes—and we agree on this—that, at least in the version that has reached us of *On the Heavens*, the last two stages are somehow integrated. Apart from that, we believe, following Jaeger, that the idea of several Unmoved Movers as the ultimate principles of celestial motion, separated by the physical bodies that they move (strictly, celestial spheres), is a mature conception of Aristotle, developed in *Metaphysics*, Λ, 8.

astronomical-theological ideas during this germinal doctrinal period, which also coincide with the general perspective of the Academy during the same period. His theological vision is mainly scattered in passages of the third book (which offers a cosmotheology), which can be read in fragments 21, 23, 24, and 26 of Valentin Rose's edition.[19] The first of these (Fr. 21) states that stars are to be counted among the gods. The second (Fr. 23) presents stars as beings bestowed with life, sensory perception and intelligence. The third fragment (Fr. 24) raises the issue of whether stars are moved by nature, by force, or voluntarily, and finally concludes that they must move voluntarily: they cannot be moved by nature since everything that is naturally moved either moves upward or downward; they cannot be moved by force since there is not a force powerful enough to move stars against their nature; their motion must originate, then, from an act of will. In the context of an enumeration of possible divine principles, the fourth fragment (Fr. 26) mentions a deity that rules the movements of the world by a sort of backward circular motion, or motion upon itself (*replicatione quadam*). This allusion has led some authors to relate the fragment to the rewinding spheres of Λ, 8, which has caused such additional difficulties. The passage from *De natura deorum*, I, 13, 33 reads: "Aristotle, in the third book of his *On Philosophy*, creates much confusion by dissenting from his master Plato. For now he ascribes all divinity to mind, now he says that the world itself is a god, now he sets another god over the world and ascribes to him the part of ruling and preserving the movement of the world by a sort of backward rotation."[20] Jaeger attempts to see in *On Philosophy*[21] the embryo of the very Primer Mover from *Metaphysics*, Λ, and interprets the expression *replicatione quadam* in a metaphysical rather than physical way as "a kind of backwards turning," concluding that: "The God to whom the world is subordinated [Fr. 26] is the transcendental unmoved mover, who guides the world as its

19 V. Rose, *Aristotelis qui ferebantur librorum fragmenta* (Leipzig: Teubner, 1886).

20 Fragment 26 of *On Philosophy* [=F 26 R3] in Barnes, *Complete Works of Aristotle*, vol. 2, p. 2396; Cf. also Rose, 26; Gigon 25, 2. The fragment of Cicero, *De natura deorum*, I, 13, 33, reads: "Aristotelesque in tertio de philosophia libro multa turbat a magistro suo Platone [non] dissentiens; modo enim menti tribuit omnem divinitatem, modo mundum ipsum deum dicit esse, modo alium quendam praeficit mundo eique eas partis tribuit, ut replicatione quadam mundi motum regat atque tueatur…." *Non* in the Latin text (a magistro suo Platone [non] dissentiens) is an insertion to Renaissance editions made at Manutius Family press, then approved by Rose but removed by Jaeger.

21 To the mentioned passages, we can add (favouring Jaeger) Fr. 16, which undoubtedly refers to an unmoved principle but is a testimony of Simplicius in reference to *On the Heavens* and not to *On Philosophy*.

final cause, by reason of the perfection of his pure thought."[22] Von Arnim and Guthrie disagree with Jaeger and consider—and we think them to be right—that this early work does not offer any clear account of the Prime Mover. Guthrie points to a further problem: the entire Ciceronian discourse is delivered by an Epicurean, which cannot but affect the representation of the young Aristotle's *On Philosophy*. Rather than directly or simply quoting *On Philosophy*, it would seem that Cicero draws from it ideas that resonate with his eclectic education. The reference to the *replicatione quadam* is certainly somewhat obscure and ambiguous, as is the rest of the passage. In fact, *On Philosophy* does not even hint at the idea that the stars are fixed to the spheres that move them; on the contrary, it would suggest that stars are themselves animated and therefore responsible for their own motion. In conclusion, what remains of the work does not provide philosophical novelty (thus, we dismiss Jaeger's hypothesis) but adopts the astronomical theology of late Platonic developments, disseminated not only in the Academy but also in the Lyceum: stars are the only gods, their superiority manifested in their regular eternal motion. Some further traces from this period can also be found in a passage from *On the Heavens* that indicates "the heaven is alive and contains a principle of motion" (δ' οὐρανὸς ἔμψυχος καὶ ἔχει κινήσεως ἀρχήν),[23] an idea reinforced in *On the Heavens*, II, 292a18–27 and 292b3–4, according to which stars are animated, and their erratic translations are attributed to celestial souls. Other ideas that set the stage for the second moment begin to appear during this period: that ether is the only element of superlunary physics; that the world is eternal, indestructible, and ungenerated; that stars are rational living beings with a certain degree of immutability.

2. *Physical Moment*: The initial *psychic moment* was followed by another, also developed during the period of his travels, recorded especially in *On the Heavens*, I. With this begins the systematic articulation of his cosmological considerations using ideas and materials derived from a religious tradition.[24] Notably, during this period, Aristotle abandons

---

22 Jaeger, *Aristotle*, p. 139.

23 *On the Heavens*, II, 2, 285a30–33; English trans. Guthrie, p. 143.

24 The traditional division of Aristotle's works suggests three periods. The first period (368–347) coincides with his first academic stay in Athens. The lost dialogues *Grylos* or *Rhetoric*, *Eudemian Ethics* or *On the Soul*, and *Protrepticus* almost certainly belong to this period. It is very likely that the treatise *On Philosophy*, of which only some fragments have been recovered, also belongs to this period. Authoritative Aristotle interpreters such as Düring and Berti consider that much of his most important works belong to this period, during which, and diverging with his master Plato, Aristotle had already outlined most of

previous views regarding ether (many of his pre-Socratic predecessors held ether to be the residence of the deity, if not the deity itself, though these reflections lacked any systematic physical connotation).[25] In *On the Heavens*, Aristotle provides dynamic grounds for his consideration: ether is the *divine* element par excellence because of the two simple motions—rectilinear and curved—the ethereal curved motion is the only one that can be *eternally* realised. Natural rectilinear motion, expressed in sublunary physics in the distribution of the four elements among concentric strata (earth, water, air, fire), operates between two insurmountable limits: the centre of the Earth and the outermost layer of the sublunary realm. Ether, on the other hand, moves in circles, thus surpassing the limits of what has beginning and end. Moreover, what defines

---

his original ideas, among them the notion of the Unmoved Movers. The oldest books of *Metaphysics* (Δ, N and book Λ), *Physics* (at least books I–IV and VII), *On the Heavens*, *On Generation and Corruption*, and book IV of *Meteorology* would belong to this period. The second period (348–335) coincides with his travelling (Assos, Mytilene, Pella), and the first outlines of his biological works almost certainly belong to this period: *On the Gait of Animals*, *Parts of Animals*, *Parva Naturalia*, *Movement of Animals*. It is quite likely, however, that these treatises, as we know them today, were finished during the third period, when he returned to Athens and headed the Lyceum. The *Eudemian Ethics*, *Politics*, and part of *Meteorology* would also belong to the second period. There is agreement, in general, that most of Aristotle's production is concentrated in the third period (335–323) during his second stay in Athens, the work of the mature philosopher. This includes, among many other works from the surviving *corpus*, the final version of the remaining biological works, probably book VIII of *Physics*, and the final versions of *On the Heavens*, *Metaphysics*, *Rhetoric*, and *Nicomachean Ethics*. The dating of *On the Soul* is uncertain, but it seems to have been written between the period of travelling and the period of his mature thought. However important this chronology may be to understand the evolution of Aristotelian thought, it is important to keep in mind that the classification of his works by periods remains tentative and approximate. Apart from the fact that his works were later edited by the tradition of his school, the evidence points to the fact that Aristotle made changes to many of his works up to his final days.

25 Identifying heaven with a creative deity (often a paternal god) who rules the cosmic order is thought to date back to the cultural influence of the Indo-European peoples that invaded Greece, Asia Minor and Mesopotamia between 2300 and 1900 BCE. Studies on religious vocabulary recognise in the Indo-European *deiwos*, meaning "luminous" or "heaven," the common origin of terms that convey deity in different languages: Sanskrit, *deva*; Irianian, *div*; Lithuan, *diewas*; Old German, *tivar*; Latin, *deus*; Greek, *dios*, from which διός and Ζεύς derive; see M. Eliade, *A History of Religious Ideas*, vol. 1: *From the Stone Age to the Eleusinian Mysteries*, trans. W. R. Trask (Chicago, IL: University of Chicago Press, 1978), p. 189. The sacred nature of the heavens and cosmic phenomena is a frequently recurring notion in Classic Greek literature, to the extent that the vast celestial space is often portrayed as Zeus's dwelling and even sometimes identified with him (see for example Aeschylus, Fr. 70; Euripides, Fr. 877; or Heraclitus Fr. 120).

this moment (or aspect) of Aristotle's thought is precisely his confidence in φύσις, for him a principle of autonomous motion, as identified in *Physics*, II, 1, 192b20: "But ... nature is a principle or cause of being moved and of being at rest in that to which it belongs primarily, in virtue of itself and not accidentally." (ὡς οὔσης τῆς φύσεως ἀρχῆς τινὸς καὶ αἰτίας τοῦ κινεῖσθαι καὶ ἠρεμεῖν ἐν ᾧ ὑπάρχει πρώτως καθ' αὑτὸ καὶ μὴ κατὰ συμβεβηκός).[26] Considered thus, celestial motion in general could rest on the nature of ether. Aristotle never fully expressed himself on this issue in such a clear physical manner, however, and this remains an unsolved point of his doctrine. Perhaps it is more appropriate to consider that while Aristotle may have contemplated the possibility of attributing the cause of celestial motion to the ethereal φύσις, he became convinced by the doctrine of the immobility of the Prime Mover, which, without refuting the previous idea, enriches it by adding a metaphysical dimension.

3. *Noesis-Noetic Monoarchetic Moment.* This third moment coincides with Aristotle's intellectual maturity and perhaps the second and third moments are simultaneous. It is represented mainly by chapters 6–7–9–10 of *Metaphysics*, Λ, and *Physics* VII and VIII, in which Aristotle elaborates his famous doctrine of the Prime Mover, the ultimate and unique guarantor of the motion of the Whole.[27] We are referring—if we add other central texts from *Metaphysics*, for example book Δ—to his most significant metaphysical speculations as regards substance, principles and causes, the corollary of which is the need for a metaphysical principle to account for physical motion. Alongside these formulations, which privilege what we might call—using a somewhat unsuitable term—a monotheistic line, are two ideas not easily reconciled with this perspective: 1) the possibility of more than one mover, introduced in *On Generation and Corruption*, II, 10, 336b27–337a7, and 2) an ecliptic motion of the Sun in the opposite direction to that of the fixed stars (so as to account for the processes

26 English trans. R. P. Hardie and R. K. Gaye in Barnes, *Complete Works of Aristotle*, vol. 1, p. 329.

27 According to Jaeger, this concurs with the main theological lines of *On Philosophy* and *Metaphysics*, Λ (except for chapter 8), as well as *On the Heavens*, II, which supports the idea that these are works from the same period (travelling period, 348–335). The idea that ether naturally moves in circles (*On the Heavens*, I) is not found in the works of this period. Consequently, *On the Heavens* would be—for Jaeger—a work from the post-monoarchetic phase in Aristotle's theology. We have opted to place it later in its definitive conception, but with the understanding that these two moments, the *psychic* and the *monoarchetic*, could overlap to some extent.

of generation and corruption in φύσις), found in *On the Heavens*, II, 3, 286b3–286b9.[28]

4. *Noesis-Noetic Oligoarchetic Moment*. In this fourth moment, represented by *Metaphysics*, Λ, 8, and taking up the issues presented in the mentioned passages of *On Generation and Corruption* and *On the Heavens*, Aristotle notices that in light of what were the recent astronomical advances at the time, his two main propositions on the immobility of the first mover—namely, that celestial circular motion is eternal, and that such motion can only be produced by an intelligible immaterial mover—had to be reconsidered; he then increases the number of movers to 55. We have already covered the reasons for this addition, though it is well worth remembering that the escalation of movers has two reasons: 1) circular motions in different directions—and even contrary in some cases—can be seen in the heavens (which requires *physica ratione* several movers); and 2) the *amount of driving force* of each mover—their velocities's magnitudes, so to speak—(even if we interpret their causality as exclusively final) is different for each circular motion, requiring a different mover for each eternal motion.

Clearly, an evolutionary interpretation of Aristotle's thought resolves some issues, though it opens others. In its favour, and historically and philologically supported, there is the fact that Aristotelian works as we know them derive from an editing process: quite often, passages on related topics written in different periods were assembled as an ahistorical unit. The order of works that we have today comes (and this is no minor point) from this heterogenous *traditio*'s work; from thence also the contradictions. However, if we can explain some of these textual ambiguities through chronology, we are still left with some interesting questions: what was, then, the definitive opinion of Aristotle?; how many divine principles does he actually consider there to be: one? forty-seven? forty-nine? or fifty-five?; are these principles gods?; are these θεοί the traditional gods?; how do they act upon the world, as a final cause or an efficient one?;[29]

---

28 It is worth considering that Jaeger believes (*Aristotle*, pp. 138–144) that the timeline of the theological sections of *Metaphysics*, Λ, except for chapter 8, coincides with the time when Aristotle wrote *On Philosophy*: for Jaeger, it was believed that the idea of a Prime Mover was congruent with the animation of stars; according to this scheme, a single metaphysical principle is responsible for the motion of the Whole, even if immanent souls—we must deem them eternal as well—cause the motion of the celestial bodies.

29 Jean Paulus, in his "La théorie du Premier Motor chez Aristote" (*Revue de Philosophie*, New ser., 4, no. 3 [May–Jun. 1933], pp. 259–294; no. 4–5 [Jul–Oct. 1933], pp. 395–424), considers that the frequently signalled contradiction between the Prime Mover of *Physics* and that of *Metaphysics* is just apparent (the first work seems to introduce it as an *efficient cause* of motion, and the second, as a *final cause*). In line with this thesis, Paulus regards the

what is the relationship between the mythical theological language—which Aristotle places great value on—and the physical-metaphysical language that offers "the philosophical form of divinity"?

Critics are still trying to find a solution to these problems. They mostly agree that it is paramount to distinguish Aristotle's early and late thought in order to explain the contradictions: Aristotle indeed seems to have modified his ideas with time, and this ought not to surprise us of such a tenacious inquirer. The chronological aspect is, then, very significant, as the essential appraisal of Aristotelian theology is at stake. And not only this, but also the ultimate meaning of his theory of principles and causes. As Philip Merlan states, the discussion is not so much about whether we are talking of a monotheist (Λ, 6–7 and Λ, 9–10) or a polytheist theology (Λ, 8), but rather of a monokinetic or polykinetic philosophical scheme.[30] Going a step further, what is at stake here is whether the comprehensive understanding of the motion of the Whole is monoarchetic or polyarchetic. As we will lay out in the following section, we believe that the final formulation of the Aristotelian thought is oligoarchetic.

---

attempts to clarify the ambiguity through genetic interpretation (Jaeger; von Arnim) as unnecessary, as *Metaphysics*, far from contradicting the text of *Physics*, merges with and completes it. In addition, Paulus believes that the issue of the Prime Mover is embedded in a larger problem, that of motion: according to this author, the Aristotelian god would be the ultimate guardian of change, rather than being. The movers from *Physics* in general are souls, and particularly the Prime Mover from book VIII is—under Paulus's interpretation—the soul of the first heaven. In *Metaphysics*, however, the Prime Mover becomes a separate substance and, as final cause, it implies and even demands an efficient cause to verify motion. There is no impediment to consider that the soul of the first heaven constitutes such an efficient cause. Thus, the remaining fifty-four spheres would similarly have a soul, which would be the efficient cause of their corresponding motions. The fact that Aristotle calls these substances "gods" is not an issue for Paulus since he believes that at that time any Greek would have regarded celestial bodies and, more widely, astronomical phenomena as divine. In addition, the notion of the heavens as animated was common to every academic philosopher during Plato's maturity.

30 Merlan, "Aristotle's Unmoved Movers," pp. 18–19.

CHAPTER 7

Aristotle's Meta-Astral Theology

> *Egregie <Aristoteles> ait numquam nos verecundiores esse debere quam cum de diis agitur. Si intramus templa compositi, si ad sacrificium accesuri vultum submittimus, togam adducimus, si in omne argumentum modestiae fingimur, quanto hoc magis facere debemus, cum de sideribus, de stellis, de deorum natura disputamus, ne quid temere, ne quid impudenter aut ignorantes affirmemus, aut scientes mentiamur!*
>
> Aristotle has said excellently that we should never be more reverent than when a subject deals with the gods. If we enter temples with composure, if, when we are about to approach a sacrifice, we lower our eyes, draw in our toga, if we assume every sign of modesty, how much more ought we to do so when we discuss the planets, the stars, the nature of the gods, lest in our ignorance we assert something rashly, impudently, or even lie knowingly!
>
> SENECA, *Questiones naturales*, VII, 30

∴

To analyse Aristotle's theological discourse, we must eschew some longstanding prejudices that seem today to be analytical predicates of his theory on divine principles. First, we have to emancipate it from layers of monotheism (Jewish, Muslim, and especially Christian) accumulated through centuries of its adoption as a philosophical foundation for revealed truths.[1] Second, but

1 As indicated by Pierre Duhem in his *Le système du monde. Histoire des doctrines cosmologiques*, vol. 5: *La Crue de l'Aristotélisme* (Paris: Hermann, 1917), p. 548, in the twelfth century, the first readers of *Metaphysics* thought that the doctrine of separate substances was in strong disagreement with Christian monotheism, as Aristotelian thought systematically requires that the Unmoved Movers be regarded as gods. Thomas Aquinas performed a remarkable metamorphosis of the separate substances of Aristotle in his *Sententia Metaphysicae Aristotelis*, particularly in Lessons 9 and 10 (about Λ, 8). There, Aquinas provides a quite accurate and detailed account of the Aristotelian text, though he transforms fifty-four out of the fifty-five Unmoved Movers into natural separate substances, that is, angels. In his reinterpretation,

© GERARDO BOTTERI AND ROBERTO CASAZZA, 2023 | DOI:10.1163/9789004525535_009

not less importantly, we must reinstate Aristotle as a man of his time, subject to the cultural practices (religious practices included) of the groups and community to which he belonged. This is essential regardless of his exercise of critical judgement and his conviction of the superiority of philosophical discourse (λόγος) over the mythical (μῦθος). He did not, it is important to underline, hold the mythical in contempt. On the contrary, he gave great value to myth as an expression of truth, considering mythical discourse as an early attempt at communicating knowledge within a historiological schema that—just as that of Plato at the beginning of the *Timaeus*—proposes a succession of cycles in which, through the continual oblivion of human striving towards knowledge, the same enigmas are repeatedly posed to new societies: the spirit of each epoch is called upon to answer these questions through its own philosophical work.[2]

The Aristotelian stance towards religion is analogous to that proposed by Hegel in *The Phenomenology of Spirit*, in which Philosophy and Religion are judged as complementary expressions of the true. Hegel grants to philosophical discourse the privileged place of more complete access to the Absolute on account of being substantially mediated by concept.[3] We believe that Aristotle does the same when he acknowledges the value of the (immature) truth that emanates from the mythical-theological corpus as well as from the religious practices of past generations. In *On the Heavens*, II, 1, 284a13–15 (§ 3), for example, he indicates that:

> Our forefathers [οἱ ἀρχαῖοι] assigned heaven [τὸν οὐρανόν], the upper region, to the gods [τοῖς θεοῖς], in the belief that it alone was imperishable [ἀθάνατον].[4]

Aquinas emphatically calls attention to the fact that Aristotle acknowledged the *dubitable* nature of his doctrine with respect to the number of beings corresponding to the number of movements (*Metaphysics*, Λ, 8, 1074a15–17, § 15). Incidentally, thirteenth-century astronomy held the number of celestial spheres to be nine or ten—depending on the interpretation—evidence that the number of spheres proposed by Aristotle had by this time been dismissed in scientific circles.

2 *Timaeus*, 20e–25d; *Metaphysics*, Λ, 8, 1074b11–12 (§ 15).

3 As an example, Hegel believes that the Holy Spirit is the realisation of divine love in the fraternal community—an objective expression of the Spirit in the form of Religion—and that the new Christian commandment "love thy neighbour" is a task to undertake within the community (*The Phenomenology of Spirit*, see ch. 7. C: "Revealed Religion").

4 English trans. Guthrie, p. 133.

This notion is further discussed in *Metaphysics*, Λ, 8, 1074b1–12, in which the Stagirite lays out his historiological thesis on the cyclic generation and corruption of civilisations:

> Our forefathers [τῶν ἀρχαίων] in the most remote ages have handed down to us their posterity a tradition, in the form of a myth [ἐν μύθου σχήματι], that these substances are gods [ὅτι θεοί τέ εἰσιν οὗτοι] and that the divine encloses the whole of nature [περιέχει τὸ θεῖον τὴν ὅλην φύσιν]. The rest of the tradition has been added later in mythical form with a view to the persuasion of the multitude and to its legal and utilitarian expediency; they say these gods are in the form of men or like some of the other animals, and they say other things consequent on and similar to these which we have mentioned. But if we were to separate the first point from these additions and take it alone—that they thought the first substances to be gods [ὅτι θεοὺς ᾤοντο τὰς πρώτας οὐσίας εἶναι]—we must regard this as an inspired [θείως] utterance, and reflect that, while probably each art and science has often been developed as far as possible and has again perished, these opinions have been preserved like relics until the present.[5]

As we know, Aristotle repeatedly reassessed the speculations of his predecessors, inaugurating a particular way of doing philosophy: to do philosophy through engaging with the history of philosophy. When we expose the nodal elements of Aristotelian theology through a lens of systematic understanding, it is clear (from the characterisation of the Prime Mover as θεός, in Λ, 7, 1072b4; from the ontological quasi-equalisation of the Prime Mover and the Unmoved Movers of Planetary Spheres, in Λ, 8, 1073a4; and from the statement that the ancients believed that πρώτας οὐσίας were θεούς, in Λ, 8, 1074b–12, § 15), that his theology requires multiple gods, that it is polytheistic, in line with the historical tradition of his time.[6]

---

5 *Metaphysics*, trans. Ross in Barnes, *Complete Works of Aristotle*, vol. 2, p. 1698; these affirmations are based on a sound Platonic tradition (*Timaeus*, 22a–c; *Critias*, 109d–110a; *Laws*, III, 677a) according to which truth flourishes on a cyclic basis, as entire civilisations are periodically destroyed by violent convulsions of nature or durations of social ἄνοια (a generalised lack of judgement).

6 Étienne Gilson also makes a case for a polytheistic interpretation of Aristotle, as stated in *God and Philosophy* (New Haven: Yale University Press, 1941), p. 33: "The world of Aristotle is there, as something that has always been and always will be. It is an eternally necessary and necessarily eternal world. The problem for us is therefore not to know how it has come into being but to understand what happens in it and consequently what it is. At the summit of the Aristotelian universe is not an Idea but a self-subsisting and eternal Act of thinking. Let us call it Thought: a divine self-thinking Thought. Below it are the concentric heavenly spheres,

Once again, we have to isolate Aristotle's own thought from the dense set of meanings associated to the term *polytheism*, which can refer to a conglomerate of animist, magical, mythical-religious, liturgical, festive, and social practices that governed literary and artistic representations in ancient Mediterranean culture. We cannot emphasise enough how far apart the theological thought of Aristotle is from this interpretation. In an attempt to provide a more accurate definition, we characterise his theology as a "meta-astral aitio-kinetic oligotheism." We have named it *oligotheism* because the number of gods is fifty-five, *meta-astral* because separate substances are required to account for the celestial motions from which the remaining sublunary motions derive, and *aitio-kinetic* because such divine substances mark the culmination of Aristotle's path towards revealing the causes of local motion (the first change and the foundation of generation and corruption) and provide a systematic account of the complexity of φύσις.

Strictly speaking, the Aristotelian gods are nothing like those of Greek mythology, nor does their number derive from complicated genealogies or age-old sagas. They are, rather, the fundamental divine principles used by the Stagirite to explain the motions of the real. He is fully aware that without this limited number of eternal guarantors to account for the permanence of motion and its particular ontic configuration (i.e., Aristotelian physics and cosmology), that which is patent (τὰ φαινόμενα) would not manifest.[7] We should bear in mind that the number of Unmoved Movers is consistent with "the principle of economy of thought," also known as Ockham's razor, which permeates

---

each of which is eternally moved by a distinct Intelligence, which itself is a distinct god. From the eternal motion of these spheres the generation and corruption, that is, the birth and death, of all earthly things are eternally caused. Obviously in such a doctrine, the theological interpretation of the world is one with its philosophical and scientific explanation. The only question is: Can we still have a religion?"

7 We should resist the temptation of regarding them as *spiritual entities* (the Latin word *spiritus*, close to the Greek ἄνεμος, refers to the idea of the *air*, *wind*, or *life-giving breath* that animates the body), as this line of interpretation deviates from Aristotelian thought and brings the matter closer to the Christian *modus cogitandi*. The movers can be characterised as immaterial, transphysical, or metaphysical, but must always be conceived as *intelligible* substances (οὐσίαι; during Antiquity this notion gave way to the possibility of reading the movers as *thoughts*) and, as such, overflowing with activity (ἐνέργεια). In "The Reality of the Aristotelian Separate Movers," pp. 322 and 328–329, Owens, however, draws attention to the possibility—which he then rejects—that the Unmoved Movers are simply thoughts (of another thinking substance). He wonders whether such thoughts could be the product of a celestial soul taking part in some kind of activity (in the manner of the Platonic World Soul) or the product of the Prime Mover itself.

the Aristotelian perspective. In other words, fewer movers would not make for a proper explanation of phenomena, according to Λ, 8.[8]

We have already offered a broad description of the Prime Mover that, we believe, applies to the remaining fifty-four movers. They are unmoved eternal substances whose sole activity is the νόησις. Then how should we represent them? The answer is (and a systematic approach demands it): philosophically—that is, as forms (εἶδος, μορφή) bursting with ἐνέργεια, for the act always accompanies the οὐσία.[9] The specific activity of these substances can only be contemplation; a contemplation that does not imply motion, an absolute self-contemplation that we can only imagine, though imprecisely, in a god.

Now, there is a certain similarity between the mythological gods and the separate substances of Aristotle, which lies mainly in their relation with non-divine entities, that is, that the former rule over the latter. This governing aspect is discussed at the end of book Λ of *Metaphysics*, which concludes with a verse by Homer (*Iliad*, II, 204), conveying the monoarchetic ideal of Aristotle (§ 15):

> οὐκ ἀγαθὸν πολυκοιρανίη· εἷς κοίρανος ἔστω.
>
> The rule of many is not good; let there be one ruler.[10]

We find the same notion expressed in astronomical terms within Λ, 8: the fifty-four lower movers appear subordinated to the Prime Mover, which influences the others by means of its greater strength. The magnitude of its might

---

8 It is worth remembering that the formula "save the phenomena" (σῴζειν τὰ φαινόμενα), famous in the history of science, is almost a verbatim derivation of the Aristotelian idea proposed in *Metaphysics*, Λ, 8, 1073b37 (§ 15): εἰ μέλλουσι συντεθεῖσαι πᾶσαι τὰ φαινόμενα ἀποδώσειν. The main goal of scientific models is to account for the observed, and this we see reflected in Aristotle's increase of the number of Unmoved Movers in Λ, 8 in order to match theory to phenomena. Significantly enough, the expression τὰ φαινόμενα σῴζειν is used by Plutarch in *De faciae quae in orbe lunae apparet*, 6. This is a key passage in the history of science because it registers the possibility of heliocentrism in Antiquity as well as its invisibilisation—perhaps driven by religion—as a viable cosmological framework. The expression is ironically uttered by Lucius, in reference to Aristarchus of Samos, and the fact that some wished him judged and sentenced for blasphemy when he supposed—based on the visible—that the heavens remained still and the Earth travelled along an oblique circle while rotating on its axis (ὅτι τὰ φαινόμενα σῴζειν ἀνὴρ ἐπειρᾶτο..., that man—trying to save the phenomena—supposed ...).

9 *Metaphysics*, Λ, 6, 1071b19–21; *Metaphysics*, Λ, 8, 1074a35–36.

10 Cited in *Metaphysics* (Λ, 10, 1076a3), English trans. Ross in Barnes, *Complete Works of Aristotle*, vol. 2, p. 1700.

is hard to determine though. The same idea had been stated in *On the Heavens*, II, 9, 291a34–291b10 (§ 28), in which Saturn is depicted as the most dominated planet (μάλιστα κρατεῖται) being the closest to the sphere of the fixed stars;[11] also replicated with absolute clarity in *On Generation and Corruption*, II, 10, 337a20–22 (§ 18):

> And if the circular movements are more than one, they must all of them, in spite of their plurality, be in some way subordinated to a single principle [πάσας δέ πως εἶναι ταύτας ἀνάγκη ὑπὸ μίαν ἀρχήν].[12]

It is precisely in the characterisation of this μία ἀρχή where philosophical thinking meets its greatest barrier.[13] Any attempt to reach its essence implies traversing the boundary between physics and metaphysics; that is, to glimpse the philosophical face of god, to comprehend—as far as is allowed to humankind—the ultimate enigma. The path followed by Aristotle is twofold: on the one hand, he examines *motion* and the journey towards its causes, the Prime Mover representing the final point of arrival; on the other, he examines the *entity qua entity* (ontology), which soon evolves into a systematic study of the *substance* (ousiology) and reaches its culmination in the understanding of the *divine substance* (theology).[14] Physics and metaphysics are seen to have a common basis, revealing themselves merely as perspectives, not as conflicting orientations.

At this point, for the sake of clarity, we will retrace Aristotle's journey towards his noesis-noetic god.[15] But first, we should remember that in the

11 Philip Merlan supports the idea that the Unmoved Movers of Planetary Spheres maintain a relation of "subordination" to the Prime Mover, and avoids considering chapter Λ, 8 in terms of monotheism/polytheism in favour of a monokinetic/polykinetic reading. For Merlan, a plurality of Unmoved Movers is neither in contradiction with the rest of book Λ, nor incompatible with any fundamental doctrine of the Aristotelian system (see "Aristotle's Unmoved Movers," p. 28). In addition, he believes that the main type of relation among the Unmoved Movers is one of priority or posteriority, not one in which each would stand as a species of a genus. They form a series, just like numbers.

12 English Trans. Joachim in Barnes, *Complete Works of Aristotle*, vol. 1, p. 552.

13 According to Jaeger, within the Peripatetic school there were those who advocated for the existence of a single principle of motion, despite the doctrine of multiple Unmoved Movers put forward in Λ, 8 (*Aristotle*, pp. 340).

14 Vigo, *Aristóteles: una introducción*, pp. 148–149.

15 David Bradshaw ("A New Look at the Prime Mover," pp. 16–20) develops an interesting—though somewhat debatable—interpretation of the scope and meaning of the noesis-noetic nature of the Prime Mover. Built on the Aristotelian description of the intellect as "having no limitations" (*On the Soul*, III, 4 and 5) as well as the fact that the intellect is "equal to itself" when it reaches actuality (*On the Soul*, III, 4, 429b5), he postulates that

physical-metaphysical discourses of *Physics*, VIII and *Metaphysics*, Λ, we must differentiate between two coexisting planes in order to avoid the error—often seen—of asserting the existence of a first moving principle in a *temporal sense* (there is no such principle for Aristotle—i.e., something that was "first in time"—since motion and time are eternal).

In order to understand these two planes, we need to consider two heterogeneous meanings of the concept "priority": an immediate moving priority in the *physical* sense (the cause "in contact" with the sensible thing that it moves) and the ultimate moving priority in the *logical-ontological* sense (the "separate" cause, the Prime Mover and the Unmoved Movers of Planetary Spheres). Notably, the succession of specific causally-linked beings, always different to one another, demands an infinite *physical* causal chain to receive the imprint of ontic diversity, in other words, a succession of causes and effects with unrepeated constituent links that, by necessity, form an infinite *succession*. The uniqueness of the entities (οὐσίαι) is caused by matter rather than form, while form is the only reality bound—so to speak—to the first and finite causation of the Prime Mover in the capacity of both formal and final cause. This does not mean that the Prime Mover is the cause of each specific ontological form in the universe but that the conditions for the actuality of the real derive from it. From a gnoseological (scientific) perspective, what matters to Aristotelian physics and metaphysics is this "finite-cause-effect-succession-that-refers-back-to-the-Prime

the Prime Mover thinks of itself just as the agent intellect does. This interpretation allows "some degree of multiplicity" (this expression is not Bradshaw's but ours) for the Prime Mover, thus building a bridge between the supreme ἀρχή and the actuality of the entities that have matter. According to Bradshaw, the νοῦς at its highest actuality would be the only reality embracing all actual natural forms; the thought of itself would contain the totality of the forms that in turn give form to sensible entities. Its perpetual ἐνέργεια would be the origin of the actuality of each form. Therefore, the noesis-noetic activity of the Prime Mover would not only enjoy the status of *kinetic cause*, as it promotes the passage from potency to act in those entities that have matter, but would also perform the function of *"the" formal cause* (again this expression is ours), being the source of all that is subject to change (under such ἀρχή, the forms would constitute an absolute unity). We believe that this interpretation, which transposes the human concept of thinking to the divine order, mistakenly imbues Aristotle's theology with pantheism. No element within the Aristotelian vision would lead us, we believe, to coherently affirm the idea of a single intellect that contains all forms of possible entities and would be identified with the formal cause of all. Strictly speaking, we can only affirm that the self-thought entailed in the actuality of the Prime Mover directly causes the motion of the sphere of the fixed stars. In addition, this motion is always local, circular, and identical to itself, constituting an ontological plenitude *per se*, at least in the sensible order. This first motion is the necessary condition for the dynamic polymorphic actualisation of forms in the sensible order, but not in the sense of a kind of original reservoir of archetypes, a notion more Platonic than Aristotelian.

Mover," not the condition of uniqueness caused by matter, in view of the unknowable nature of the ultimate material substratum of each being. Notably, we cannot but acknowledge that, for Aristotle, the series of movers in contact with one another in the sensible plane is infinite both backwards and forwards. However, there can *only* be an infinite series if the analysis concentrates on the individual entities comprising each mover-moved link in a causal chain subject to temporality.[16] It is in this alternative sense that Aristotle posits that the causal series cannot be infinite because, without the unbroken actuality of the first *logical-ontological* cause, there would not be "actual-realities-that-are-potential-realities-of-something-else," those we experience through the senses. Here the series is limited to just three agents: the Prime Mover, the local motion of the first heaven, and (if we symbolically reduce for a moment all astronomical complexity to the action of the first heaven) the ever-changing sublunary realities. The previously-mentioned absence of the "principle of inertia" in Aristotelian thought forces us to think that the *actuality* of the highest metaphysical principle is fully unfolded in the actuality of celestial motion, which, in turn, unfolds in the actuality of the present, in the "being" of that which will soon cease to be. In this sense, the

16 Keen reader of Aristotle, Thomas Aquinas finds a way around the problem posed by the eternity of the world by reformulating, in his *secunda via*, the Aristotelian proof of a necessary first efficient cause without allusion to whether or not the world has a temporal beginning (*Summa Theologiae*, I, Q. 2, A. 3, answer): "The second way is from the nature of the efficient cause. In the world of sense we find there is an order of efficient causes [*causarum efficientium*]. There is no case known (neither is it, indeed, possible) in which a thing is found to be the efficient cause of itself; for so it would be prior to itself, which is impossible. Now in efficient causes it is not possible to go on to infinity, because in all efficient causes following in order, the first is the cause of the intermediate cause, and the intermediate is the cause of the ultimate cause, whether the intermediate cause be several, or only one. Now to take away the cause is to take away the effect. Therefore, if there be no first cause among efficient causes, there will be no ultimate, nor any intermediate cause. But if in efficient causes it is possible to go on to infinity, there will be no first efficient cause, neither will there be an ultimate effect, nor any intermediate efficient causes; all of which is plainly false. Therefore it is necessary to admit a first efficient cause [*aliquam causam efficientem primam*], to which everyone gives the name of God"; this English version from the *The Summa Theologiæ of St. Thomas Aquinas*, second and revised ed., 1920, translated by Fathers of the English Dominican Province (Online Edition from New Advent, 2017). He further elaborates on the matter of eternity or the temporal beginning of the world in *Summa Theologiae*, I, Q. 46, A. 2, in which he concludes that the first efficiency of God is so regardless of whether or not the world has a beginning, since God *may have created* either an eternal or time-limited world. Cf. J. A. Castello Dubra, "Creación, cambio y eternidad del mundo en Tomás de Aquino," in J. G. J. Ter Reegen, L. de Boni, M. Costa (comps.), *Tempo e Eternidade na Idade Média* (Porto Alegre: EST Edições, 2007), pp. 102–108.

entire series is actual. Without the ever-acting presence of the Prime Mover, there would be no motion or, in other words, if the moving action of the Prime Mover were to stop, the universe would stop at once.[17]

When we schematise Aristotle's ideas on this point, we notice that what he first analyses in entities having matter is change, pointing to the existence of "contact" between causes and effects and the *simultaneous* action of the mover and the moved thing. Now, based on the evidence that entities (with matter) are forever subject to change, and acknowledging that matter implies *potency* (δύναμις) and that what is, potentially, can only come to be in actuality through an actualisation of its potentiality, Aristotle concludes that there must be an entity completely deprived of materiality-potentiality, which as such is the most real of all existing things (the only stable permanent thing), as well as intelligible and best—revealing here a Platonic influence.[18] There must, then, be a foundation for motion in which the act occurs without potency and, from a systematic perspective, this source is required to be a substance without potentiality (pure act). He arrives at the conclusion that such a substance is necessary since the verification of motion relies on it, and that its action has no beginning in time, despite being primary.

The Prime Mover is outside of time; motion and time issue from its immaterial energetic substance but are not in contact with it. Such is the necessary conclusion from the Aristotelian theory of change, as only the attraction produced by this immaterial substance on the first heaven can account for this massive celestial motion that has no beginning—essential to sustain the dynamics of the universe—and that is visible to the naked eye every night.

That said, how can we reconcile the ideas that, on the one hand, change and motion are eternal and, on the other, there is a first mover? The answer requires distinguishing between a *space-time plane* and a *suprasensible plane*, which is precisely what Aristotle does. This means that motion is explained by virtue of the metaphysical plane, which provides the sensible plane with its ultimate foundation.[19] The succession of entities in the sublunary domain is

---

17 Despite Aristotle's misinterpretation of the necessary conditions for local motion in nature, we cannot but acknowledge his *brilliant intuition* on the need of a *continuous motion* as the first condition to support all existing things. Classical physics explains this continuity through the "law of conservation of linear momentum," which governs our understanding of all natural phenomena.

18 For Aristotle any form is intelligible, but only the highest object of intellection—the Prime Mover, whose sole "determination" is being νόησις, so to speak—is so in an eminent sense.

19 The Prime Mover, though it exists by and of itself, shares (being the essence of things) its immaterial quality with every abstract form. Therefore, we should bear in mind that the metaphysical proposal of Aristotle is broader than the issue of the Unmoved Movers. It

without beginning or end, in a process akin to the successive generations of life in a living species, like our own, in which no entity is identical to a previous entity.[20] Something different happens, however, with the eternal motion of the first heaven as well as with the ecliptic motion of the Sun and the stars, whose revolutions result from perfect circular motions, whose beginning and end are at every moment identical and *now*.[21] Still, this infinite succession of specific entities does not imply an infinite regression of movers in the physical plane. Undoubtedly, the sphere of the fixed stars is the first physical mover to the extent that if its rotation were to cease, sublunary becoming would stop. But we have to separate the motion of this sphere from its consequences: the ontic change (generation and corruption) is guaranteed by the typical multi-directional local motion of the sublunary domain, whose origin or necessary

---

is well known that for Aristotle matter is practically synonymous with potentiality or the capacity for change.

20 The infinite nature of the parent-child series is analysed in *Physics*, III, 6, 206a25–30. In *On the Heavens*, II, 10, 336b26–34, the unlimited continuity of species is presented as contributing to the perfection of the universe.

21 We should be mindful that, as indicated by Guthrie ("The Development—I," p. 167), the circular celestial motion (κύκλῳ φορά) is a *sui generis* motion, so much so that it can barely be called motion if we consider that Aristotle defines κίνησις as ἐντελέχεια τοῦ κινητοῦ ἀτελής (the act of that which is incomplete). The regular motion of the heavens rests on an exceedingly dense platform of actuality, with minimum potentiality (local only), which explains its stability and predictability. Strictly speaking, the condition of the perfectly homogeneous celestial spheres, the vehicles for the regular revolution of the stars on their axes, is such that their motion can be identified with rest. Their motions exhibit no beginning or end and lack local potentiality, thus constituting a unique mode of actuality. In a way, Ross (*Aristotle*, p. 144) intuits this idea and maintains that the rotation of the celestial sphere is the closest sensible approximation to the eternal, unchanging, pure act: the life of god (the same is suggested by Bradshaw in "A New Look at the Prime Mover," p. 6). The sublunary entities, on the other hand, whose frail actuality is constantly threatened by the tempestuous sea of potentiality, constitute a perfect contrast. In this respect, in *Laws*, x, 898a–b, Plato also compares the rotation of the sphere on its axis with the living plenitude of the divine intellect: "Now of these two motions, the one that moves always in one place must necessarily move around some centre, being an imitation of circular things turned on a lathe, and it must in every way have the greatest possible kinship and resemblance to the revolution of Intelligence [νοῦς] ... Surely, if we said that moving according to what is the same, in the same way, in the same place, around the same things, toward the same things, and according to one proportion and order characterised both Intelligence and the motion that moves in one place—speaking of them as images of the motions of a sphere turned on a lathe—we'd never appear to be poor craftsmen of beautiful images in speech ... On the other hand, wouldn't the motion that never moves the same, nor according to what is the same, nor in the same place, nor around the same things, nor toward the same things, nor in one place, nor in regularity, or order, or some proportion, be akin to complete lack of Intelligence? [ἄνοια]"; English trans. Pangle, p. 296.

condition is the circular, regular, and eternal local motion of the first heaven and, beyond this, the Prime Mover. Such motion causes the *never-ending* succession of specific entities, of which, in their substantial (hylomorphic) realisation, no two are equal, whether prior or posterior.

However, the regular eternal motion of the first heaven, a motion-motor that operates on other entities in the physical plane but is itself without beginning in time, demands its own principle of motion, as everything that moves does so by virtue of something. This mover is transphysical and its actuality cannot be related with or compared to the actuality of any entity that has matter, whether superlunary or sublunary. In the metaphysical plane, the Prime Mover and the other Unmoved Movers introduced by Aristotle are not successively ordered but constitute, each individually, the principle of motion of one specific sphere. Furthermore, the number of Unmoved Movers cannot be conceived as infinite, as this is not necessary to their being according to the Aristotelian concepts of motion and change. The number of transphysical principles of motion will be, ultimately, the minimum required to explain observable phenomena; that is, the planetary motions causing generation and corruption in the sublunary world.

The motion of the first heaven, for its part, guarantees the infinitude of time and of all motion, as well as the finitude of space, without which the circular motion of the Whole—revealed by the diurnal rotation of the firmament—would not be possible. This is so because no mobile thing can travel through infinite space in a finite amount of time. In other words, the Prime Mover is first because it has always moved the first heaven on a regular basis and will forever do so, and every motion in the universe depends on it, even if we cannot speak of a first motion *in time*, since time, like motion, is eternal. More precisely, for Aristotle, "time is not movement, but only movement in so far as it admits of enumeration, in terms of a before and after."[22]

Thus, Aristotle approaches the specific relation between the Prime Mover and the first moved thing by revoking the doctrine of the mover-moved contact, for there cannot be contact between the most exalted sensible entity (the first heaven) and its intelligible principle. A complete lack of contact does not, however, mean a lack of relation. In fact, this relation is of the greatest significance to Aristotelian thought. It is one of admiration or love or desire on the part of the first heaven towards the Prime Mover, which, separate, transcendent, and indifferent to the world, moves all of reality by virtue of the

22 *Physics*, IV, 11, 219b2; English trans. R. P. Hardie and R. K. Gaye in Barnes, *Complete Works of Aristotle*, vol. 1, p. 372.

attracting plenitude and goodness of its being. This indifference to the world on the part of Aristotle's god (a loved god) was to become—by means of a remarkable conceptual metamorphosis—the love of the Christian God (a loving God) in the context of the extraordinary integration of Aristotle into the core of the Judeo-Christian message, most particularly through the refined conceptual development of European Scholasticism from the twelfth to the fourteenth century. Specifically, the medieval distinction between the *primum movens* (first mover) and the *primum mobile* (first moved) comes to the aid of those of us who try to understand the relation between the Prime Mover and the world. The *primum movens* (the noesis-noetic θεός of Aristotle that resides at the apex of reality) moves, in its capacity of final cause, thanks to its ontological plenitude, attracting the *primum mobile* (the first heaven or sphere of the fixed stars). For its part, the motion of the *primum mobile* is the result of the greatest possible force and regularity, which explains why the motion of the fixed stars is the fastest and most regular of all celestial motions, both according to Aristotle and to the medieval cosmological *traditio*. This first-moved (along with the other spheres moved by other Unmoved Movers) moves all other existing things in the capacity of efficient cause. Only in this sense can we say that the Prime Mover moves all things, for, in truth, it exists apart and comes into contact with nothing, not even the first heaven. Considering that the first moved (together with the totality of celestial spheres) *efficiently* moves the lower strata of reality, we can legitimately affirm that the Aristotelian god is the origin of all motion and, broadly speaking, of all that is. This is because the evanescent act of corruptible entities depends, in an ontological sense, on the motion of the heavens, guaranteed by the absolute act of, firstly, the Prime Mover and, in a subsidiary manner, the remaining Unmoved Movers. The ontological dependence of corruptible entities on the Prime Mover should not, however, be mistaken for the Christian *creatio*. In addition, the first motion, that of the fixed stars, is the expression of a fully materialised form that, in perpetual pursuit of its end, becomes the efficient cause of every specific being, without which they would never come to be as they are. But, unlike the Christian God, the Prime Mover (represented as the final and formal cause of the motion of the first heaven) does not contain the archetypes or—in Aristotelian terms—the forms of specific beings. This is because in the Aristotelian view each entity comprises its own end, which has not been assigned by a transphysical source but, if anything, has as its direct cause another entity belonging to the same species (biological succession being an example). In order to understand the specificities of causation in Aristotelian physics and metaphysics, the analysis has to distinguish efficient causes (the first being the motion of the first heaven) from formal causes (the Prime Mover being the first and formal cause

of the motion of the first heaven and the indirect cause of the becoming of specific entities through the efficient cause of the first heaven).

Regarding the type of causation involved with these immaterial entities, in her influential "Que fait le premier moteur d'Aristote?,"[23] Sarah Broadie clears the way for linking the activity of the Prime Mover to astronomy, physics, and biology (at the expense of the traditional view that focuses on the noesis-noetic character of the Prime Mover without elaborating on its cosmological role). Broadie recognises two different types of noetic activities: the *contemplative*, a theoretical type that operates independently of the physical world; and the *kinetic*, a practical type with space-time consequences. For this philosopher—and in this we agree—the main activity of the Prime Mover is to stir the sphere of the first heaven into motion.[24] Broadie believes that the notion according to which the central activity of the Prime Mover would be contemplation—as expressed in *Metaphysics*, Λ, 7—is a somewhat artificial exegetical construct rather than an Aristotelian postulate. Moreover, she favours the idea that the Prime Mover is, above all, a kinetic agent (hence an efficient cause), a notion that—she claims—can be traced throughout the entire Aristotelian corpus, particularly in his biological works.

Broadie highlights that the connecting point between the intelligible and the sensible is "the specificity of circular motion," namely, the fact that it is an eternal motion, thus eluding any possibility of being experienced, as we can perceive motion but not its eternal nature. According to her, however, the identification of the circular motion of the first heaven with the self-thinking intellect and, at the same time, of this intellect with a pure rational soul (by analogy with human beings) makes it easier to regard the self-moving astral souls as efficient causes. Under this interpretation, the eternal circular motion of the body of the sphere would cause the visible motion of the stars much in the same way that the invisible soul of a person causes their visible motions. It should be noted that such a reading of the Unmoved Mover(s) is akin to Platonic thought in so far as it relates the νοῦς to circular motion[25] and designates the souls of stars as the cause of celestial motions. Yet, we know that Aristotle roundly rejected the Platonic identification of the circular motion with the intellect and the soul.[26] The transcendence of the mover as proposed by Broadie is very similar to the notion of immanence; as she puts it: "there

23 S. Broadie, "Que fait le premier moteur d'Aristote?," *Revue Philosophique de la France et de l'Étranger* 183, (1993), pp. 375–411.

24 Broadie, "Que fait le premier moteur d'Aristote?," pp. 375–376.

25 *Timaeus*, 34a, 37a, 40a–b; *Laws*, x, 897d ff.

26 *On the Soul*, I, 3, 406b26–407b12.

cannot be a mover as such without the sphere."[27] Although we agree with her interpretation in so far as it takes the circular motion of heavens as that which binds together physics and metaphysics, Broadie's view seems to us unsuitable (for reasons that we shall elaborate in chapter 8), if we are to understand the moving immanence of the spheres as the presence of souls in the astral bodies, or simply in terms of efficient causality.

The rotational motion of the first heaven, Broadie claims, provides the "permanent conditions" (but not "the structures or forms") of all that happens in the sublunary domain,[28] so that all that we know (intelligise) about the Mover is that it eternally moves the sphere. To locate it outside the moved sphere would make for a distorted picture of the Prime Mover's kinetic activity. We can say little of the intimate unfathomable relation between sphere and mover, she concludes. The role of the Aristotelian god as kinetic guarantor, however, offers a different case, an issue that she highlights repeatedly and which has been taken up practically unanimously by recent critics.

At this point, it is worth outlining the significant milestones of the Aristotelian argument contained in *Physics* and *Metaphysics*. In Book VII, chapter 1 of *Physics*, the Stagirite concludes that "everything that is in motion must be moved by something" (πᾶν ἀνάγκη τὸ κινούμενον ὑπό τινος κινεῖσθαι),[29] a principle that applies universally to every "in-contact" cause as well as to the relation between transcendental substances and the celestial spheres (which, also being moved things, are moved by some cause).

We note that this principle in no way affirms whether or not there is motion: we know that there is always motion between in-contact causes but this conclusion does not apply to transcendent causes (Unmoved Movers) which are apart from, or outside of, all motion.

Moreover, in *Physics*, book VIII, Aristotle develops the concept of "first mover" at least in two different registers. On the one hand, he points to the properties of certain provisional or relative first movers—human or animal souls—that are apparent autonomous origins of motion (we have analysed the example of the person that moves the hand, which moves the stick, which moves the stone). Concerning human beings (and we can generalise to the animal soul in general), he tells us that "in the man, however, we have reached a mover that is not so in virtue of being moved by something else" (οὗτος δ'

27 Broadie, "Que fait le premier moteur d'Aristote?," p. 401 (our trans.).

28 Broadie, "Que fait le premier moteur d'Aristote?," p. 396.

29 *Physics*, VII, 1, 242a 14–15; English trans. R. P. Hardie and R. K. Gaye in Barnes, *Complete Works of Aristotle*, vol. 1, p. 407.

οὐκέτι τῷ ὑπ' ἄλλου κινεῖσθαι),[30] which makes the human a *first mover* in this sense. Yet, as the soul is subject to accidental changes of location, it cannot be the first mover in an absolute sense.[31] Aristotle takes a critical attitude towards the Platonic notion of the soul as a principle of motion, which he conveys in *Metaphysics*, Λ, 6, 1072a1–4:

> But again Plato, at least, cannot even say what it is that he sometimes supposes to be the source of movement—that which moves itself; for the soul is later, and simultaneous with the heavens, according to his account.[32]

For the Stagirite, this interpretation has obvious limits. Plato's World Soul cannot be a principle of motion because the demiurge has shaped it from pre-existing materials (the Same, the Different and the Same-Different) mixed in a receptacle. Moreover, nowhere in Timaeus's myth is the origin of the demiurge and the elements properly explained (still, it should be noted that Aristotle's is not the only possible interpretation of this myth). Regardless, once we dismiss the possibility that the soul is a principle in a supreme sense, the task remains of investigating the first mover in an absolute sense.

Aristotle refers to such a mover in *Metaphysics*, Λ, 6–10, but also in a significant passage of *Physics*, VIII, 6, 259b19–27 (§ 13), in which, based on the analysis of relative first movers (animal souls), he concludes that there must be such a principle:

> Hence we may be sure that if a thing belongs to the class of unmoved things which move themselves accidentally [i.e. by the soul], it is impossible that it should cause continuous motion. So the necessity that there should be motion continuously requires that there should be a first mover

30 *Physics*, VIII, 5, 256a9; English trans. R. P. Hardie and R. K. Gaye in Barnes, *Complete Works of Aristotle*, vol. 1, p. 427.

31 Change in general must be caused by a self-moving thing (the soul) or by the activity of an unmoved thing (the Unmoved Movers). The soul, however, cannot be the source of change in an absolute sense for two reasons: it is subject to accidental changes of location (it moves with the body that it animates) and, according the Aristotle, it generates and corrupts at the same time as the human body. Thus, it cannot be eternal nor postulated as the cause of continuous eternal change. Neither can the *successive* action of souls account for change in general, as the potential condition of that which is not eternal would interrupt the motion. There must, then, be a general cause of motion that also embraces the successive generation and corruption of non-absolute Unmoved Movers (souls) to account for the eternal continuous nature of change in general.

32 English trans. Ross in Barnes, *Complete Works of Aristotle*, vol. 2, p. 1693.

> that is unmoved even accidentally, if, as [25] we have said, there is to be in the world of things an unceasing and undying motion, and the world is to remain self-contained and within the same limits; for if the principle is permanent, the universe must also be permanent, since it is continuous with the principle.[33]

This key passage of *Physics* makes a clear distinction between a first mover in a relative sense and the first mover in an absolute sense; only the latter corresponding to the metaphysical principle of motion, which is for Aristotle divine. The argumentations of *Physics* and *Metaphysics* clearly complement each other, and the alleged divergences often signalled between them in the critical tradition do not in fact appear necessary at all.

For clarity, we will now outline both approaches, convinced that they are profoundly consistent with each other. The argumentation in the *Physics*, along with some additional ideas introduced in *On Generation and Corruption*, can be summarised as follows:

> In the world, we observe successive changes that, we must assume, belong to some infinite series, in the sense that the succession of causes and effects never ceases and the component entities are always individual in a strict sense, not repeated. The infinite nature of these contingent series of entities (for example, parent-child chains in any species) must, however, also refer back to a necessary foundation that exists in a state of permanent actuality; otherwise, the series can at some point cease to be. The foundation of successive universal motion should, then, consist of at least one unmoved substance or, if more than one, a finite number.
>
> Therefore, a study of unmoved substances should be conducted, as there are some unmoved substances that do not always exist (souls) and these may be the origin of limited causal series (for example, of the person that moves the hand that moves the stick that moves the stone) but not of the *continuity* of the first heaven's motion. In order to substantiate continuous motion, we must postulate an immaterial substance that cannot move, not even accidentally (the soul is therefore excluded as it moves accidentally along with the body).
>
> An entirely unmoved immaterial substance is what causes the continuous motion of the heavens, the first foundation of successive and contingent sublunary motions. To this motion, we must add the cyclical

33 English trans. R. P. Hardie and R. K. Gaye in Barnes, *Complete Works of Aristotle*, vol. 1, p. 433.

> motion of the Sun along the ecliptic (caused by another unmoved substance similar to the previous but hierarchically beneath it). Together these cause the cyclical processes of generation and corruption in nature.

In turn, the argumentation of *Metaphysics* can be summarised as follows:

> The local circular motion of the first heaven—which we know through reasoning and observation—has always existed and will always exist.
>
> Where there is something moved there is also something which moves, yet that which is both in motion and moves another is necessarily intermediate. We must seek, still, the principle from which the motion arises.
>
> A motion such as that of the first heaven (regular and eternal) must emanate from a principle whose identity is always pure act (completely lacking potentiality). If this principle were to have a residue of potentiality, the eternity of motion would not be imaginable, as it would not manifest permanently—which is in fact what experience shows. This principle must lack potentiality (be immaterial) and cause movement simply as that which is loved and is intelligible.
>
> Thus, by necessity there must first be an eternal immaterial substance that *acts* always and moves without being moved. Aristotle calls it "god." It must be a substance because the first heaven is a substance and only a substance can move another substance.
>
> The only possible action of this substance is thought, the thought of the noblest of entities: itself. Every mutable being aspires to reach its own and proper end through action, because everything that is potentially something else aspires to the act; but this condition is only possible in as much as the celestial local motion, the moving cause par excellence in the background of becoming, persists. Therefore, all of reality hangs on the pure act that is the Prime Mover, whose plenitude is the ultimate origin (and draw) of eternal motion.

The comparison of these two arguments reveals that the basic set of doctrinal elements of both treatises (*Physics* and *Metaphysics*) is one in the same; the mode of exposition, though, is not. In *Physics*, the analysis centres on the motion of moving series, mainly around the problem of mechanical or efficient causation, as may be expected in a treatise addressing change in the sensible order. Although the main subject is motion, broadly speaking, Aristotle nonetheless anticipates here the necessity of a cause existing beyond the sensible order, one completely unmoved, which he compares to the role of the human

soul as regards the actions that it engenders in the body, though never strictly identifying it with the soul as such. In *Metaphysics*, the explanatory framework for causation—unlike that of *Physics*—is deeply onto-theological and coherent with the general orientation of Greek thought in that it assimilates the celestial to the divine. This is significant as it links the argument of *Metaphysics*, Λ with the pre-Socratic and Platonic efforts to understand the philosophical foundation of reality in religious terms. In the case of Aristotle, however, the deification of the principal and ultimate cause of everything assumes a philosophical rather than religious or liturgical character. This explains why the Stagirite regards traditional deities as a primitive or pre-philosophical way of understanding the unmoved intelligible principles.[34] The consequences of Aristotle's onto-theological proposal echo across every order of reality: that which is divine-actual becomes the ultimate foundation of the actual in general, so its divine nature is, to some extent, transmitted to all the other entities. One could say, without betraying Aristotle, that for any contingent entity *to be in act* is to absorb the most intimate part of the divine, the *act* of the necessary being, as properly interpreted in Thomas Aquinas's metaphysics—this is, however, leaving the differences between the Aristotelian naturalism and the Judeo-Christian creationism aside, and rejecting of course any pantheist interpretation of the Aristotelian thought.[35]

---

34 *Metaphysics*, Λ, 9, 1074b1–14.

35 As indicated by Jaeger in his *Theology of the Early Greek Philosophers*, trans. from the manuscript by E. S. Robinson for The Gifford Lectures, 1936, (Oxford: Clarendon Press, 1947), pp. 32–33), we should strip the references to the divinity, so frequent in the thought of the first philosophers, from any religious connotation. Regarding the first philosopher (Anaximander) who named the principle of all things "god," Jaeger tells us: "We have no right, for instance, to complain that Anaximander's god is not a god one can pray to, or that physical speculation is not true religion" (p. 32). The fact that he called this principle god is more likely attributable to a limitation in the proto-scientific language (rather than to a pious feeling) that, Jaeger suggests, had not yet found, in the dawn of Greek though, another way to celebrate the ontological dignity that provided the foundation for all things.

CHAPTER 8

The Animation of Celestial Bodies

> Pourquoi est-ce que les planètes ne parlent pas?
> Premièrement parce qu'elles n'ont rien à dire, deuxièmement parce qu'elles n'en ont pas le temps, troisièmement parce qu'on les a fait taire.
>
> Why is it that the planets don't speak?
> Firstly, because they've got nothing to say; secondly,
> because they don't have time;
> thirdly, because they've been silenced.
>
> JACQUES LACAN, *The Seminar of Jaques Lacan*, Bk II, seminar of 25 May 1955

∴

Connected with the issue of the evolution of Aristotelian thought is a question that conditions any astronomical interpretation; that is, whether stars are imbued with souls. The question is above all relevant to Aristotle's mature stage of cosmological thinking, by which time his astronomical system has already been outlined (*Metaphysics*, Λ, 8). The problem is both historical and systematic. *Historical* because the idea of the first heaven as a divine entity (due to its excellence) that moves by virtue of its soul, which is κινήσεως ἀρχή, was already expressed in *Phaedrus*, 245e and *Laws*, X, 896b, leading us to reconsider some Platonic elements present in his thought. *Systematic* because if we regard the ethereal spheres of Λ, 8 as animated beings, we are confronted with the issue of whether fifty-five independent souls seek to reach, in some sense, their corresponding movers, and *eo ipso* the problem of where those souls would reside (in the stars?, the spheres?, somewhere else?).[1]

In *Metaphysics*, Λ, Aristotle introduces his final explanation on the ultimate causes of nature's unceasing motion. He justifies the perpetual motion of all existing things by introducing eternal Unmoved Movers that are directly responsible for the local motion of the stars and indirectly responsible for all change

1 Essentially, this is the soul-related approach to the relation between celestial bodies and Unmoved Movers adopted by Mugnier in *La théorie du Premier Moteur*.

© GERARDO BOTTERI AND ROBERTO CASAZZA, 2023 | DOI:10.1163/9789004525535_010

in the sublunary realm. These unique immaterial entities—metaphysical realities, unlike any other entity within Aristotelian thought—have always represented a considerable interpretive challenge for two main reasons. On the one hand, they are transphysical entities, alien to sensible perception; on the other, Aristotle offers little explanation as to the way in which these immaterial movers constitute the cause of physical motions observed in celestial bodies. He states that the first among these principles of motion—the Prime Mover, which causes the daily motion of the first heaven—moves without being moved, and "produces motion by being loved" (κινεῖ δὴ ὡς ἐρώμενον).[2] The meaning of this expression is rather obscure, making it difficult to decide whether we should read it literally or simply as an analogy or metaphor for an action unlike any known phenomena.

Consolidated during the twentieth century, the traditional interpretation—strongly influenced by this image of the Prime Mover as an object of love—proposes the Prime Mover as final cause of the motion of the first heaven. However, in recent decades, further developments have questioned this conception, supporting the thesis that the Unmoved Movers must be understood as the efficient causes of the motion of celestial spheres.[3] Enrico Berti, a strong advocate of this interpretation, considers that the condition of final cause must, nevertheless, be attributed to the Prime Mover, but as the final cause of itself, not of celestial motion.[4] This is not a new debate; the ancient exegesis had already considered this issue and proposed similar though not identical solutions. In fact, Theophrastus (*c.*371–*c.*287), Aristotle's disciple and friend, questioned the nature and role of the Unmoved Movers and suggested identifying the first cause of motion with the thinking activity (διάνοια) of stars. He thus subscribed to the hypothesis of efficient causality in so far as, from this perspective, the prime and noblest mover would be the cognitive activity of the celestial body's own soul and, hence, an internal principle of motion. Such an interpretation would see the motion of celestial bodies as a voluntary act of the soul, animating through the desire born of the faculty of thought.[5] Similarly, Cicero (*On the Nature of Gods*, II, 16, 44), interpreting the Aristotelian cosmology, affirms:

---

2 *Metaphysics*, Λ, VII, 1072b3; English trans. Ross in Barnes, *Complete Works of Aristotle*, vol. 2, p. 1694.

3 E. Berti, "Unmoved mover(s) as efficient cause(s) in *Metaphysics* Λ 6," pp. 427–451; Berti, *Ser y tiempo en Aristóteles*; A. Kosman, "Aristotle's Prime Mover," in *Self-motion*, pp. 135–153; L. Judson, "Heavenly Motion and the Unmoved Mover," in *Self-motion*, pp. 155–171; and M. L. Gill, "Aristotle on Self-Motion," also in *Self-motion*, pp. 15–34.

4 Berti, "Unmoved mover(s) as efficient cause(s) in Metaphysics Λ 6," p. 463.

5 Theophrastus, *Metaphysics*, II, 8–9.

> Nor yet can it be said that some stronger force compels the heavenly bodies to travel in a manner contrary to their nature, for what stronger force can there be? It remains therefore that the motion of the heavenly bodies is voluntary [*restat igitur, ut motus astrorum sit voluntarius*].[6]

Although several passages in the Aristotelian corpus represent celestial bodies as animated beings,[7] the motions manifested by these bodies do not appear to be the consequence of voluntary acts, as Theophrastus and Cicero believe. About this, and with respect to the motion of the first heaven, Aristotle clarifies that:

> A third supposition is equally inadmissible, namely that it is by the constraint of a soul that it endures for ever: for such a life as the soul would have to lead could not possibly be painless or blessed.[8]

This statement does not deny that the sphere of the first heaven is an animated body, it simply emphasises that its soul cannot *strive*, as would be the case in a volitional act such as that performed by the Platonic soul, always striving towards that which it desires.

That Aristotle compares the motion of the heavens with the act of thinking and knowing—in so far as he affirms that the Prime Mover moves as that which is desired (τὸ ὀρεκτὸν) and intelligible (τὸ νοητόν)[9]—implies that the sphere of the fixed stars, moved by the Prime Mover, is an intelligent entity bestowed with a rational (divine) soul.[10] We may reasonably extend this condition to the other celestial spheres responsible for the specific motions of the stars, which, Aristotle believes, are equally animated by Unmoved Movers, one for each sphere. However, the idea that the origin of motion stems from the thinking activity of the souls of celestial bodies, understood as a voluntary act towards an end (an efficient causality), is in fact a Platonic interpretation, not Aristotelian. For Plato, the World Soul, instilled by the demiurge in his creation, penetrates, encompasses, moves, and orders the totality of all existing things. Like the human soul, this soul of the Whole is a principle of motion (as it is a self-moving mover), but unlike the human soul, it is also—at least according to the Aristotelian interpretation that takes the myth literally—the ultimate

---

6 Cicero, *De Natura Deorum*, trans. H. Rackham, Loeb Classical Library (London: W. Heinemann; Cambridge, MA: Harvard University Press, 1933), p. 165.
7 *On the Heavens*, II, 2, 285a30; II, 12, 292a18–21.
8 *On the Heavens*, II, 1, 284a23; English trans. Guthrie, p. 135.
9 *Metaphysics*, Λ, 7, 1072a26.
10 *On the Heavens*, II, 1, 284a2–284b6.

principle of universal motion, as it embraces and governs all things without being externally conditioned. In consequence, the Platonic World Soul voluntarily arranges and sets into motion all things, thus becoming a sort of first mover that constitutes a principle of motion in nature,[11] and—by analogy with Aristotle's categories—an efficient cause of motion. Its "will"—please allow us the anachronism—dominates not only the harmonic order of stars (also considered animated living things by Plato)[12] but also the becoming of sublunary entities. Moreover, as the body of the world is formed by the interaction of the elements, the *anima mundi* is the ultimate source of proportion and harmony that vibrates, finely combining identity and alterity, within all things.[13]

Against this interpretation, Aristotle counters that the first mover is a *separate unmoved* entity that—bound to his conception of motion—must necessarily possess the ultimate principle explaining the perennial motion of an unborn world.[14] The Stagirite also denies the possibility that the Unmoved Movers are the souls of celestial bodies. Souls, as the principle of life in the human or any other animated being, are consigned to the body and so move along with it, if only accidentally, while the Unmoved Movers of the heavens must be unmoved, even accidentally.[15] Moreover, Aristotle states explicitly that the Platonic World Soul cannot be the principle that he seeks, since in the Platonic conception the soul that animates the body of the world has been created by the demiurge *together with* the universe, thus it cannot be a first principle in an absolute sense.[16]

In order to fully comprehend the notion of a cosmos in need of Unmoved Movers such as that proposed by Aristotle, it is essential to bear in mind that—even if Aristotle was able, with his remarkable intuition, to notice that local motion is the origin of all things—the train of thought that led him to confirm a Prime Mover as necessary to explain φύσις belongs to a theoretical framework that lacks the *principle of inertia* and the *principle of conservation of momentum* that today rule the understanding of universal dynamics. Aristotle is convinced that "everything that is in motion must be moved by

11 *Timaeus*, 34c.

12 Plato describes the stars as animated entities in *Timaeus*, 38e and *Laws*, x, 898d–e and 899a–b. Similarly, in *Epinomis* (setting aside the issue of its authenticity), stars have souls (νοῦν ἔχειν, 982c) and are animated (ἔμψυχα, 983a).

13 *Timaeus*, 36e–37c.

14 *Physics*, VIII, 5–6.

15 *Physics*, VIII, 6, 259b1–31.

16 *Metaphysics*, Λ, 6, 1071b37–1072a2.

something" (ἅπαν τὸ κινούμενον ὑπό τινος ἀνάγκη κινεῖσθαι).[17] Consequently, any given motion becomes the endpoint of a *causal chain* formed by a myriad of simultaneous sequential movers that move and are moved *in actuality* (each entity in each present is in contact with the first heaven through this sequence of movers that *instantly* transmit the actuality of celestial motion, which, in turn, is received from the Prime Mover).[18] For Aristotle, motion is not a property that can be transmitted from one body to another; Descartes would claim the same almost twenty centuries later.[19] Therefore, if something moves, it is because of an actual causal action on the part of an entity to which Aristotle attributes the condition of mover. This principle, though incorrect from the perspective of our current physics, is a crucial step in his deductive path towards the Prime Mover. If every motion is an element in a simultaneous and actual causal chain, the sequence of elements must be finite, even if we cannot know each term of the mover-moved series. Aristotle elaborates on this aspect of necessary finitude in *Metaphysics*, α, 2, 994a1–19, asserting that its determining reasons are both phenomenal and logical. Every order of causation is—in its conceptual proposition (material, formal, motor, or final) and as experience shows—an ordered event, not so much in time as in being, since one thing derives from another. It is precisely from this perception of phenomena that we draw the very notion of cause. Given that this successive order, with respect to specific motions, implies a first, middle, and last term of the causal process, if the causal series were infinite none of its constituents could be distinctively designated as middle, first, or last. Moreover, as we call the first term "cause" (whether in a strict or absolute sense), if the series were infinite, then there would be neither a cause of any given motion nor the possibility of knowledge of that motion, since to know something is to know its cause.[20] In other words, if the series were infinite, the starting point of reasoning (that is, that every motion is "caused") would become invalid, because none of the terms in the series could be called a "cause" in an absolute sense. If, then, the series comprising every motion is assumed to be finite (a postulate

17 *Physics*, VII, 1, 241b34; English trans. R. P. Hardie and R. K. Gaye in Barnes, *Complete Works of Aristotle*, vol. 1, p. 407.

18 We believe that "the immediate nature of the actuality of the becoming Whole" is what positions Aristotle's philosophy as—just as he meant it to be—the hinge and mid-position that betters the Parmenidian notion of absolute rest and Heraclitus's elusive flow.

19 As regards the motion of each substance in nature, Descartes argues: "… if it is stopped in some place, it will never depart from that place unless others chase it away; and if it has once begun to move, it will always continue with an equal force until others stop or retard it"; *Le Monde, Ou Traité De La Lumière*, bilingual ed., English trans. M. S. Mahoney (New York: Abaris Books, 1979), p. 61.

20 *Physics*, I, 1, 184a12; *Metaphysics*, α, 2, 993b23; *Posterior Analytics*, II, 11, 94a20–24.

also present in Thomism), there must be a principle, a first mover that, as such, is unmoved; otherwise, there shall be a prior mover causing its motion. In addition, we should keep in mind that this reasoning regarding the first causes of local motion explains and is inherent to any type of change in nature, not only a change of place, since Aristotle understands that any change in nature (qualitative, related to substance, increase or decrease) begins with a local motion.[21]

Once we accept that unceasing local motion (principle of every change in nature) requires at least one Unmoved Mover, the problem lies in understanding how such a singular entity causes the motion of the first moved thing. The answer to this question must apply not only to the action of the Prime Mover on the sphere of the fixed stars but to the action of all Unmoved Movers comprising the Aristotelian astronomical system.

Before expounding our own interpretation of the matter, however, we need to outline the difference between the notions of "unmoved mover" and "rational soul of the celestial bodies." The Prime Mover is evidently an eternal unmoved entity separated from sensible entities.[22] Therefore, it must be different and apart from the soul of the sphere it moves, and also from the thinking activity of that sphere, both of these being aspects integral to the sphere itself as a rational, existing, living thing in the sensible order. The Prime Mover is precisely the supreme good and object of thought of this sphere. It is the content of its cognitive activity, which is its ultimate goal and good, and in which it wishes to remain and always remains, without any further end.[23]

Moreover, it is reasonable to think that the soul of the sphere of the fixed stars is a pure intellectual soul, for why would such a soul have need of any of the other faculties that characterise the rational human soul? What would be the function of the nutritive, perceptual, or motor faculty in the celestial bodies? They do not increase or decrease because they are eternal and unchanging in their substance. From the very shape of the sphere of the fixed stars and other celestial spheres it can also be assumed that they—like celestial bodies—possess neither limbs nor organs with which to move themselves.[24] In addition, if the only activity of such bodies—considered gods—consists in contemplation, it is reasonable to think that they do not require or desire anything apart from the intellectual faculty because, contrary to the perceptual

21 *Physics*, VIII, 7, 260a26–260b7; *On Generation and Corruption*, II, 10336a19.

22 *Metaphysics*, Λ, 7, 1073b4.

23 *Metaphysics*, Λ, 7, 1072a25–1072b13.

24 *On the Heavens*, II 11, 291b11–18.

faculty, which depends on the body, the intellect can comprehend of its own accord.[25]

Separated from any other faculty, the intellect is impassive and unmixed, a condition necessary to receive in unmitigated fashion the forms of objects. Aristotle affirms that the intellect is nothing but pure possibility before the act of comprehending, in which the intellect is realised as thought and, as such, becomes the object or, more precisely, the form of the object thought.[26] The intellect, Aristotle tells us, is the form of the forms (ὁ νοῦς εἶδος εἰδῶν; *On the Soul*, III, 8, 432a2). This mutual implication and identification between the form (εἶδος) and the intellect—or knowledge—in act (ἐνέργειαν ἐπιστήμη) is the reason why the Stagirite states that the human soul is, in some way, all existing things; because the forms themselves and the intellect-in-act that thinks them are one.[27] For the human soul, this identity between object and thought is never complete, as the objects of sensible knowledge are indeed forms abstracted from the material substratum. This substratum, a constituent part of the object itself, is unknowable by virtue of its undefined nature.[28] By contrast, when a purely rational soul such as that of the first heaven thinks of the Prime Mover, this latter (an entity existing without matter) is one and the same as the thought that thinks it. Aristotle refers to the identity between the act of thinking and its object in the following passages:

> Thought is itself thinkable in exactly the same way as its objects are. For in the case of objects which involve no matter, what thinks and what is thought are identical for speculative knowledge and its object are identical.[29]
>
> And thought thinks itself because it shares the nature of the object of thought; for it becomes an object of thought in coming into contact with and thinking its objects, so that thought and object of thought are the same. For that which is capable of receiving the object of thought, i.e. the substance, is thought. And it is active when it possesses this object. Therefore the latter rather than the former is the divine element which

25 *On the Soul*, III, 4, 429b5–10.

26 *On the Soul*, III, 5, 430a14–20.

27 *On the Soul*, III, 7, 491a1; III, 8, 431b21.

28 *Metaphysics*, Z, 10, 1036a9.

29 *On the Soul*, III, 4, 430a3–5; English trans. Smith in Barnes, *Complete Works of Aristotle*, vol. 1, p. 683.

> thought seems to contain, and the act of contemplation is what is most pleasant and best.[30]

We thus must distinguish two different orders of reality that are, nonetheless, from a certain perspective the same. First, there is the order of impassive, unaltered, and immaterial eternal entities that lack magnitude, have no parts, and unceasingly act. The Prime Mover falls into this category, as do the other Unmoved Movers deemed indispensable to explain the totality of planetary motions in the Aristotelean schema.[31] Then, there is the order of the rational souls of celestial spheres, whose sole activity is the pure theoretical intellection of their objects, their Unmoved Movers. If we accept this distinction, the notion of the Unmoved Mover corresponds to the condition of final cause in a way similar—though not identical—to the schema of analysis proposed by Aristotle to account for the motion of living beings.[32]

In the case of beings endowed with reasoning, happiness is the ultimate end of every existence. For ethereal living things, as well as human beings, this end is the contemplation of the good that is proper to each. In the case of a celestial body, this end must be understood as already realised and consisting in the identification of pure understanding (which constitutes the soul of each celestial body) with its corresponding Unmoved Mover (which is its own Good) and entails for each the happiness proper to a god.

Among Arab thinkers, Ibn Rushd (or Averroes) also agrees that the soul of celestial bodies is entirely and only occupied with the faculty of comprehending. He interprets the Prime Mover in a way similar to the one we have exposed: as final and formal cause. However, Averroes also attributes to the Prime Mover the condition of efficient or motor cause since—with respect to Aristotle's quadripartite classification—he combines three causes in one. The material cause is excluded because the Prime Mover is immaterial, lacking potentiality. For Averroes, this intelligible entity, the object of thought of the soul of the first heaven, would achieve actuality simply by being thought by the soul of the sphere. Its immaterial condition removes any need for embodiment in a material substratum such as that required for the objects of human appetite. For illustration, Averroes provides the image of a work yet in an artisan's mind. As a form, in thought, the work is but an efficient cause of the artisan's labour, and it is exactly *through* that labour that the form reaches full actuality

30 *Metaphysics*, Λ, 7, 1072b20–24; English trans. Ross in Barnes, *Complete Works of Aristotle*, vol. 2, p. 1695.

31 *Metaphysics*, Λ, 8, 1074a1–22.

32 *On the Soul*, III, 10.

as form in matter, which is the final cause of the labour.[33] In the case of celestial bodies, and unlike the case of human beings, the Aristotelian affirmation that understanding and the intelligible are one and the same must be conceived, according to Averroes, as if the desire for the object (efficient cause) and the actuality of the object (final and formal causes) are identical. This motion of the unfinished, which we saw in the artisan's labour, is therefore not possible for celestial bodies.

In our opinion, Averroes's interpretation of the nature of the Prime Mover and of the soul of the sphere—the sphere of the fixed stars in this case—is correct, but we believe he is in error when he considers the Prime Mover simply as an efficient cause. If the Prime Mover—a self-existing reality and final cause, being the object of appetite of the first heaven—were to be merged with its presence in thought, this transcendent entity would then become the very soul of the heavens, a notion that falls again into Plato's causal interpretation of motion. That is, the being-in-act of the Prime Mover would not be distinguishable from the thought of the soul of the first heaven, and, in result, this soul would become a being much like the World Soul of the *Timaeus*.[34] This interpretation, like that of Theophrastus, prioritises the activity of thought of the celestial souls over the entities themselves, turning the Unmoved Movers into mere products of the thinking activity of the divine soul of each sphere. If we seek fidelity to the Aristotelian texts, we have to assume, conversely, that the

33 *Ibn Rushd's Metaphysics: A Translation with Introduction of Ibn Rushd's Commentary on Aristotle's* Metaphysics *Book Lām*, English trans. C. Genequand, Islamic Philosophy, Theology and Science, Texts and Studies, vol. 1, ser. ed. H. Daiber (Leiden: E. J. Brill, 1986), §§ 1593–1598.

34 In *Laws*, x, 896b, Plato defines the soul as a "principle of motion" (ἀρχὴ κινήσεως), while in 898d–e he concludes that the stars must be endowed with souls both good and divine (899a–b). The celestial order is matched (898c) to the regularity typical of the intellect (νοῦς). These ideas are evidently present in several instances of Aristotelian thought, despite not being systematised in the same way that we find in Plato. Moreover, Aristotle inherits from Plato the idea that spontaneous motion is prior to and nobler than transmitted motion (894b–d), though he criticises the idea, introduced in the *Timaeus*, that the World Soul may offer an explanation for universal motion. In *Metaphysics*, Λ, 6, 1071b31–1072a3, the Stagirite argues that this soul was fashioned by the demiurge, who, for his part, was suddenly introduced in the myth with no further explanation and as an obvious pedagogical device. In contrast, Aristotle attempts to—and eventually does—deduce the characteristics that the first mover should possess, thus improving upon the Platonic stance, which was solely based on a definition (895c–e). He criticises the Platonic notion, arguing that if things were as Plato suggests, every animated being would have a beginning in time. The motion of the world cannot, however, have a beginning in time, because if that were the case it would have come into being from pure potentiality. Every self-moving being has constant accidental motion, while the supreme principle must not move at all, least of all accidentally (see Jaeger, *Aristotle*, p. 342).

Prime Mover is a primary entity and that the understanding of the first heaven actualises itself in thinking the Prime Mover and not the other way around. The same would apply to the other Unmoved Movers.

Among recent interpretations, Berti also assigns to the Prime Mover a soul that, being an efficient cause, produces the motion of the heavens in a continuous and voluntary way.[35] However, his interpretation of the Aristotelian texts differs from Platonic-like conceptions such as those of Theophrastus, Averroes, or Cicero, as Berti believes—and we agree with this—that the Prime Mover is an entity separate from the body of the stars. The Aristotelian principle of motion understood as a soul in Berti's proposal is not equivalent to the World Soul of the *Timaeus*. In contrast to the Platonic construct, for Berti, the Prime Mover moves outside the body of the world, having a transcendent nature; nor can it be identified with the Platonic demiurge because, unlike the latter and similarly to the World Soul, the Prime Mover eternally and continually supports the dynamics of the universe through a voluntary act.

Having made this distinction between the Unmoved Movers and the souls of the spheres, we must now consider the specific issue of the causal relation that determines the motion of celestial spheres. Aristotle attributes these motions to the "desiring" condition of the souls of celestial bodies in as much as the Prime Mover "produces motion by being loved" (κινεῖ δὴ ὡς ἐρώμενον; *Metaphysics*, Λ, VII, 1072b3). We believe that the difficulty in reconciling the love of the stars with the condition of the Unmoved Movers as final causes resides in the Platonic—rather than the Aristotelian—interpretation of love (ἔρως) as the cause of desire. Within the Theory of Forms, and once the feeling of love is elevated as a pursuit of absolute knowledge, Platonic love is but the desire to always possess the Good, to contemplate the Idea of Good, which, for humankind, remains always an unsatisfied desire due to the strict separation between the sensible realm (home to incarnate souls) and the intelligible world of Ideas.[36] This explains why the Platonic *Eros* is not a θεός but a δαίμων. We find a similar stance in Aristotle: the most elevated expression of love matches a pursuit of knowledge that promises to be the pinnacle of pleasure and happiness. If such a state of knowledge constitutes the end pursued by the most real and authentic human desire, if it is what is most loved, then the difference between Plato and Aristotle lies in the fact that for Aristotle this is an achievable condition. The final cause that moves human desire is such because it constitutes a realisable end. Therefore, if the ultimate end and virtue pursued

35 E. Berti, "Unmoved mover(s) as efficient cause(s) in Metaphysics Λ 6," p. 465.

36 *The Symposium*, 201d–212a.

by human beings is to possess the good, this cannot be the Idea of the Good. Against this Platonic doctrine Aristotle says, in the *Nicomachean Ethics*:

> And similarly with regard to the Idea; even if there is some one good which is universally predicable of goods or is capable of separate and independent existence, clearly it could not be achieved or attained by man; but we are now seeking something attainable.[37]

Each entity has its own end, which is its own good and, in the case of human beings and gods, leads to happiness and contentment. The end of human life is this virtuous condition, a state achievable through philosophical effort. Aristotle identifies this state of plenitude with contemplative, immobile activity, not with production and motion.[38] According to Aristotle, happiness is pleasure, and:

> But of pleasure the form is complete at any and every time. Plainly, then, pleasure and movement must be different from each other, and pleasure must be one of the things that are whole and complete.[39]

This end, which makes human beings wiser,[40] is not unreachable but, rather, untenable.

In some passages of his treatises on ethics, Aristotle compares the human and the divine condition, showing their radical difference. Given the composite nature of human beings nothing can be always pleasant, so happiness cannot be lasting, whereas the simple nature of gods enables them to enjoy forever a single pleasure, such that their happiness is permanent.[41] This reading of the divine nature is verified in the celestial bodies, or gods from Aristotle's perspective, as they are without internal conflict, despite being composed of both body and soul.[42] In the case of human beings, conflict and tension is an inherent condition derived not from the body-soul duality but from the compound nature of the body. It is in the body where antagonistic tendencies manifest and ultimately give rise to inclinations in conflict with its constituent

---

37 *Nicomachean Ethics*, I, 6, 1096b33–36; English trans. Ross (rev. Urmson) in Barnes, *Complete Works of Aristotle*, vol. 2, p. 1733.

38 *Nicomachean Ethics*, X, 8, 1178b7.

39 *Nicomachean Ethics*, X, 4, 1174b7; English trans. Ross (rev. Urmson) in Barnes, *Complete Works of Aristotle*, vol. 2, p. 1856.

40 *Nicomachean Ethics*, X, 8, 1179a29 and ff.

41 *Nicomachean Ethics*, VII, 14, 1154b22–32.

42 *On the Heavens*, I, 3, 270a19–23.

elements. Within the human body, the elements struggle to reach their natural place, each different from the other.

This tendency explains why all living things eventually perish and why, during their existence, they never arrive at a stable state in which to remain and last.[43] Ethereal bodies, on the contrary, have a simple nature. Within them, there are no opposing elements that may struggle, and they are always and forever in their natural place in motion along their circular path, eternal living things in a state of perpetual enjoyment—actual gods. In Aristotle's words:

> For if the nature of anything were simple, the same action would always be most pleasant to it. This is why God always enjoys a single and simple pleasure; for there is not only an activity of movement but an activity of immobility, and pleasure is found more in rest than in movement. But "change in all things is sweet", as the poet [Euripides] says, because of some vice.[44]

Gods's natural activity and immobility correlates perfectly with the state of self-thinking thought typical of the Unmoved Movers of *Metaphysics*, Λ, 6, but also with the thinking activity of celestial souls, though they have motion. The supposed circular motion of stars does not contradict the Aristotelian statement that the activity implied in a state of complete happiness agrees with stillness rather than motion. That passage refers, it would seem, to a kind of motion that characterises a change, such as generation or corruption, or, in the case of local motion, to some finite process. It does not seem to refer to the circular motion of stars that, according to *Physics*, is identical to stationariness.[45] This activity—which is the perfect happiness and ultimate end of both human and divine beings, pursuing nothing outside of itself—is of a contemplative nature, exempt of effort or strain.[46]

If celestial bodies are gods and their activity and happiness consist of pure contemplation, such activity must be rooted in the love that, according to Aristotle, the soul of the first heaven feels for the Prime Mover. Understanding the nature of this love should help us to understand the way in which the Prime Mover causes the daily motion of the sphere of the first heaven; and this regardless of whether the expression "moves as that which is loved" is interpreted in

---

43 *On the Heavens*, II, 6, 288b13–17.

44 *Nicomachean Ethics*, VII, 14, 1154b25–27; English trans. Ross (rev. Urmson) in Barnes, *Complete Works of Aristotle*, vol. 2, p. 1825.

45 *Physics*, VIII, 9, 265a33–265b3.

46 *Nicomachean Ethics*, X, 8, 1178b7–33.

a literal sense, or as allegory or metaphor. It is also true that our only access to this interpretive line is our knowledge of the condition of human. As we have already pointed out, however, this understanding permits only a first approximation, after which a "quantum leap" is necessary to imagine the unique condition experienced by the divine soul of stars, which can be similar, but never identical, to that of the human soul. This fact becomes more evident when we remember that astral souls inhabit ethereal bodies, and that ether is of a completely different nature from the sublunary elements. Moreover, their state of happiness and enjoyment is simple and permanent, which is inherently impossible for the human condition as conceived by Aristotle.

In the Aristotelian treatises that address the feeling of love, it is clear that for human beings, the loved and desired—the ultimate end of our yearnings—is a purely contemplative and self-sufficient life.[47] Nevertheless, the Aristotelian discussion of these topics deals mainly with the vicissitudes of the human condition, always in need of external elements to ensure the well-being that allows the mind to concern itself with pure knowledge. Among the external conditions that are part of this context, human happiness appears to be particularly subject to relations with other individuals; not a physical or utilitarian relation but one marked by an intellectual and cognitive love that may be synthesised in the notion of "friendship" (φιλία). Although Aristotle frequently uses the notion of ἔρως in a broad sense, love as ἔρως is, properly speaking, more related to desire and the pleasure of the senses,[48] while φιλία denotes a more elevated feeling of love, connected with virtuous living and contemplation.[49] This is the sense (love as φιλία) chosen by Aristotle to theorise about profound love, and is the one we must attend to in order to understand the condition of celestial motion.

At the end of the *Eudemian Ethics*, Aristotle links this two-fold condition of human beings: on the one hand happiness seems to be determined by contemplative isolation, yet on the other hand, and at the same time—as experience amply demonstrates—we are in need of external elements (mainly friends) to achieve a happiness that is complete or lasting. At this point, Aristotle departs from theoretical speculation and argues through evidence; namely, that unlike gods, human beings need to share their lives with others. But why do they need community? To understand this particularly human condition, says Aristotle,

47 *Nicomachean Ethics*, x, 8.

48 *Eudemian Ethics*, VII, 12, 1245a25–27; *Nicomachean Ethics*, IX, 5, 1167a4.

49 Owens ("The Reality of the Aristotelian Separate Movers," p. 324) points out, as does Alexander of Aphrodisias (Fr. 31), that the pleasure proper to separate substances, or Unmoved Movers, is not the type of pleasure that follows affections but the pleasure of the act of thinking itself.

we must realise that the act and end of human life is to perceive and to know,[50] and consequently that friendship, which is expressed as a shared life, is more than anything else perception and knowledge experienced or held in common with another. In this reasoning Aristotle seeks to understand why even the wisest person has need of friendship.

What, then, is the relation between that pure contemplation that pursues self-sufficiency, and friendship? When Aristotle posits knowledge as the purpose of life, to the extent that in a certain way to live is to know, he does not refer to an abstraction of knowledge, something that would be, in each person, the same thing. The knowledge to which Aristotle makes reference—the knowledge identified with life itself—is, rather, specific to each person; otherwise, he would be implicitly accepting that one person's life is the same as another's, which does not seem to be the case.[51] In the context within which the sentiment of friendship is explained, such affirmations appear logical, since a person cannot choose to attach themselves to someone else's life—this being the act through which friendship is expressed—without being, first, someone themselves; i.e., without having autonomously forged their own life. Aristotle repeatedly states that in order to love others, one must first love oneself (φιλαυτία). This self-referential act is also an act of self-knowledge through which a person affirms their own existence, which is why the knowledge of each person, introduced as the ultimate end of human desire, primarily implies self-knowledge and φιλαυτία. If a person seeks to know, but does not perceive their own self other than in the act of knowing—to the extent that only by perceiving the object of knowledge do they perceive their own identity—then they desire above all to know themselves. Aristotle asserts, in this respect, that:

> If, then, of such a pair of corresponding series there is always one series of the desirable, and the known and the perceived are in general constituted by their participation in the nature of the determined; so that to wish to perceive one's self is to wish oneself to be of a certain definite character,—since, then, we are not in ourselves possessed of each of such characters, but only by participation in these qualities in perceiving and knowing—for the perceiver becomes perceived in that way and in that respect in which he first perceives, and according to the way in which and the object which he perceives; and the knower becomes known in the same way—therefore it is for this reason that one always desires to

50 *Eudemian Ethics*, VII, 12, 1244b24–25; *Nicomachean Ethics*, IX, 9, 1170a19.

51 *Eudemian Ethics*, VII, 12, 1244b23–1235a1.

> live, because one always desires to know; and this is because he himself wishes to be the object known.[52]

The key to the φιλία may thus be found in these considerations regarding self-consciousness of our own existence, encoded in self-knowledge as well as in the Aristotelian intuition that a friend is in a sense another self. An individual comes to know themselves as good in the community of thought with their like, because for Aristotle, unlike Descartes, "we can contemplate our neighbours better than ourselves."[53] The Aristotelian god does not have this need of friends because a god has within itself the possibility to be its own object of knowledge without mediation: a god thinks itself and needs nothing outside itself. This act of thinking is the consummate end and the archetype of the completely happy contemplative life that human beings can only achieve fleetingly and in community with others. We notice our own existence when we perceive and know, to the extent that through this act we perceive and know ourselves. This condition, however, depends on the exteriority of the known object, a quality that prevents us from reducing everything to the fundamental unity of the self. It is precisely this attempt to unify the knowledge of oneself—that cannot be realised outside sensible perception—that drives one to look for other selves, in whom one recognises oneself, and to join together in *amicitia*—to use the Roman expression—which is but common perception and knowledge.[54] Gods are free from these human constraints that place the self somehow outside the self. It can be said that gods live with their backs turned on the world; they "see" only themselves and are in need of nothing. This is why Aristotle regards their knowledge as simple and one. Being fully realised entities, they are the self-reference of the universal order, in as much as they remain always in their proper and ultimate end. All causes become one in them, even the efficient cause (as efficient cause of their own actuality, which is eternal). The Aristotelian θεός cannot, hence, be the producer of anything but its own thinking, and certainly cannot be responsible for voluntary acts in the natural order. Universal harmony is not caused by the action of a singular entity on the totality of things; it is the name that we give to the ultimate end (final cause) proper to each entity, in so far as it can be achieved and in whose plenitude only the Prime Mover and the stars (also divine) remain indefinitely.

---

52 *Eudemian Ethics*, VII, 12, 1245a1–10; English trans. Solomon in Barnes, *Complete Works of Aristotle*, vol. 2, p. 1973.

53 *Nicomachean Ethics*, IX, 9, 1169b33; English trans. Ross (rev. Urmson) in Barnes, *Complete Works of Aristotle*, vol. 2, p. 1849.

54 *Eudemian Ethics*, VII, 12, 1245a29–37.

If the feeling of love moves the heavens, it must be love in the sense of the Aristotelian φιλαυτία, a human pre-condition to love others and fundamentally to know oneself. The Aristotelian god (the Prime Mover and the object of the first heaven's love) is self-thinking thought, an activity that easily correlates with the Stagirite's considerations on self-love. This condition must be assigned to the other Unmoved Movers of *Metaphysics*, Λ, 8 as well, being those responsible for the specific motions of the stars. An additional issue arises, however, in that Aristotle also attributes the status of divine entities to celestial bodies in circular motion. If we accept that, as gods, they can only love themselves in a purely contemplative activity, then we should assume that their corresponding Unmoved Movers constitute a kind of immutable and simple image, an exalted divine and stationary register of their sameness, presented as an object that exists by itself and transcends the theoretical contemplation of the soul of its corresponding sphere. A similar interpretation is proposed by Theokritos Kouremenos,[55] who reasonably believes that each Unmoved Mover, as an immaterial intelligible object, contains, so to speak, the parameters of the motion of each sphere; that is, the corresponding period, direction of rotation, and orientation of their axes, as if such data were the very essence of these bodies. The statement that a love such as φιλαυτία moves the celestial bodies is coherent, under this interpretation, with the Aristotelian affirmation that, with respect to the souls of the spheres in the act of knowing their object, understanding and the intelligible are one and the same, given that the essence of each sphere is nothing but its state of motion, as defined by the parameters mentioned.

If we remove from the gods the possibility of acting and producing, what are they left with, asks Aristotle, but pure contemplation?[56] We could also ask: if we deprive the gods of friends, as Aristotle does, what are they left with but love for themselves, the condition of their self-sufficiency?[57] The contemplation of their pure simple sameness and their self-love is the nature of what the Stagirite calls θεός. The Aristotelian god attains, in this self-contemplation of its simplicity, its own end, unaware of all else. The vicissitudes of generation and corruption are alien to it, both from a cognitive and productive perspective in a voluntary sense. This image has nothing in common with the Platonic demiurge, creator of the universe and identifiable (in a metaphorical sense)

55 Theokritos Kouremenos, *Heavenly Stuff: The Constitution of the Celestial Objects and the Theory of Homocentric Spheres in Aristotle's Cosmology* (Stuttgart: Franz Steiner Verlag, 2010), p. 42.

56 *Nicomachean Ethics*, X, 8, 1178b20.

57 *Eudemian Ethics*, VII, 12, 1244b7–10.

with a plurality of ideal archetypes.[58] The demiurge of the *Timaeus* contains, so to speak, all the Ideas of which things are but imitations, a plurality to be imagined only in the realm of its thinking activity.[59] In Plato's words, the demiurge created the world "looking to the eternal" (πρὸς τὸ ἀίδιον ἔβλεπεν), as an image (εἰκών) of the eternal Ideas.[60] The Soul instilled in the world by the demiurge may be understood as an efficient cause in so far as it encompasses all materiality and causes the motion of the body of the world in its entirety; just as the living soul causes its organism to function. Fire, earth, water, and air are brought together and linked by the harmony of numbers and reason,[61] which has its source in the soul and penetrates the whole from the centre to the outermost reach of the heavens.[62] Like the Aristotelian gods, the World Soul is self-sufficient[63] but, unlike them, its thinking activity is not manifested in the sensible order as pure simple circular motion. On the contrary, we may say that it animates, knowingly and intentionally, the universal dynamic flow comprised in the generation and corruption of all things.[64]

---

58 In his translation of the *Timaeus*, Conrado Eggers Lan suggests the possibility of equivalence between the demiurge—the best of beings, creator of all things—and the set of ideal archetypes—intelligible beings that always exist. This suggestion is supported by an ambiguous Platonic expression, which appears to mix both terms by claiming that the World Soul has been "brought into being by the most excellent of things intelligible and eternal" (τῶν νοητῶν ἀεί τε ὄντων ὑπὸ τοῦ ἀρίστου ἀρίστη γενομένη τῶν γεννηθέντων); *Timaeus*, 37a1, English trans. Cornford, p. 94. See the discussion on pp. 118–119, note 62, of Eggers Lan's translation (Buenos Aires: Colihue, 1999). The meaning of the quotation is somewhat unclear, however. Moreover, in another more explicit utterance, Plato seems to differentiate the demiurge from the set of Ideas: "whenever a demiurge looks to that which is always unchanging, and uses it for model in fashioning the form and quality of his work, all that he thus accomplishes must be good" (ὅτου μὲν οὖν ἂν ὁ δημιουργὸς πρὸς τὸ κατὰ ταὐτὰ ἔχον βλέπων ἀεί, τοιούτῳ τινὶ προσχρώμενος παραδείγματι, τὴν ἰδέαν καὶ δύναμιν αὐτοῦ ἀπεργάζηται); *Timaeus*, 28a6, English trans. modified from Cornford, p. 22.

59 We certainly agree with a non-literal interpretation of Timaeus's myth. The text (29a–b) only affirms that the demiurge fashioned the world in the image of Ideas (eternal and comprehensible by rational discourse and understanding) but does not elaborate as to where the demiurge and the Ideas come from or how they come to be. Instead, we prefer to think of the demiurge as a poetic resource that allows Plato to account for the causal passage from the intelligible to the sensible, from the World of Ideas to phenomena. The causal versatility of the demiurge is *de facto* demonstrated by the literary effectiveness of the work, allowing Plato to cover the problems arising from φύσις, without any loss to the scientific value of its doctrines on account of the mythical nature of the presentation.

60 *Timaeus*, 29a–b.

61 *Timaeus*, 31b–32c.

62 *Timaeus*, 36e–37a.

63 *Timaeus*, 34b.

64 *Timaeus*, 33c–d.

CHAPTER 9

Aristotle's System in Perspective

All'alta fantasia qui mancò possa;
ma già volgeva il mio disio e 'l velle,
sì come rota ch'igualmente è mossa,
l'amor che move il Sole e l'altre stelle.

Here power failed the lofty phantasy;
but already my desire and my will were revolved,
like a wheel that is evenly moved,
by the Love which moves the sun and the other stars.

DANTE ALIGHIERI, *The Divine Comedy*, Paradiso, Canto XXXIII

∴

An evaluation of Aristotle's astronomical system in its entirety requires some further considerations, both systematic and historical. In the first place, as we saw in the Introduction, the lack of a single definitive text articulating Aristotle's ideas on planetary motions in an integrated manner means that any interpretation aiming at completeness must enter into risky territory. We have undertaken our own interpretation undeterred, with full awareness of this, because we see in the attempt at a unified view of this unsolved issue (the number, nature, and function of the Unmoved Movers according to *Metaphysics*) a possibility to glimpse its *meaning*, even if we may not decipher it completely. Faced with the fragmented character of Aristotle's discourse on the non-physical principles animating celestial motions, we can only accept our limitations and formulate cautious but—as much as possible—well-founded hypotheses.

Concerning the Stagirite's scientific knowledge, as reflected in his astronomical considerations, it is perhaps worth noting that despite the obvious and enormous distance between Aristotelian and classical physics, they nonetheless share profound affinities. For example, the way that Aristotle understood the governing action of the first heaven over the universe resembles our own understanding of the solar system (its ordering, cohesion, and structure), though he was without the benefit of Copernicus's advances. For Aristotle, the dominant element in motion, generally, is the diurnal motion originating in

© GERARDO BOTTERI AND ROBERTO CASAZZA, 2023 | DOI:10.1163/9789004525535_011

the sphere of the fixed stars at the periphery of the system; our contemporary interpretation of the solar system also has a single dominant force, though now at the centre of the system, exerted by the gravitational force of the sun (the primary pole of the mass-distance equation).

The general notion of "force field" that characterises the different kinds of interactions between bodies, and especially that of a gravitational field, is not a mere abstraction or explanatory device of our physics, but represents what philosophers would call a "substantial entity," a οὐσία. It is the means by which the sun "touches" or exerts its influence on each celestial body, arranging their specific motions. Thomas Kuhn has likewise observed a proximity between the physical-mathematical causal explanation involved in our contemporary notion of "field" and the idea of formal cause in the Aristotelian explanation.[1] Whatever equation guides the physicist's work, and whether or not it is correct—in our case, the Law of Universal Gravitation—this becomes the "form" that provides the explanation of phenomena. Similarly, the complete form-actuality of the Prime Mover and other Unmoved Movers is the explanation of cosmic phenomena and, indirectly, of φύσις in general.

In mathematical physics, just as in Aristotelian metaphysics, the explanatory precedence of a formula-law and its predictive capacity (despite its limitations in the context of theories involving statistical and non-deterministic causality) constitute an explanatory teleological framework comparable to Aristotle's notions of formal and final cause. According to physics as understood today, the local intensity of the attractive force of the gravitational field within the solar system determines the direct orbital speed of each star, as if it pulled them. In Aristotelian astronomy, the most salient diurnal motion (as seen from Earth) is caused by the "mass of ether" that we have called the *System of the First Heaven*, which literally drags—and also pushes—each star. The governing action of the first heaven over celestial motions (mediated by the ethereal spheres that compose the *System of the First Heaven*) has, then, in terms of an efficient cause *per se*, characteristics akin to the action of the sun through its gravitational field. The group of the first planetary spheres (or more specifically their velocity spectrum), led by the first heaven, is then analogous both as formal and efficient cause to a gravitational field; it is the means by which the Prime Mover "communicates" to each planet both its *raison d'être* and the ultimate end of its motions, bringing order and systematic unity to the Whole.

1 T. S. Kuhn, *The Essential Tension: Selected Studies in Scientific Tradition and Change* (Chicago/London: University of Chicago Press, 1997), pp. 21–30.

This proposed view of Aristotle's ideas runs contrary to what can be read in some philosophical literature which sees Platonic rather that Aristotelian thought as the germinal antecedent of the physical-mathematical approach that characterises our current understanding of phenomena. This standpoint is understandable to some extent: Plato's work, unlike that of Aristotle, overflows with references to Pythagorean mathematics—mainly algebraic—to account for the behaviour of nature. In contrast, Aristotle not only refutes several Pythagorean notions but also explicitly denies that a parade of numbers and abstract geometric shapes can penetrate the essence of the physical. His cosmology is distinctively different. However, the basis of sublunary φύσις in Aristotle's astronomical system is, as we have seen, eminently mathematical, not because it exalts *more platonico* the ratio between numbers or mystical harmonic proportions, but because it admits geometrical abstraction as an instrument for understanding celestial phenomena. This branch of mathematics, valued by Aristotle as the legacy of his master Eudoxus, and later systematised and axiomatised by Euclid (i.e., geometry), is the most enduring and significant Greek contribution to today's science.[2]

Anyone who delves deeply into the perspectives of both Plato and Aristotle soon notices fundamental differences between them and will discover a greater proximity of Aristotle's way of doing science to the scientific tradition, broadly speaking, from Descartes to Einstein. The Platonic cosmology, with its beautiful harmonic heavens, constitutes a very ambitious mode of approaching scientific explanation (shared by Kepler, among others), by which the ἐπιστήμη is extended to celebrate the mythical-musical character of reality. In contrast, Aristotle's thought operates, we believe, in a more cautious manner, but its strength and originality stand like a giant—to use an image popular since the Middle Ages—from whose shoulders our own science is able to look upon and penetrate this complex world. Our very frame of thought and way of doing science are heirs to his distinctive approach to problems. He is in many ways the culmination of what Heisenberg refers to when he says:

---

2 The astronomer and geometer Eudoxus is a remarkable character in the history of Greek mathematics and we believe that he is a key historical influence in Aristotle's geometrical conception of the cosmos. Surprisingly enough, Plato's scientific works do not refer to Eudoxian astronomic theories, despite the fact that he was a prominent scholar in the Academy, while Aristotle seemed fascinated by both the man and his thought. The *Timaeus*, a sort of updated academic encyclopedia and Plato's scientific final testimony, is undoubtedly the result of Plato's affinity with Pythagorean thought—one might also hypothesise, reading between lines, that it holds a subtle dismissal of Eudoxian ideas too. Indeed, the theories of Eudoxus followed a different branch within the Academy, one less Pythagorean, so to speak, and—as we would now call it—more *scientific*, in the current sense of this word.

> … what always distinguished Greek thought from that of all other peoples was its ability to change the questions it asked into questions of principle and thus to arrive at new points of view, bringing order into the colourful kaleidoscope of experience and making it accessible to human thought.… Whoever delves into the philosophy of the Greeks will encounter at every step this ability to pose questions of principle, and thus by reading the Greeks he can become practised in the use of the strongest mental tool produced by Western thought.[3]

This way of approaching phenomena continues in our scientific methods of today. And given that the observation of celestial phenomena leads naturally to a speculation of causes, it is no surprise that Aristotle, with his extraordinary mind absorbed in contemplation of the nightly celestial parade over the Aegean, his thought framed by the intuition of a unique and continuous heaven, would regard the contemplative life—the deepest and ultimate understanding of such *order*—as a sublime experience.[4]

In connection with this standpoint, and from an *epistemological* perspective, the celestial system of Aristotle is an explanatory *model* that we feel justified in calling "scientific." The same is true for the simpler models of Eudoxus and Callippus. The sheer grandeur of the Aristotelian system derives from its idealising pursuit, typical of the few scientific schemas capable of sustaining through to their ultimate consequences the extra-scientific convictions that animate them.[5]

---

3 W. Heisenberg, "Classical Education, Science and the West," in *The Physicist's Conception of Nature*, pp. 51–67, trans. A. J. Pomerans, (London: Hutchinson & Co, 1958), pp. 52–53; originally published as *Das Naturbild der heutigen Physik* (Hamburg, 1955).

4 This approach to phenomena is not, finally, so different from the typical thought experiment of classical physics and is similar to the mathematisation of nature (the current mode, if we may say, of theoretical contemplation). Everything suggests that, in *On the Heavens*, II, 8, Aristotle abstracts from his primary concern the specific planetary motions as if they belonged to a secondary order superimposed on the central phenomena of diurnal motion shared by all celestial bodies. In a similar manner, when Galileo considers how objects fall, he puts aside the question of air friction, which allows him to understand the fundamental laws governing falling bodies.

5 This characteristic is also found in the polyhedral hypothesis of the young Johannes Kepler (1571–1630), who studied the problem of interplanetary distances in his *Mysterium cosmographicum* (1596) and proposed a fascinating cosmological schema based on two fundamental ideas: that God is an exalted geometer; and that God manifested his work (the universe) by choosing the simplest and most perfect three-dimensional figures—namely, the sphere and the five regular solids of Plato (tetrahedron, hexahedron, octahedron, dodecahedron, icosahedron). Kepler realises that the intervals between the six visible planets are only five (Mercury–Venus–Earth–Mars–Jupiter–Saturn) and then posits that one may explain the

On another level, even the most superficial analysis of the historical significance of Aristotle's system cannot but acknowledge its persistent influence in the Western philosophical-astronomical tradition well into modernity. The homocentric systems developed, mainly, in the fourth century BCE—especially from the Alexandrine Greek period—gave way, regarding their kinematic aspects, to eccentric systems (incorporating epicycles and epicyclets, equants and deferents) commonly known as "Ptolemaic." But these transmutations did not affect their essential physical-metaphysical aspects (the search for the ultimate *causes* of celestial motions), which continued to be treated in Aristotelian terms well into the seventeenth century. We can point, for example, to the role assigned to the Unmoved Movers of Planetary Spheres by some medieval authors (both Arab and Latin) and thinkers of the Renaissance who would occasionally insert them into non-Aristotelian metaphysical frameworks and adapt them to tackle problems such as the relationship between God and the natural world (as understood in creationist terms), or the function and inner hierarchy of the Intelligences or legions of angels.[6]

A brief overview of the *fortune* of Aristotle's astronomical system reveals how the matter of the *causes* of celestial motions was reformulated repeatedly, especially during the Middle Ages. Due to the syncretism typical of historical processes, the Platonic idea that stars are driven by their souls merged with

measure of these intervals by interleaving the five perfect solids, provided that 1) each planet is located on a sphere and has a circular orbit, and 2) the regular solids are perfectly inscribed in spheres while at the same time circumscribing smaller spheres, also perfectly. The Maker would have thus tessellated the cosmic space in the simplest and most beautiful way. This polyhedral hypothesis reveals the extent to which scientific progress has embedded within it pre-existing philosophical and religious practices and conceptions.

6 We should bear in mind that the notion of a causal link between angelic powers and celestial motions gave way to more physical perspectives in various different contexts, both in the Arab and Roman worlds. Alpetragius—the Latinised name of al-Bitrūjī—(*c.*1150–1204) proposed a new astronomical notion that rejected much of the Ptolemaic model and located the Prime Mover in the periphery of a (spherical) universe. According to his interpretation, the Prime Mover's moving force is transmitted directly towards the centre, weakening with distance from the periphery. Consequently, he believed that the tides were the last visible effects of this universal cosmic force. His ideas influenced the astronomy of Robert Grosseteste (*c.*1175–1253) and were absorbed by the group of heterodox thinkers whose doctrines were condemned in 1270 and 1277 by Étienne Tempier, bishop of Paris. Jean Buridan (1295–1358), for his part, followed in the steps of Alpetragius, arguing that the "impetus" (or propelled force) with which God set the celestial spheres into motion is preserved due to a lack of resistance (or friction) in the celestial realm. However, Nicolas Oresme (1323–1382) assumed that there is in fact resistance in the heavens—while at the same time subscribing to the traditional notion of the Intelligences moving the celestial bodies—and that the set of visible celestial motions results from a combination of such "resistance" and the force created by the plenitude of the Intelligences.

the Aristotelian notion that certain finite separate substances would produce the celestial motions through attraction, thus giving new meaning to both doctrines in light of Judeo-Christian angelology. This process also led to some remarkable conceptual alchemies, often marked by a certain medieval "principle of symmetry," which tended to harmonise different orders of reality and heterogeneous sources within the tradition.[7]

This issue had already been extensively discussed by Greek commentators. John Philoponus (490–570), for instance, rejected the idea that celestial spheres are moved by souls or angels, considering that the force driving the stars is directly imprinted on them by God. This conjecture becomes problematic, however, if we consider one of the central notions of Aristotle's theory of motion: that there must be contact between mover and moved thing.[8]

In turn, the Arabic tradition inherited the Neoplatonic metaphysical matrix and forged a cosmological schema in which the Unmoved Movers of Planetary Spheres—conceived as Intelligences, though fewer in number (reduced, in the best-known schemas from fifty-four to ten)—were depicted as the producers of nine planetary spheres through a process of emanation. By way of a series of hypostases, the spiritual, astral, and hyletic worlds in this conception were perfectly interwoven and united. Synthesising Al-Farabi's (870–950) ideas, Rafael Ramón Guerrero says:

> The intelligent One, in thinking of itself as intelligence, opens space to something other than itself as the content of knowledge. This is the first emanated intellect, the first created being, the possible being that receives its existence from the necessary Being. This first intellect, which is also one, contains plurality as it can think of itself as something distinct from the first Being. Beginning with this first emanation, the process is as follows: the first emanated intellect, upon knowing the first Being, gives rise to a second intellect and, in knowing itself, produces the sphere of the first heaven, endowed with a body, which is its matter, and a soul,

7 We are referring to the persistence—in graphs and schemas of illuminated manuscripts—of *rotae* by which different strata of medieval culture appear related to others. For example, in an illustration of the Weissenburg 70 codex (131v) of the Herzog August Bibliothek Wolfenbüttel, within the image of a wheel the seven Our Father petitions are linked to the Seven Gifts of the Holy Spirit and the seven Beatitudes in the Sermon of the Mount. See other similar examples in R. Casazza, *Iconology of the Medieval and Renaissance Iconography of voluntas* (MA Thesis, The Warburg Institute, University of London, 1995). Cosmological schemas in illuminated manuscripts also share this characteristic, typical of medieval iconography.

8 See M. D. Boeri, "'Entre motor y movido debe haber contacto': una dificultad en la teoría aristotélica del movimiento (*Física*, 266b27–267a12)," *Revista Latinoamericana de Filosofía*, 24, 2 (1998), pp. 251–262.

> which is its form. In turn, the second intellect gives rise to a third intellect and to a new sphere. The process is repeated until it reaches the tenth intellect, which completes the separate intellects, and the ninth sphere, that of the moon, the last of the celestial bodies.[9]

This is essentially the same cosmological schema adopted by Avicenna (980–1037).[10] It is adopted in good part by both the Arabic and Christian traditions in so far as it identifies the Intelligences with angels, resulting in metaphysical problems within medieval Scholasticism regarding, especially, the principle of individuation with respect to angels (by definition *unmoved, immaterial, natural substances*). Being spiritual substances, they lack matter, which is, for Aristotle, exactly what makes it possible to differentiate between individuals of the same species. (For Thomas Aquinas, any particular entity is such due to its *materia signata quantitate*, its "matter as determined by quantity," which is precisely its substantial principle of individuation).

The reception of Aristotle's cosmology in the Middle Ages was not uniform. There were naturalistic readings alongside more metaphysical interpretations, and many theologians endeavoured to achieve a synthesis of the two. This can be seen in a host of pre-Copernican versions of the cosmic structure developed between the thirteenth and sixteenth centuries, the main exponents consisting of nine or ten spheres. In the former, the *primum mobile* is considered the ninth sphere, which has a 24-hour revolution period and pulls the other eight spheres through their diurnal motions. The eighth sphere, that of the firmament, moves one degree per century in the opposite direction from the zodiac's trajectory along an oblique circle, which accounts for the precession of the equinoxes. Other schemas introduce a tenth sphere between the sphere of the firmament and the *primum mobile*. This extra sphere, which becomes through its position sphere number nine (also called the crystalline sphere), is attributable to biblical sources—specifically to Genesis, in which we read that God placed the "waters above" on the firmament.[11] Other versions of a

9 R. R. Guerrero, *El pensamiento filosófico árabe*, Historia de la filosofía 8 (Madrid: Cincel, 1985), p. 99; our translation.

10 See for example the excellent study by S. F. Afnan, *Avicenna: His Life and Works* (London: Allen and Unwin, 1958); esp. "Introduction," pp. 9–38, and "Chapter 4: Metaphysics," pp. 106–135.

11 Genesis 1: 6–8 (King James Version): "And God said, 'Let there be a firmament in the midst of the waters, and let it divide the waters from the waters'. And God made the firmament, and divided the waters which were under the firmament from the waters which were above the firmament: and it was so. And God called the firmament Heaven. And the evening and the morning were the second day."

more astronomical nature used this ninth sphere to explain the "trepidation" noticed by the Arab astronomer Thebit (or Thābit, 9th c.), a small cyclical motion of the celestial poles in the North-South direction, supposedly identifiable with the nutation of the axis of the Earth (a kind of wobble of the axis of the Earth accompanies the precession of the equinoxes, changing the latitude of the poles).

The thirteenth century gave birth to a significant tradition of treatises about the celestial sphere whose main representatives were Johannes de Sacrobosco (whose work was commented by Robertus Anglicus, Michael Scott and Cecco d'Ascoli, among others) and Robert Grosseteste. In Sacrobosco's *De sphaera mundi* (chapters I and II), he takes up the schema of nine spheres and designates the outermost sphere as responsible for the regular 24-hour westward motion, which he deems, following Plato, a "rational" motion due to its resemblance to the rational motion of the microcosmos of the human being in its path from the Creator, through the realm of creatures, and back to the Creator where they find rest. The opposite eastward motion of the remaining eight spheres (the precession of the firmament and the direct motions of the planets) he considers "irrational" or "sensual," comparing them to the microcosmic motion of humans when, rising from corruptible things towards the Creator, they then descend again towards the impure world. (In both cases the emphasis is on the return to the same point, as in celestial motion). The first motion in this interpretation—in contrast to the Aristotelian proposal (§ 25)—is initiated at the Arctic and Antarctic poles, and its fundamental plane is the equinoctial circle, while that of the second motion is the ecliptic plane.

Grosseteste also opts for a schema of nine spheres in his *De sphaera* (a remarkable manual of positional astronomy). Thomas Aquinas, for his part, in his *Summa Theologiae* (First Part, question 68, art. 4), taking up ideas from classical patristics, refers to the existence of three heavens: a luminous heaven called "empyrean"; a transparent one called "aqueous or crystalline"; and a third one called "sidereal," which is partly transparent and partly luminous.[12]

12 *Summa Theologiae*, I, Q. 68, A. 4—"Utrum sit unum caelum tantum": "Respondeo: Et secundum hoc, ponuntur tres caeli. Primum totaliter lucidum, quod vocant *empyreum*. Secundum totaliter diaphanum, quod vocant caelum *aqueum* vel *cristallinum*. Tertium partim diaphanum et partim lucidum actu, quod vocant caelum *sidereum*; et dividitur in octo sphaeras, siclicet in sphaeram stellarum fixarum, et septem sphaeras planetarum; quae possunt dici octo caeli" ("Whether there is only one heaven?": "I answer that ... in this body there are three heavens; the first is the empyrean, which is wholly luminous; the second is the aqueous or crystalline, wholly transparent; and the third is called the starry heaven, in part transparent, and in part actually luminous, and divided into eight spheres. One of these is the sphere of the fixed stars; the other seven, which may be called the

This last is divided into eight celestial spheres: the sphere of the fixed stars and the seven spheres of the wandering planets. For Thomas and his contemporaries, the addition of the crystalline heaven completes the typical medieval schema of nine concentric spheres.

In his *Sententia libri Metaphysicae Aristotelis*, Thomas examines chapter Λ, 8 of *Metaphysics* line by line in order to shed light on the Stagirite's obscure astronomical and theological thought. There, as elsewhere, he seeks balance and convergence among disparate elements, some of which are decidedly inconsistent with the Christian worldview. Beginning with the idea that an Unmoved Mover causes movement as would an object desired or intellected, Thomas clarifies and broadens the Aristotelian idea that the first heaven must be animated. If it were otherwise, argues Thomas, the motion of the first heaven would be impossible, since only that which is animated is capable of desire or intellection (Lesson 8, no. 1). In the *intelligentia* (the intellectual act of the first heaven's soul), whose object is the most intelligible (God), we find complete happiness. The first mover is then characterised, following Aristotle, as an intelligent and intelligible substance identified with God (Lesson 8, no. 7). Thomas understands this *primum movens* as a being whose act is *optima et sempiterna vita* or, in Aristotelian terms, *a substantia sempiterna et immobilis, separata a sensibilibus*, a definition that is imprecise in its reference to the everlasting but not entirely unsuitable to the Christian God (Lesson 8, no. 12). He celebrates the statement that fullness of life is found in God, clearly expounded in the Prologue of the Gospel of John, and affirms that if *Deus est ipsa vita* (Lesson 8, no. 9), then it is understandable that the ancient thinkers, following Aristotle, had regarded God as an *animal sempiternum et optimum* (Lesson 8, no. 9).

Continuing his review of the Aristotelian text, Thomas then admits the possibility of other immaterial eternal substances (those to which Aristotle attributes the planetary motions) but struggles with the question of the number of these substances. He introduces, in consequence, a subtle shift that Christianises Aristotle's doctrine: he maintains that the Prime Mover is God, but as to the multiple *substantiae immateriales naturaliter sempiternae* (Lesson 9, no. 5) that move the stars, these are not to be seen as gods but as angelic powers that move, *virtute infinita,* the desiring souls of their respective celestial spheres (his argument on this point is not explicit but adopts the form

seven heavens, are the spheres of the planets"; English trans. by the Fathers of the English Dominican Province (New Advent online ed.).

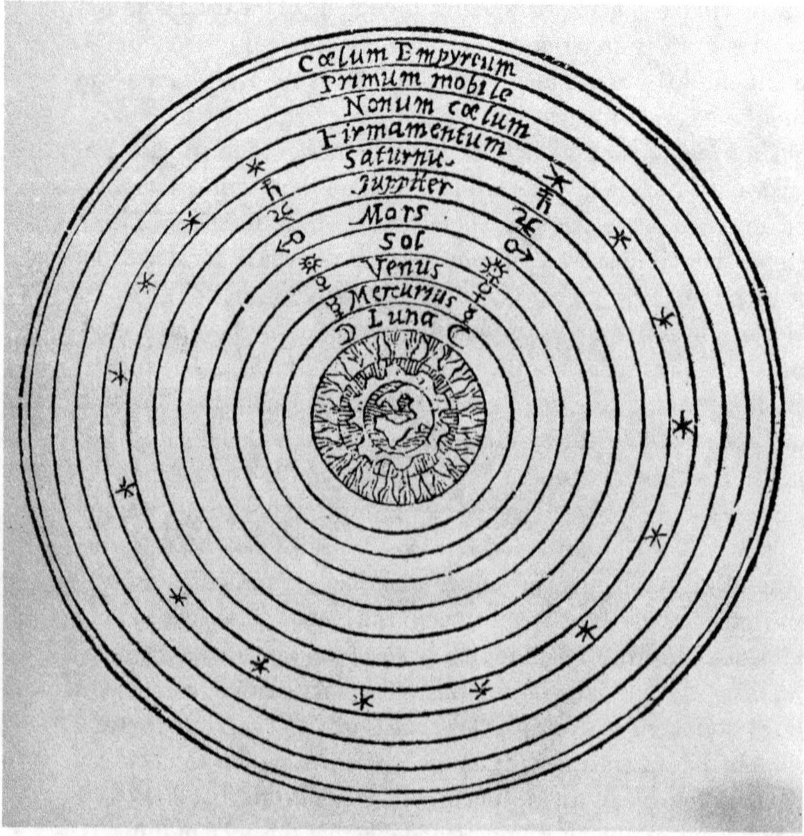

FIGURE 48 The celestial spheres of the medieval cosmos according to an engraving by late-Renaissance astronomer Christopher Clavius.

The relative positions of the Sun, Venus, and Mercury follow the Alexandrine order and differ from those proposed by Plato, Eudoxus, Callippus, and Aristotle, which are based on the Egyptian order. Moreover, the Aristotelian sphere of the fixed stars has been extended to include four heavens, the *firmamentum*, the *nonum coelum*, the *primum mobile* and the *coelum empyreum*. Based on Hipparchus's discovery of precession and Ptolemy's planetary models, medieval astronomy proposes a different explanation of celestial motions than that represented in the Greek homocentric models. The sphere responsible for diurnal revolution is the first moved thing (*primum mobile*) which has westward self-motion on a 24-hour revolution period. This motion "accidentally" accompanies all the other inner spheres, which also move by themselves but in the opposite direction, each having its particular period of return. The ninth heaven completes a revolution every one hundred years, advancing one degree—which constitutes a miscalculation of the precession of the equinoxes (36,000 years) attributed by some to Ptolemy—and dragging the eighth sphere, which has its own trepidation (a motion intended to solve a variation of precession in its velocity and direction relative to a middle position); see Antonio Durán Guardeño, "Ciencia y Renacimientos: Thābit ibn Qurra y Gerolamo Cardano," in eds. José Ferreirós and Antonio Durán Guardeño, *Matemáticas y matemáticos* (Sevilla: Universidad de Sevilla, 2003), p. 45. The *coelum empyreum* is a symbolic representation of a spiritual sphere that contains the legions of angels; Christopher Clavius, *In sphaeram Ioannis de Sacro Bosco commentarius*, Rome, 1585. Sala del Tesoro, Biblioteca Nacional.

of an *ad litteram* comment).[13] The angelic beings move—thanks to their purely spiritual nature—with a constant force and eternally through infinite time,[14] and their plurality is not an issue given that the Thomist doctrine allows for an additional principle of individuation for angels (the *actus essendi*), regardless of their immateriality.[15] Connected with this is a point regarded, within the Thomist tradition, as the greatest philosophical achievement of human reason—an achievement of Aquinas himself, according to that tradition—and one that partially unveils the mystery of Creation: that there is no distinction in God between *essentia* and *esse*. Put in other words, there is, then, such a distinction between these two metaphysical principles in every created entity. The Aristotelian distinction between matter and form was found by Thomas to be inadequate for spiritual substances and pure forms, such as angels. These, he believed, were composed of two metaphysical principles: *essentia* (providing their limits), and *esse* or *actus essendi* (granted by God in creation). Thus, as compound beings, the simplicity of angels is not absolute, and as created beings, though closer to the Divine principle than any other created being on

13 *Sententia libri Metaphysicae* (Turin, 1950), Bk. XII, Lesson 9, no. 5: "Unde manifestum est quod necesse est, quot sunt lationes astrorum, tot esse substantias, quae sunt naturaliter sempiternae, et secundum se immobiles et sine magnitudine, propter causam supra assignatam, quia scilicet movent tempore infinito, et sic per consequens virtute infinita." ("It is clear, then, that there must be as many substances as there are motions of the stars, and that these substances must be by nature eternal and essentially immovable and without magnitude, for the reason given above (1076:C 2548–50), i.e., because they move in infinite time and therefore have infinite power"; *Commentary on the Metaphysics of Aristotle*, trans. J. P. Rowan (Chicago: Regnery, 1961).

14 *Summa theologiae*, I, Q. 110, A. 3—"Utrum corpora obediant angelis ad motum localem": Respondeo: Natura autem corporalis est infra naturam spiritualem. Inter omnes autem motus corporeos perfectior est motus localis ... Et ideo natura corporalis nata est moveri immediate a natura spirituali secundum locum. Unde et philosophi posuerunt suprema corpora moveri localiter a spiritualibus substantiis. Unde videmus quod anima movet corpus primo et principaliter locali motu." ("Whether bodies obey the angels as regards local motion?": "I answer that ... corporeal nature is below the spiritual nature. But among all corporeal movements the most perfect is local motion.... Therefore the corporeal nature has a natural aptitude to be moved immediately by the spiritual nature as regards place. Hence also the philosophers asserted that the supreme bodies are moved locally by the spiritual substances; whence we see that the soul moves the body first and chiefly by a local motion"; trans. Fathers of the English Dominican Province (New Advent online ed.).

15 P. Faitanin, "La individuación de las sustancias separadas según Tomás de Aquino—I," *Aquinate*, no. 1,6 (2011), pp. 20–49. According to Faitanin, angels are separate spiritual forms immediately created by God. In the creative act of God, these forms are illuminated by his Truth, his Love, his Goodness, and his Wisdom, and *eo ipso*, are individuated by the limitation of their own essences. Of course, the discussion about the principle of individuation of angels—an issue well known for its complexity within Thomist and Scholastic traditions generally—exceeds this brief account.

account of their ontological perfection, they are nevertheless infinitely far from the absolute excellence of the Creator.

Next, Thomas initiates an exposition of general astronomical statements based largely—as he tells us—on the authority of Simplicius and incorporating additional knowledge of his time. He begins by noting that the discovery of the displacement of the poles was made by Hipparchus only in the 2nd century BCE, and therefore was not known during Aristotle's time (the displacement is attributed, today, to the precession of the axis of the Earth). Consequently, an extra sphere was necessary to account for this motion (Lesson 9, no. 6). He then criticises the Ptolemaic models (which he calls "Pythagorean") for violating the fundamental principles of Aristotelian cosmology: the orbital eccentricity of these models, he says, would generate voids between the celestial spheres. Using a non-historical mode of criticism, and not knowing a key characteristic of the models (that they are essentially *geometrical*), he dismisses them on account of not being *physical models*, just as Aristotle criticised those of Eudoxus and Callippus.[16]

Regarding the number of Unmoved Movers, Thomas asserts that this was decided by Callippus and Aristotle, who together completed and corrected the Eudoxian system.[17] He considers the number to be a contingent (and mistaken) product of the astronomy of their time (his own position it that there are only nine material spheres).[18] Also, Thomas repeatedly emphasises the distinctive quality of the Prime Mover with respect to other principles of sphere motion, something which Aristotle does not do (he considers them different only in terms of hierarchy and motor force, not regarding their ontological structure),

16 *Sententia libri Metaphysicae*, Bk. XII, Lesson 10, no. 3.

17 Ibid, Bk. XII, Lesson 10, no. 12: "Fuit autem Calippus, ut Simplicius dicit, cum Aristotele Athenis conversatus, cum eo ea quae ab Eudoxo inventa fuerant, corrigens et supplens." ("Now Callippus, as Simplicius tells us, was associated with Aristotle at Athens when the discoveries of Eudoxus were corrected and supplemented by him"; *Commentary on the Metaphysics*, English trans. Rowan). Then, in Bk. XII, Lesson 10, no. 14: "Sed praeter has ponebant [Aristoteles et Calippus] alias, quas vocabant revolventes." ("But in addition to these spheres they posited others, which they called revolving spheres"; trans. Rowan). Aquinas may have got the idea that Callippus and Aristotle worked together from *Simplicii Commentarius in IV libros Aristotelis de Caelo*, ed. S. Karsten (Utrecht: Kemink and Son, 1865), 493, 5, in which Simplicius affirms that Callippus had studied with Polemarchus, had arrived to Athens after Eudoxus's time as head of the Academy, and had lived with Aristotle in Athens, where together they corrected and broadened the discoveries made by Eudoxus.

18 The crystalline heaven referred to in the above-quoted passage from the *Summa Theologiae* (I, Q. 68, A. 4—"Utrum sit unum caelum tantum") and to which medieval thinkers attributed the diurnal revolution of the stars, is regarded as the ninth material sphere, unlike the empyrean heaven, which is purely spiritual.

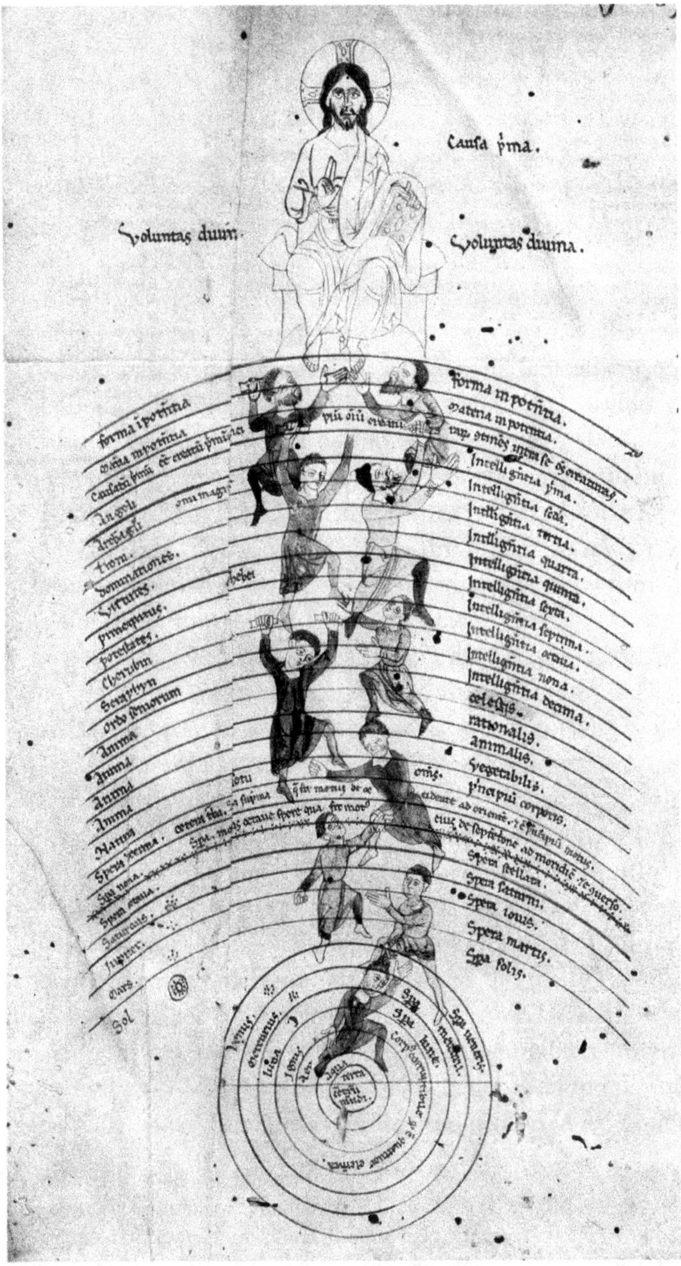

FIGURE 49
A Neoplatonic cosmological schema designed in the North of Italy at the beginning of the twelfth century and included in an anonymous treatise on the fate of the soul (Paris, Bibliothèque Nationale, Ms. Lat. 3236A, f. 90r). The arcs represent spheres, while the ascending men indicate the return to the Divine Unity, represented by Christ in Majesty (*creator omnium—deus—causa prima*). Beardless men are seen in the lower strata of the path while the more spiritual bearded men appear in the upper strata. The schema reflects the doctrine of the ten intelligences emanated from the Divine Unity, as proposed by Arab philosophers. In the medieval Christian environment, however, these intelligences were identified with the Judeo-Christian legions of angels, as shown in the illumination.

and recognises that speculation about the number of Unmoved Movers leaves room for doubt (… *de hoc … potest esse dubitation.* Lesson 9, no. 8).[19]

Medieval reception of Aristotle's astronomical system reached its culmination—in terms of its strength, novelty, and consequences—in the philosophical movement known as "radical" or "heterodox" Aristotelianism (or by the less appropriate name of "Latin Averroism"), emerging principally from the Faculty of Arts of the University of Paris between approximately 1240 and 1290.[20] The key ideas of this intellectual movement—whose main representatives were Liberal Arts professors including Siger of Brabant and Boetius of Dacia—have reached us largely through indirect sources, and were framed within the broader intellectual process of the rediscovery of Aristotelian works in the West from the late twelfth century. The reception of Aristotle at the time was such that his doctrines were viewed as the highest expression of human reason and his work generated—especially in the newly emerging universities—a deep reformulation of the meaning and orientation of intellectual activity. The dominant current of this reformulation sought to fit together Christian revealed truth with what appeared to them, in these new texts, as philosophical truths, in order to provide a rational foundation for theology. Some thinkers, however, perhaps without measuring its consequences, gave free rein to Aristotle's philosophical truth, viewing it, in a sense, as independent of the truths of faith (a position known as the "doctrine of double truth"). The result was the development of an Aristotelianism that strayed beyond acceptable limits of the time, generating an immediate reaction from ecclesiastical authorities. A series of theses condemned in 1270 and 1277 by Étienne Tempier, bishop of Paris, show the extent to which Aristotelian cosmology—interpreted under the influence of Arabic Necessitarianism—posed a challenge to some of the cornerstones of Christian thought. The condemned theses addressed a variety of issues, among them the eternity of the world and the plurality of the Unmoved Movers, the astral determinism of human conduct, the doctrine of the eternal return, the mediating role of the Intelligences in the act of creation, and the supremacy of philosophical life. What follows is a review of the main theses condemned by the bishop (thereby forbidding their teaching)

---

19 *Sententia libri Metaphysicae*, Bk. XII, Lesson 10, no. 30: "Relinquitur igitur quod primum movens immobile sit unum, non solum ratione speciei, sed etiam numero." ("It remains, then, that the first unmoved mover is one not only in its intelligible structure but also in number"; *Commentary on the Metaphysics*, trans. Rowan).

20 See P. Mandonnet, *Siger de Brabant et l'averroïsme latin au XIIIme siècle*, 2 vols. (Louvain: Institut Supérieur de Philosophie de l'Université, 1908–1911), esp. vol. 1, pp. 231–233; see also K. Flasch, *Aufklärung im Mittelalter? Die Verurteilung von 1277* (Mainz: Dieterich, 1989).

concerning astronomical matters—quite eloquent, all—stemming from this radical interpretation of Aristotelian astronomy:

> On the eternity of the world and
> the plurality of the Unmoved Movers
>
> 91—*Quod ratio philosophi demonstrans motum coeli esse aeternum non est sophistica; et mirum quod homines profundi hoc non vident.* (That the reasoning of the Philosopher proving that the motion of the heaven is eternal is not sophistic, and that it is surprising that profound men do not perceive this. [80])[21]
>
> 66—*Quod plures sunt motores primi.* (That there is more than one prime mover.[99])
>
> On the eternal return
>
> 6—*Quod redeuntibus corporibus coelestibus omnibus in idem punctum, quod fit in XXX sex milibus annorum, redibunt idem effectus, qui sunt modo.* (That with all the heavenly bodies coming back to the same point after a period of thirty-six thousand years, the same effects as now exist will reappear. [92])
>
> 21—*Quod nichil fit a casu, sed omnia de necessitate eveniunt, et, quod omnia futura quae erunt, de necessitate erunt, et quae non erunt, impossibile est esse, et quod nichil fit contingenter, considerando omnes causas.* (That nothing happens by chance, but everything comes about by necessity, and that all the things that will exist in the future will exist by necessity, and those that will not exist are impossible, and that nothing occurs contingently if all causes are considered. [102])

21 English translations of the theses are from "Condemnation of 219 Propositions," trans E. L. Fortin and P. D. O'Neill, in *Medieval Political Philosophy: A Sourcebook*, ed. Ralph Lerner and Muhsin Mahdi, (New York: Free Press, 1963), pp. 335–54; the numbers in square brackets are those given by Mandonnet in the abovementioned Louvain 1908 edition of *Siger de Brabant*. Thomas Aquinas dedicates Question 46 of the First Part of his *Summa Theologiae* to this matter concluding—without threatening the act of creation—that God may have created an eternal world. We know through faith, he adds, that the world was created over time, so the Aristotelian conclusion is incorrect though without faults of logic.

On the determinism of the will

132—*Quod orbis est causa voluntatis medici, ut sanet.* (That a sphere is the cause of a doctor's willing to cure. [155])

133—*Quod voluntas et intellectus non moventur in actu per se, sed per causam sempiternam, scilicet corpora coelestia.* (That the will and intellect are not moved in act by themselves but by an eternal cause, namely, the heavenly bodies. [153])

161—*Quod effectus stellarum super liberum arbitrium sunt occulti.* (That the effects of the stars upon free choice are hidden. [156])

162—*Quod voluntas nostra subiacet potestati corporum coelestium.* (That our will is subject to the power of the heavenly bodies. [154])

Somewhat after these heterodox thinkers, Dante Alighieri (1265–1321) portrays the relation between the Intelligences and the celestial spheres in his *Divine Comedy*, in the memorable Canto XXVIII of Paradiso, as Beatrice and Dante arrive at the limit of the physical world and view the legions of angels that move the spheres. In Dante's schema, the Seraphim move sphere nine, the *primum mobile* and outermost sphere of the system; the Cherubim move sphere eight, the sphere of the fixed stars; the Thrones move Saturn; the Dominions, Jupiter; the Virtues, Mars; the Powers, the Sun; the Principalities, Venus; the Archangels, Mercury; and the Angels, the Moon.[22] Gustave Doré's (1832–1883) celebrated illustrations of the legions of angels disappearing into a luminous point (a metaphor for the divine plenitude) depicts the profound intuition recovered by this tradition: the true reality is the spiritual reality, while the sensible is governed by an intelligibility that characterises the supersensible plane and expresses itself in the astral order—the replication of the celestial domain in the sublunary domain being but a degraded imitation.

More than a century after Nicolaus Copernicus's *De revolutionibus orbium coelestium* (1543), we still find Aristotelian—i.e., geocentric—representations that, following the medieval tradition, construct hierarchical arrangements of separate angelic substances with celestial bodies, and of celestial bodies with sublunary entities. A well-known example is Athanasius Kircher's *Musurgia*

22 See the discussion (in Spanish) on this in A. Gangui, *Poética astronómica—El cosmos de Dante Alighieri* (Buenos Aires: Fondo de Cultura Económica, 2008), pp. 112–118.

FIGURE 50 The Celestial Rose. Dante and Beatrice gaze upon the legions of angels in the empyrean heaven, beyond the last heaven (Paradiso, Canto XXXI) in an engraving by Gustave Doré.

universalis (1650).[23] Reinterpreting concepts introduced by Nicholas of Cusa, mainly in his *De coniecturis* (*On Conjectures*), Kircher proposes a U-P figure (so named because it combines figures U and P of Cusa) to represent a physical-cosmological arrangement of the universe. In it, legends point to the

23 For a study on the scope of the relation between the angelic powers and the sublunary world, see R. Casazza, "Las figuras universal (U) y paradigmática (P) del *De coniecturis* de Nicolás de Cusa, a la luz de la interpretación cosmológica (figura U-P) de Athanasius Kircher," in *Identidad y alteridad en Nicolás de Cusa. Actas del II Congreso Cusano de Latinoamérica* (Buenos Aires: Biblos, 2010), pp. 201–223.

influence of each angelic choir (Seraphim, Cherubim, Thrones, Dominions, Virtues, Powers, Principalities, Archangels, and Angels) on the various celestial spheres (Immobilia, Firmament, Saturn, Jupiter, Mars, the Sun, Venus, Mercury, and the Moon, respectively). In turn, the celestial spheres exert their influence on the ontological strata of the sublunary domain (the composite, that which is, that which vegetates, that which feels, human beings, fire, air, water, and earth).

The astronomical thought of Aristotle has undergone many transformations throughout the centuries, including variations that have misrepresented essential aspects. His ideas and their various interpretations continue to permeate our conceptions of reality even today, especially, as commented by Bachelard and Sloterdijk, that of a cosmic sphericity that seems to govern every aspect of our existence.[24] Similarly, the relation between planets and spiritual forces is still present, *mutatis mutandis*, in deep religious and philosophical doctrines, for example in the Kabbalah, some formulations of Christian and Muslim mysticism, or Rudolf Steiner's Anthroposophy. We see it also in eclectic conceptual pastiches of a more dubious nature, such as in self-help and New Age literature. Despite the transformations and bastardisations of the doctrinal core, the effort of classical Greek astronomical thought has left a strong and fundamental mark in philosophy: the search for a systematic and comprehensive explanation—built on a sound heuristic foundation and integrating principles and ends, causes and effects, motion and rest—of every strata of its object of study. Just by raising our eyes to the sky on a clear night, the beautiful intuition of those philosophers—that is, that the celestial order emanates from the intelligible order of its principles—challenges us with all its enigmas and compels us to once again take up the task of explaining (or at least trying to explain) the cosmic symphony of the universe in philosophical terms. Therefore, as in that brief time and space of ancient Greek intellectual life with all its breadth, philosophy must again, we believe, look to the sky for inspiration, orientation and peace. Despite the general bewilderment of the present, we find hope in the fact that the formidable celestial revolution is there, always, before our eyes and proclaiming its truths, so that sooner or later, as we search for answers to some of the many questions that distress us, we will notice its simplest message: "Know, children, that you are part of a cosmos!"

---

24 Gaston Bachelard, *The Poetics of Space* (Boston: Beacon Press, 1994), and Peter Sloterdijk, *Spheres*, 3 vols. (Cambridge: MIT Press, 2011–2014–2016).

Appendices

APPENDIX 1

Astronomical Fragments

The following fragments are meant to provide the reader with a bilingual selection of the main Aristotelian texts on cosmological and astronomical issues. The order and arrangement of the selected texts, as well as their thematic divisions and titles, were determined according to the main theses proposed in this work.

The source of the Greek texts is The Thesaurus Linguae Graecae, developed by the Packard Humanities Institute and the Perseus Project & Others, Diogenes (3.1.6) version, designed by P. J. Heslin (1999–2007). The source of each English text is indicated above its entry. We have occasionally introduced additional paragraph breaks to better match with the Greek texts. Line numbers in the English texts are approximate.

This appendix aims at juxtaposing the texts in the two languages to favour their comparison by readers who may wish to see longer segments than those presented in the main text, as well as segments that are not quoted. It does not aim at a systematic analysis of the original texts. Scholars interested in more philological aspects will wish to consult the critical canonical editions of Aristotle's works that offer an adequate critical apparatus.

Aristotle's Cosmology

§ 1—"Heaven" has many meanings
On the Heavens, I, 9, 278b9–21

§ 2—The divinity and eternity of the heavens
On the Heavens, I, 9, 279a18–279b3

§ 3—The ungendered incorruptible nature of the heavens
On the Heavens, II, 1, 284a3–18

§ 4—The empirical evidence of the eternity and immutability of the first heaven
On the Heavens, I, 3, 270b1–270b16

§ 5—The perfection of the celestial sphere
On the Heavens, II, 4, 287b15–21

§ 6—The spherical nature of the heavens
On the Heavens, II, 4, 286b10–26

§ 7—The proof of the finitude of the celestial sphere derived from its circular motion
On the Heavens, I, 5, 271b26–272a6

© GERARDO BOTTERI AND ROBERTO CASAZZA, 2023 | DOI:10.1163/9789004525535_012

§ 8—The regularity of the motion of the celestial sphere
On the Heavens, II, 6, 288a13–288a27

§ 9—The contiguity of celestial spheres
On the Heavens, II, 4, 287a2–12

§ 10—The spherical nature of stars and the impossibility of their axial rotation
On the Heavens, II, 8, 290a7–290b12

§ 11—The emergence of heat due to the interaction between the sphere of ether and the spheres of fire and air
Meteorology, I, 3, 340b6–341a2

The Prime Mover

§ 12—The unicity and eternity of the Prime Mover
Physics, VIII, 6, 259a6–259a20

§ 13—The need for the existence of a Prime Mover
Physics, VIII, 6, 259b20–260a18

§ 14—Atlas, Zeus, and the perpetuity of heavenly motion
Movement of Animals, 2, 699a12–700a6

The Unmoved Movers of Planetary Spheres

§ 15—The number of the Unmoved Movers
Metaphysics, Λ, 8, 1073a3–1074b14

§ 16—The need for the existence of other motion(s) apart from the continuous motion of the sphere of the fixed stars to explain generation and corruption
On the Heavens, II, 3, 286a3–286b9

§ 17—The eternal causes of the divine things that we perceive
Metaphysics, E, 1, 1026a6–22

§ 18—The subordination of the Unmoved Movers of Planetary Spheres to the Prime Mover
On Generation and Corruption, II, 10, 337a15–25

§ 19—The reasons why a single body has many motions in the case of planets and many bodies have a single motion in the case of the sphere of the fixed stars
On the Heavens, II, 12, 291b28–293a14

Theophrastus and Plotinus's objections to the doctrine on the plurality of the Unmoved Movers

§ 20—Doubts (of Theophrastus) about the multi-substance theology of Aristotle
Theophrastus, *On First Principles*, I, 1–II, 8

§ 21—The shortfall (according to Theophrastus) in the discourse of astrologers about the first movers
Theophrastus, *On First Principles*, IV, 27–28

§ 22—Criticism (of Plotinus) to the impossibility of individuation (due to the absence of matter) of Aristotle's Unmoved Movers
Plotinus, *Enneads*, V, 1, 8–9

The Astronomical System of Aristotle

§ 23—The influence of the superlunary on the sublunary
Meteorology, I, 2, 339a19–339a32

§ 24—The simultaneity and contiguity in the mover-moved series
Physics, VII, 2, 243a32–243a40; 243a11–243a18

§ 25—The force produced by the attraction of the Prime Mover on the circumference of the sphere (rather than on its axis)
Physics, VIII, 10, 267a21–267b9

§ 26—The simultaneous pulling and pushing (on the surface of a sphere) in the rotation produced by the Unmoved Movers
Physics, VII, 2, 244a2–6

§ 27—The lineal velocity increase (from centre to periphery) in the circular motion of the spheres that carry the stars
On the Heavens, II, 8, 289b1–290a7

§ 28—The influence of the motion of the first heaven on the zodiacal revolution of the planets
On the Heavens, II, 10, 291a32–291b10

§ 29—The influence of ecliptic obliquity on generation and corruption processes
On Generation and Corruption, II, 10, 336a23–336b14

§ 30—On how the cyclicality of natural processes mirrors the circular motions of heaven
On Generation and Corruption, II, 10, 336b26–337a7

§ 31—The continuity and eternity of circular motion
Physics, VIII, 8, 264b9–19

Aristotle's Cosmology

§ 1—"Heaven" has many meanings
On the Heavens, I, 9, 278b9–21
English Source: *On the Heavens*, trans. W. K. C. Guthrie (1939).

[278b9] Εἴπωμεν δὲ πρῶτον τί λέγομεν εἶναι τὸν οὐρανὸν καὶ ποσαχῶς, ἵνα μᾶλλον ἡμῖν δῆλον γένηται τὸ ζητούμενον.

Ἕνα μὲν οὖν τρόπον οὐρανὸν λέγομεν τὴν οὐσίαν τὴν τῆς ἐσχάτης τοῦ παντὸς περιφορᾶς, ἢ σῶμα φυσικὸν τὸ ἐν τῇ ἐσχάτῃ περιφορᾷ τοῦ παντός· εἰώθαμεν γὰρ τὸ ἔσχατον καὶ τὸ ἄνω μάλιστα καλεῖν οὐρανόν, ἐν ᾧ καὶ τὸ θεῖον πᾶν ἱδρῦσθαί φαμεν. Ἄλλον δ' αὖ τρόπον τὸ συνεχὲς σῶμα τῇ ἐσχάτῃ περιφορᾷ τοῦ παντός, ἐν ᾧ σελήνη καὶ ἥλιος καὶ ἔνια τῶν ἄστρων· καὶ γὰρ ταῦτα ἐν τῷ οὐρανῷ εἶναί φαμεν. Ἔτι δ' ἄλλως λέγομεν οὐρανὸν τὸ περιεχόμενον σῶμα ὑπὸ τῆς ἐσχάτης περιφορᾶς· τὸ γὰρ ὅλον καὶ τὸ πᾶν εἰώθαμεν λέγειν οὐρανόν.

[278b9] Let us first establish what we mean by *ouranos*, and in how many senses the word is used, in order that we may more clearly understand the object of our questions.

In one sense we apply the word *ouranos* to the substance of the outermost circumference of the world, or to the natural body which is at the outermost circumference of the world; for it is customary to give the name of *ouranos* especially to the outermost and uppermost region, in which also we believe all divinity to have its seat. Secondly we apply it to that body which occupies the next place to the outermost circumference of the world, in which are the moon and the sun and certain of the stars; for these, we say, are in the *ouranos*. We apply the word in yet another sense to the body which is enclosed by the outermost circumference; for it is customary to give the name of *ouranos* to the world as a whole.

§ 2—The divinity and eternity of the heavens
On the Heavens, I, 9, 279a18–279b3
English Source: *On the Heavens*, trans. W. K. C. Guthrie (1939).

[279a18] Διόπερ οὔτ' ἐν τόπῳ τἀκεῖ πέφυκεν, οὔτε χρόνος αὐτὰ ποιεῖ γηράσκειν, οὐδ' ἐστὶν οὐδενὸς οὐδεμία μεταβολὴ τῶν ὑπὲρ τὴν ἐξωτάτω τεταγμένων φοράν, ἀλλ' ἀναλλοίωτα καὶ

[279a18] Wherefore neither are the things there born in place, nor does time cause them to age, nor does change work in any way upon any of the beings whose allotted place is beyond the outermost

ἀπαθῆ τὴν ἀρίστην ἔχοντα ζωὴν καὶ τὴν αὐταρκεστάτην διατελεῖ τὸν ἅπαντα αἰῶνα. (Καὶ γὰρ τοῦτο τοὔνομα θείως ἔφθεγκται παρὰ τῶν ἀρχαίων. Τὸ γὰρ τέλος τὸ περιέχον τὸν τῆς ἑκάστου ζωῆς χρόνον, οὗ μηθὲν ἔξω κατὰ φύσιν, αἰὼν ἑκάστου κέκληται. Κατὰ τὸν αὐτὸν δὲ λόγον καὶ τὸ τοῦ παντὸς οὐρανοῦ τέλος καὶ τὸ τὸν πάντα χρόνον καὶ τὴν ἀπειρίαν περιέχον τέλος «αἰών» ἐστιν, ἀπὸ τοῦ «αἰεὶ εἶναι» τὴν ἐπωνυμίαν εἰληφώς, ἀθάνατος καὶ θεῖος). Ὅθεν καὶ τοῖς ἄλλοις ἐξήρτηται, τοῖς μὲν ἀκριβέστερον τοῖς δ' μαυρῶς, τὸ εἶναί τε καὶ ζῆν.

Καὶ γάρ, καθάπερ ἐν τοῖς ἐγκυκλίοις φιλοσοφήμασι περὶ τὰ θεῖα, πολλάκις προφαίνεται τοῖς λόγοις ὅτι μασι περὶ τὰ θεῖα, πολλάκις προφαίνεται τοῖς λόγοις ὅτι τὸ θεῖον ἀμετάβλητον ἀναγκαῖον εἶναι πᾶν τὸ πρῶτον καὶ ἀκρότατον· ὃ οὕτως ἔχον μαρτυρεῖ τοῖς εἰρημένοις. Οὔτε γὰρ ἄλλο κρεῖττόν ἐστιν ὅ τι κινήσει (ἐκεῖνο γὰρ ἂν εἴη θειότερον) οὔτ' ἔχει φαῦλον οὐδέν, οὔτ' ἐνδεὲς τῶν αὑτοῦ καλῶν οὐδενός ἐστιν. [279b] Καὶ ἄπαυστον δὴ κίνησιν κινεῖται εὐλόγως· πάντα γὰρ παύεται κινούμενα ὅταν ἔλθῃ εἰς τὸν οἰκεῖον τόπον, τοῦ δὲ κύκλῳ σώματος ὁ αὐτὸς τόπος ὅθεν ἤρξατο καὶ εἰς ὃν τελευτᾷ.

motion: changeless and impassive, they have uninterrupted enjoyment of the best and most independent life for the whole *aeon* of their existence. Indeed, our forefathers were inspired when they made this word, *aeon*. The total time which circumscribes the length of life of every creature, and which cannot in nature be exceeded, they named the *aeon* of each. By the same analogy also the sum of existence of the whole heaven, the sum which includes all time even to infinity, is *aeon*, taking the name from ἀεὶ εἶναι ("to be everlastingly"), for it is immortal and divine. In dependence on it all other things have their existence and their life, some more directly, others more obscurely.

In the more popular philosophical works, where divinity is in question, it is often made abundantly clear by the discussion that the foremost and highest divinity must be entirely immutable, a fact which affords testimony to what we have been saying. For there is nothing superior that can move it—if there were it would be more divine—and it has no badness in it nor is lacking in any of the fairness proper to it. [279b] It is too in unceasing motion, as is reasonable; for things only cease moving when they arrive at their proper places, and for the body whose motion is circular the place where it ends is also the place where it begins.

§ 3—The ungendered incorruptible nature of the heavens
On the Heavens, II, 1, 284a3–18
English Source: *On the Heavens*, trans. W. K. C. Guthrie (1939).

[284a3] Διόπερ καλῶς ἔχει συμπείθειν ἑαυτὸν τοὺς ἀρχαίους καὶ μάλιστα πατρίους ἡμῶν ἀληθεῖς εἶναι λόγους, ὡς ἔστιν ἀθάνατόν τι καὶ θεῖον τῶν ἐχόντων μὲν κίνησιν, ἐχόντων δὲ τοιαύτην ὥστε μηθὲν εἶναι πέρας αὐτῆς, ἀλλὰ μᾶλλον ταύτην τῶν ἄλλων πέρας· τό τε γὰρ πέρας τῶν περιεχόντων ἐστί, καὶ αὕτη τέλειος οὖσα περιέχει τὰς ἀτελεῖς καὶ τὰς ἐχούσας πέρας καὶ παῦλαν, αὐτὴ μὲν οὐδεμίαν οὔτ’ ἀρχὴν ἔχουσα οὔτε τελευτήν, ἀλλ’ ἄπαυστος οὖσα τὸν ἄπειρον χρόνον, τῶν δ’ ἄλλων τῶν μὲν αἰτία τῆς ἀρχῆς, τῶν δὲ δεχομένη τὴν παῦλαν.

Τὸν δ’ οὐρανὸν καὶ τὸν ἄνω τόπον οἱ μὲν ἀρχαῖοι τοῖς θεοῖς ἀπένειμαν ὡς ὄντα μόνον ἀθάνατον· ὁ δὲ νῦν μαρτυρεῖ λόγος ὡς ἄφθαρτος καὶ ἀγένητος, ἔτι δ’ ἀπαθὴς πάσης θνητῆς δυσχερείας ἐστίν, πρὸς δὲ τούτοις ἄπονος διὰ τὸ μηδεμιᾶς προσδεῖσθαι βιαίας ἀνάγκης, ἣ κατέχει κωλύουσα φέρεσθαι πεφυκότα αὐτὸν ἄλλως· πᾶν γὰρ τὸ τοιοῦτον ἐπίπονον, ὅσῳπερ ἂν ἀϊδιώτερον ᾖ, καὶ διαθέσεως τῆς ἀρίστης ἄμοιρον.

[284a3] Therefore we may well feel assured that those ancient beliefs are true, which belong especially to our own native tradition, and according to which there exists something immortal and divine, in the class of things in motion, but whose motion is such that there is no limit to it. Rather it is itself the limit of other motions, for it is a property of that which embraces to be a limit, and the circular motion in question, being complete, embraces the incomplete and finite motions. Itself without beginning or end, continuing without ceasing for infinite time, it causes the beginning of some motions, and receives the cessation of others.

Our forefathers assigned heaven, the upper region, to the gods, in the belief that it alone was imperishable; and our present discussion confirms that it is indestructible and ungenerated. We have shown, also, that it suffers from none of the ills of a mortal body, and moreover that its motion involves no effort, for the reason that it needs no external force of compulsion, constraining it and preventing it from following a different motion which is natural to it. Any motion of that sort would involve effort, all the more in proportion as it is long-lasting, and could not participate in the best arrangement of all.

§ 4—The empirical evidence of the eternity and immutability of the first heaven
On the Heavens, I, 3, 270b1–270b16
English Source: *On the Heavens*, trans. W. K. C. Guthrie (1939).

[270b1] Διότι μὲν οὖν ἀΐδιον καὶ οὔτ' αὔξησιν ἔχον οὔτε φθίσιν, ἀλλ' ἀγήρατον καὶ ἀναλλοίωτον καὶ ἀπαθές ἐστι τὸ πρῶτον τῶν σωμάτων, εἴ τις τοῖς ὑποκειμένοις πιστεύει, φανερὸν ἐκ τῶν εἰρημένων ἐστίν.

Ἔοικε δ' ὅ τε λόγος τοῖς φαινομένοις μαρτυρεῖν καὶ τὰ φαινόμενα τῷ λόγῳ· πάντες γὰρ ἄνθρωποι περὶ θεῶν ἔχουσιν ὑπόληψιν, καὶ πάντες τὸν ἀνωτάτω τῷ θείῳ τόπον ἀποδιδόασι, καὶ βάρβαροι καὶ Ἕλληνες, ὅσοι περ εἶναι νομίζουσι θεούς, δῆλον ὅτι ὡς τῷ ἀθανάτῳ τὸ ἀθάνατον συνηρτημένον· ἀδύνατον γὰρ ἄλλως. Εἴπερ οὖν ἔστι τι θεῖον, ὥσπερ ἔστι, καὶ τὰ νῦν εἰρημένα περὶ τῆς πρώτης οὐσίας τῶν σωμάτων εἴρηται καλῶς. Συμβαίνει δὲ τοῦτο καὶ διὰ τῆς αἰσθήσεως ἱκανῶς, ὥς γε πρὸς ἀν-θρωπίνην εἰπεῖν πίστιν· ἐν ἅπαντι γὰρ τῷ παρεληλυθότι χρόνῳ κατὰ τὴν παραδεδομένην ἀλλήλοις μνήμην οὐθὲν φαίνεται μεταβεβληκὸς οὔτε καθ' ὅλον τὸν ἔσχατον οὐρανὸν οὔτε κατὰ μόριον αὐτοῦ τῶν οἰκείων οὐθέν.

[270b1] From what has been said it is clear why, if our hypotheses are to be trusted, the primary body of all is eternal, suffers neither growth nor diminution, but is ageless, unalterable and impassive.

I think too that the argument bears out experience and is borne out by it. All men have a conception of gods, and all assign the highest place to the divine, both barbarians and Hellenes, as many as believe in gods, supposing, obviously, that immortal is closely linked with immortal. It could not, they think, be otherwise. If then—and it is true—there is something divine, what we have said about the primary bodily substance is well said. The truth of it is also clear from the evidence of the senses, enough at least to warrant the assent of human faith; for throughout all past time, according to the records handed down from generation to generation, we find no trace of change either in the whole of the outer most heaven or in any one of its proper parts.

§ 5—The perfection of the celestial sphere
On the Heavens, II, 4, 287b15–21
English Source: *On the Heavens*, trans. W. K. C. Guthrie (1939).

[287b15] Ὅτι μὲν οὖν σφαιροειδής [15] ἐστιν ὁ κόσμος, δῆλον ἐκ τούτων, καὶ ὅτι κατ' ἀκρίβειαν ἔντορνος οὕτως ὥστε μηθὲν μήτε χειρόκμητον ἔχειν παραπλησίως μήτ' ἄλλο μηθὲν τῶν ἡμῖν ἐν ὀφθαλμοῖς φαινομένων. Ἐξ ὧν γὰρ τὴν σύστασιν εἴληφεν, οὐδὲν οὕτω δυνατὸν ὁμαλότητα δέξασθαι καὶ ἀκρίβειαν ὡς ἡ τοῦ πέριξ σώματος [20] φύσις· δῆλον γὰρ ὡς ἀνάλογον ἔχει, καθάπερ ὕδωρ πρὸς γῆν, καὶ τὰ πλεῖον ἀεὶ ἀπέχοντα τῶν συστοίχων.

[287b15] Our arguments have clearly shown that the universe is spherical, and so accurately turned that nothing made by man, nor anything visible to us on the earth, can be compared to it. For of the elements of which it is composed, none is capable of taking such a smooth and accurate finish as the nature [20] of the body which encompasses the rest; for the more distant elements must become ever finer in texture in proportion as water is finer than earth.

§ 6—The spherical nature of the heavens
On the Heavens, II, 4, 286b10–26
English Source: *On the Heavens*, trans. W. K. C. Guthrie (1939).

[286b10] Σχῆμα δ' ἀνάγκη σφαιροειδὲς ἔχειν τὸν οὐρανόν· τοῦτο γὰρ οἰκειότατόν τε τῇ οὐσίᾳ καὶ τῇ φύσει πρῶτον. Εἴπωμεν δὲ καθόλου περὶ τῶν σχημάτων, τὸ ποῖόν ἐστι πρῶτον, καὶ ἐν ἐπιπέδοις καὶ ἐν στερεοῖς. Ἅπαν δὴ σχῆμα ἐπίπεδον ἢ εὐθύγραμμόν ἐστιν ἢ περιφερόγραμμον. Καὶ τὸ μὲν εὐθύγραμμον [15] ὑπὸ πλειόνων περιέχεται γραμμῶν, τὸ δὲ περιφερόγραμμον ὑπὸ μιᾶς. Ἐπεὶ δὲ πρότερον [τῇ φύσει] ἐν ἑκάστῳ γένει τὸ ἓν τῶν πολλῶν καὶ τὸ ἁπλοῦν τῶν συνθέτων, πρῶτον ἂν εἴη τῶν ἐπιπέδων σχημάτων ὁ κύκλος. Ἔτι δὲ εἴπερ τέλειόν ἐστιν οὗ μηδὲν ἔξω τῶν αὑτοῦ λαβεῖν δυνατόν, ὥσπερ [20] ὥρισται πρότερον, καὶ τῇ μὲν εὐθείᾳ πρόσθεσίς ἐστιν ἀεί, τῇ δὲ τοῦ κύκλου οὐδέποτε, φανερὸν ὅτι τέλειος ἂν εἴη ἡ περιέχουσα

[286b10] The shape of the heaven must be spherical. That is most suitable to its substance, and is the primary shape in nature. But let us discuss the question of what is the primary shape, both in plane surfaces and in solids. Every plane figure is bounded either by straight lines or by a circumference; the rectilinear [15] is bounded by several lines, the circular by one only. Thus since in every genus the one is by nature prior to the many, and the simple to the composite, the circle must be the primary plane figure. Also, if the term "perfect" is applied, according to our previous definition, [20] to that outside which no part of itself can be found, and addition to a straight line is always possible, to a circle never, the circumference of the circle must be a perfect

τὸν κύκλον· ὥστ’ εἰ τὸ τέλειον πρότερον τοῦ ἀτελοῦς, καὶ διὰ ταῦτα πρότερον ἂν εἴη τῶν σχημάτων ὁ κύκλος. Ὡσαύτως δὲ καὶ ἡ σφαῖρα τῶν στερεῶν· μόνη γὰρ περιέχεται μιᾷ ἐπιφανείᾳ, τὰ δ’ εὐθύγραμμα πλείοσιν· ὡς γὰρ ἔχει ὁ κύκλος ἐν τοῖς ἐπιπέδοις, οὕτως ἡ σφαῖρα ἐν τοῖς στερεοῖς.

line: granted therefore that the perfect is prior to the imperfect, this argument too demonstrates the priority of the circle to other figures. By the same reasoning the sphere is the primary solid, for it alone is bounded by a single surface, rectilinear solids by several. The place of the sphere among solids is the same as that of the circle among plane figures.

§ 7—The proof of the finitude of the celestial sphere derived from its circular motion
On the Heavens, I, 5, 271b26–272a6
English Source: *On the Heavens*, trans. W. K. C. Guthrie (1939).

[271b26] Ὅτι μὲν τοίνυν ἀνάγκη τὸ σῶμα τὸ κύκλῳ φερόμενον πεπεράνθαι πᾶν, ἐκ τῶνδε δῆλον. Εἰ γὰρ ἄπειρον τὸ κύκλῳ φερόμενον σῶμα, ἄπειροι ἔσονται αἱ ἀπὸ τοῦ μέσου ἐκβαλλόμεναι. Τῶν δ’ ἀπείρων τὸ διάστημα ἄπειρον· διάστημα δὲ λέγω τῶν γραμμῶν, οὗ μηδὲν ἔστιν ἔξω λαβεῖν μέγεθος ἁπτόμενον τῶν γραμμῶν. Τοῦτ’ οὖν ἀνάγκη ἄπειρον εἶναι· τῶν γὰρ πεπερασμένων ἀεὶ ἔσται πεπερασμένον. Ἔτι δ’ ἀεὶ ἔστι τοῦ [272a] δοθέντος μεῖζον λαβεῖν, ὥστε καθάπερ ἀριθμὸν λέγομεν ἄπειρον, ὅτι μέγιστος οὐκ ἔστιν, ὁ αὐτὸς λόγος καὶ περὶ τοῦ διαστήματος· εἰ οὖν τὸ μὲν ἄπειρον μὴ ἔστι διελθεῖν, ἀπείρου δ’ ὄντος ἀνάγκη τὸ διάστημα ἄπειρον εἶναι, οὐκ ἂν ἐνδέχοιτο κινηθῆναι κύκλῳ· τὸν δ’ οὐρανὸν ὁρῶμεν κύκλῳ στρεφόμενον, καὶ τῷ λόγῳ δὲ διωρίσαμεν ὅτι ἐστί τινος ἡ κύκλῳ κίνησις.

[271b26] The following arguments make it plain that every body which revolves in a circle must be finite. If the revolving body be infinite, the straight lines radiating from the centre will be infinite. But if they are infinite, the intervening space must be infinite. “Intervening space” I am defining as space beyond which there can be no magnitude in contact with the lines. This must be infinite. In the case of finite lines it is always finite, and moreover [272a] it is always possible to take more than any given quantity of it, so that this space is infinite in the sense in which we say that number is infinite, because there exists no greatest number. If then it is impossible to traverse an infinite space, and in an infinite body the space between the radii is infinite, the body cannot move in a circle. But we ourselves see the heaven revolving in a circle, and also we established by argument that circular motion is the motion of a real body.

§ 8—The regularity of the motion of the celestial sphere
On the Heavens, II, 6, 288a13–288a27
English Source: *On the Heavens*, trans. W. K. C. Guthrie (1939).

[288a13] Περὶ δὲ τῆς κινήσεως αὐτοῦ, ὅτι ὁμαλής ἐστι καὶ οὐκ [15] ἀνώμαλος, ἐφεξῆς ἂν εἴη τῶν εἰρημένων διελθεῖν. Λέγω δὲ τοῦτο περὶ τοῦ πρώτου οὐρανοῦ καὶ περὶ τῆς πρώτης φορᾶς· ἐν γὰρ τοῖς ὑποκάτω πλείους ἤδη αἱ φοραὶ συνεληλύθασιν εἰς ἕν.

Εἰ γὰρ ἀνωμάλως κινήσεται, δῆλον ὅτι ἐπίτασις ἔσται καὶ ἀκμὴ καὶ ἄνεσις τῆς φορᾶς· ἅπασα γὰρ ἡ ἀνώμαλος φορὰ καὶ ἄνεσιν ἔχει καὶ ἐπίτασιν καὶ ἀκμήν. Ἀκμὴ [20] δ' ἐστὶν ἢ ὅθεν φέρεται ἢ οἷ ἢ ἀνὰ μέσον, οἷον ἴσως τοῖς μὲν κατὰ φύσιν οἷ φέρονται, τοῖς δὲ παρὰ φύσιν ὅθεν, τοῖς δὲ ῥιπτουμένοις ἀνὰ μέσον. Τῆς δὲ κύκλῳ φορᾶς οὐκ ἔστιν οὔτε ὅθεν οὔτε οἷ οὔτε μέσον· οὔτε γὰρ ἀρχὴ οὔτε πέρας οὔτε μέσον ἐστὶν αὐτῆς ἁπλῶς· τῷ τε γὰρ χρόνῳ ἀΐδιος [25] καὶ τῷ μήκει συνηγμένη καὶ ἄκλαστος· ὥστ' εἰ μή ἐστιν ἀκμὴ αὐτοῦ τῆς φορᾶς, οὐδ' ἂν ἀνωμαλία εἴη· ἡ γὰρ ἀνωμαλία γίγνεται διὰ τὴν ἄνεσιν καὶ ἐπίτασιν.

[288a13] The next thing which our discussions have to explain is that its motion is regular and not [15] irregular. (I refer to the first heaven and the primary motion; in the lower regions a number of motions are combined into one.)

If it moves irregularly, there will clearly be an acceleration, climax, and retardation of its motion, since all irregular motion has retardation, acceleration, and climax. The climax [20] may be either at the source or at the goal or in the middle of the motion; thus we might say that for things moving naturally it is at the goal, for things moving contrary to nature it is at the source, and for things whose motion is that of a missile it is in the middle. But circular motion has in itself neither source nor goal nor middle. There is no absolute beginning or end or mid-point of it, for in time it is eternal [25] and in length it returns upon itself and is unbroken. If then there is no climax to its motion, there will be no irregularity, for irregularity is the result of retardation and acceleration.

§ 9—The contiguity of celestial spheres
On the Heavens, II, 4, 287a2–12
English Source: *On the Heavens*, trans. W. K. C. Guthrie (1939).

[287a2] Ἐπεὶ δὲ τὸ μὲν πρῶτον σχῆμα τοῦ πρώτου σώματος, πρῶτον δὲ σῶμα τὸ ἐν τῇ ἐσχάτῃ περιφορᾷ, σφαιροειδὲς ἂν εἴη τὸ τὴν κύκλῳ [5] περιφερόμενον φοράν. Καὶ τὸ συνεχὲς ἄρα ἐκείνῳ· τὸ

[287a2] But the primary figure belongs to the primary body, and the primary body is that which is at the farthest circumference, hence it, the body which revolves in a circle, [5] must be spherical in shape.

γὰρ τῷ σφαιροειδεῖ συνεχὲς σφαιροειδές. Ὡσαύτως δὲ καὶ τὰ πρὸς τὸ μέσον τούτων· τὰ γὰρ ὑπὸ τοῦ σφαιροειδοῦς περιεχόμενα καὶ ἁπτόμενα ὅλα σφαιροειδῆ ἀνάγκη εἶναι· τὰ δὲ κάτω τῆς τῶν πλανήτων ἅπτεται τῆς ἐπάνω σφαίρας. Ὥστε σφαιροειδὴς ἂν εἴη πᾶσα· πάντα γὰρ ἅπτεται καὶ συνεχῆ ἐστι ταῖς σφαίραις.

The same must be true of the body which is contiguous to it, for what is contiguous to the spherical is spherical, and also of those bodies which lie nearer the centre, for bodies which are surrounded by the spherical and touch it at all points must themselves be spherical, and the lower bodies are in contact with the sphere above. It is, then, spherical through and through, seeing that everything in it is in continuous contact with the spheres.

§ 10—The spherical nature of stars and the impossibility of their axial rotation
On the Heavens, II, 8, 290a7–290b12
English Source: *On the Heavens*, trans. W. K. C. Guthrie (1939).

[290a7] Ἔτι δ' ἐπεὶ σφαιροειδῆ τὰ ἄστρα, καθάπερ οἵ τ' ἄλλοι φασὶ καὶ ἡμῖν ὁμολογούμενον εἰπεῖν, ἐξ ἐκείνου γε τοῦ σώματος γεννῶσιν, τοῦ δὲ σφαιροειδοῦς δύο κινήσεις εἰσὶ καθ' αὑτό, κύλισις καὶ δίνησις, εἴπερ οὖν κινεῖται τὰ ἄστρα δι' αὑτῶν, τὴν ἑτέραν ἂν κινοῖτο τούτων· ἀλλ' οὐδετέραν φαίνεται.

Δινούμενα μὲν γὰρ ἂν ἔμενεν ἐν ταὐτῷ καὶ οὐ μετέβαλλε τὸν τόπον, ὅπερ φαίνεταί τε καὶ πάντες φασίν. Ἔτι δὲ πάντα μὲν εὔλογον τὴν αὐτὴν κίνησιν κινεῖσθαι, μόνος δὲ δοκεῖ τῶν ἄστρων ὁ ἥλιος τοῦτο δρᾶν ἀνατέλλων καὶ δύνων, καὶ οὗτος οὐ δι' αὑτὸν ἀλλὰ διὰ τὴν ἀπόστασιν τῆς ἡμετέρας ὄψεως· ἡ γὰρ ὄψις ἀποτεινομένη μακρὰν ἑλίσσεται διὰ τὴν ἀσθένειαν. Ὅπερ αἴτιον ἴσως καὶ τοῦ στίλβειν φαίνεσθαι τοὺς ἀστέρας τοὺς ἐνδεδεμένους, τοὺς δὲ

[290a7] Again, since the stars are spherical (as others assert, and the statement is consistent with our own opinions, seeing that we construct them out of the spherical body), and there are two motions proper to the spherical as such, namely rolling and rotation about an axis, the stars, if they moved of themselves, would move in one of these two ways. But to all appearances they move in neither.

If their motion were rotation, they would remain in the same place and not change their position, which would be contrary to observation and the universal consensus of men. Besides, we may take it that all have the same motion, and none of the stars appears to rotate except the sun at its rising or setting, and that not of itself but only because we view it at such a great distance: that is, our sight, when used at long range, becomes weak and unsteady. This is possibly the reason also why the fixed stars appear to twinkle

πλάνητας μὴ στίλβειν· [20] οἱ μὲν γὰρ πλάνητες ἐγγύς εἰσιν, ὥστ' ἐγκρατὴς οὖσα πρὸς αὐτοὺς ἀφικνεῖται ἡ ὄψις· πρὸς δὲ τοὺς μένοντας κραδαίνεται διὰ τὸ μῆκος, ἀποτεινομένη πόρρω λίαν. Ὁ δὲ τρόμος αὐτῆς ποιεῖ τοῦ ἄστρου δοκεῖν εἶναι τὴν κίνησιν· οὐθὲν γὰρ διαφέρει κινεῖν τὴν ὄψιν ἢ τὸ ὁρώμενον.

Ἀλλὰ μὴν [25] ὅτι οὐδὲ κυλίεται τὰ ἄστρα, φανερόν· τὸ μὲν γὰρ κυλιόμενον στρέφεσθαι ἀνάγκη, τῆς δὲ σελήνης ἀεὶ δῆλόν ἐστι τὸ καλούμενον πρόσωπον. Ὥστ' ἐπεὶ κινούμενα μὲν δι' αὑτῶν τὰς οἰκείας κινεῖσθαι κινήσεις εὔλογον, ταύτας δ' οὐ φαίνεται κινούμενα, δῆλον ὅτι οὐκ ἂν κινοῖτο δι' αὑτῶν.

Πρὸς δὲ [30] τούτοις ἄλογον τὸ μηθὲν ὄργανον αὐτοῖς ἀποδοῦναι τὴν φύσιν πρὸς τὴν κίνησιν (οὐθὲν γὰρ ὡς ἔτυχε ποιεῖ ἡ φύσις), οὐδὲ τῶν μὲν ζῴων φροντίσαι, τῶν δ' οὕτω τιμίων ὑπεριδεῖν, ἀλλ' ἔοικεν ὥσπερ ἐπίτηδες ἀφελεῖν πάντα δι' ὧν ἐνεδέχετο προϊέναι καθ' αὑτά, καὶ ὅτι πλεῖστον ἀποστῆσαι τῶν ἐχόν των [35] ὄργανα πρὸς κίνησιν.
Διὸ καὶ εὐλόγως ἂν δόξειεν ὅ τε [290b] ὅλος οὐρανὸς σφαιροειδὴς εἶναι καὶ ἕκαστον τῶν ἄστρων. Πρὸς μὲν γὰρ τὴν ἐν ἑαυτῷ κίνησιν ἡ σφαῖρα τῶν σχημάτων χρησιμώτατον (οὕτω γὰρ ἂν καὶ τάχιστα κινοῖτο καὶ μάλιστα κατέχοι τὸν αὐτὸν τόπον), πρὸς δὲ τὴν εἰς τὸ πρόσθεν [5]

but the planets do not. [20] The planets are near, so that our vision reaches them with its powers unimpaired; but in reaching to the fixed stars it is extended too far, and the distance causes it to waver. Thus its trembling makes it seem as if the motion were the stars'—the effect is the same whether it is our sight or its object that moves. [25] On the other hand it is equally clear that the stars do not roll. Whatever rolls must turn about, but the moon always shows us its face (as men call it).

Thus: if the stars moved of themselves they would naturally perform their own proper motions; but we see that they do not perform these motions; therefore they cannot move of themselves.

[30] Another argument is that it would be absurd for nature to have given them no organs of motion. Nature makes nothing in haphazard fashion, and she would not look after the animals and neglect such superior beings as these. Rather she seems to have purposely deprived them of every means of progressing by themselves, and made them as different as possible from creatures which have [35] organs of motion.

The assumption is therefore justified that [290b] both the heaven as a whole and the separate stars are spherical, for the sphere is at once the most useful shape for motion in the same place—since what is spherical can move most swiftly and can most easily maintain its position unchanged—and [5]

ἀχρηστότατον· ἥκιστα γὰρ ὅμοιον τοῖς δι' αὐτῶν κινητικοῖς· οὐδὲν γὰρ ἀπηρτημένον ἔχει οὐδὲ προέχον, ὥσπερ τὸ εὐθύγραμμον, ἀλλὰ πλεῖστον ἀφέστηκε τῷ σχήματι τῶν πορευτικῶν σωμάτων. Ἐπεὶ οὖν δεῖ τὸν μὲν οὐρανὸν κινεῖσθαι τὴν ἐν ἑαυτῷ κίνησιν, τὰ δ' ἄλλα [ἄστρα] μὴ προϊέναι δι' αὐτῶν, εὐλόγως ἂν ἑκάτερον εἴη σφαιροειδές· οὕτω γὰρ μάλιστα τὸ μὲν κινήσεται τὸ δ' ἠρεμήσει.

the least suited to progression, the latter because it least resembles bodies which are self-moving. It has no separate or projecting parts as a rectilinear figure has, and is of a totally different shape from forward-moving bodies.

Since, then, the heaven must move within its own boundaries, and the stars must not move forward of themselves, we may conclude that both are spherical. This will best ensure to the one its movement and to the others their immobility.

§ 11—The emergence of heat due to the interaction between the sphere of ether and the spheres of fire and air

Meteorology, I, 3, 340b6–341a2

English Source: *Meteorologica*, trans. H. D. P. Lee (1952).

[340b6] Τὸ μὲν γὰρ ἄνω καὶ μέχρι σελήνης ἕτερον εἶναι σῶμά φαμεν πυρός τε καὶ ἀέρος, οὐ μὴν ἀλλ' ἐν αὐτῷ γε τὸ μὲν καθαρώτερον εἶναι τὸ δ' ἧττον εἰλικρινές, καὶ διαφορὰς ἔχειν, καὶ μάλιστα ᾗ καταλήγει πρὸς τὸν ἀέρα καὶ πρὸς τὸν περὶ τὴν γῆν κόσμον. φερομένου δὲ τοῦ πρώτου στοιχείου κύκλῳ καὶ τῶν ἐν αὐτῷ σωμάτων, τὸ προσεχὲς ἀεὶ τοῦ κάτω κόσμου καὶ σώματος τῇ κινήσει διακρινόμενον ἐκπυροῦται καὶ ποιεῖ τὴν θερμότητα. δεῖ δὲ νοεῖν οὕτως καὶ ἐντεῦθεν ἀρξαμένους. τὸ γὰρ ὑπὸ τὴν ἄνω περιφορὰν σῶμα οἷον ὕλη τις οὖσα καὶ δυνάμει θερμὴ καὶ ψυχρὰ καὶ ξηρὰ καὶ ὑγρά, καὶ ὅσα ἄλλα τούτοις ἀκολουθεῖ πάθη, γίγνεται τοιαύτη καὶ ἔστιν ὑπὸ κινήσεως καὶ ἀκινησίας, ἧς τὴν αἰτίαν καὶ τὴν ἀρχὴν εἰρήκαμεν πρότερον.

[340b6] We maintain that the celestial region as far down as the moon is occupied by a body which is different from air and from fire, but which varies in purity and freedom from admixture, and is not uniform in quality, especially when it borders on the air and the terrestrial region. Now this primary substance and the bodies set in it as they move in a circle set on fire and dissolve by their motion that part of the lower region which is closest to them and generates heat therein. We are also led to the same view if we reason as follows: The substance beneath the motions of the heavens is a kind of matter, having potentially the qualities hot, cold, wet and dry and any others consequent upon these; but it only actually acquires and has any of these in virtue of motion or rest, about whose originating cause we have already spoken elsewhere.

Ἐπὶ μὲν οὖν τοῦ μέσου καὶ περὶ τὸ μέσον τὸ βαρύτατόν ἐστιν καὶ ψυχρότατον ἀποκεκριμένον, γῆ καὶ ὕδωρ· περὶ δὲ ταῦτα καὶ ἐχόμενα τούτων, ἀήρ τε καὶ ὃ διὰ συνήθειαν καλοῦμεν πῦρ, οὐκ ἔστι δὲ πῦρ· ὑπερβολὴ γὰρ θερμοῦ καὶ οἷον ζέσις ἐστὶ τὸ πῦρ. ἀλλὰ δεῖ νοῆσαι τοῦ λεγομένου ὑφ' ἡμῶν ἀέρος τὸ μὲν περὶ τὴν γῆν οἷον ὑγρὸν καὶ θερμὸν εἶναι διὰ τὸ ἀτμίζειν τε καὶ ἀναθυμίασιν ἔχειν γῆς, τὸ δὲ ὑπὲρ τοῦτο θερμὸν ἤδη καὶ ξηρόν. ἔστιν γὰρ ἀτμίδος μὲν φύσις ὑγρὸν καὶ θερμόν, ἀναθυμιάσεως δὲ θερμὸν καὶ ξηρόν· καὶ ἔστιν ἀτμὶς μὲν δυνάμει οἷον ὕδωρ, ἀναθυμίασις δὲ δυνάμει οἷον πῦρ. τοῦ μὲν οὖν ἐν τῷ ἄνω τόπῳ μὴ συνίστασθαι νέφη ταύτην ὑποληπτέον αἰτίαν εἶναι, ὅτι οὐκ ἔνεστιν ἀὴρ μόνον ἀλλὰ μᾶλλον οἷον πῦρ.

Οὐδὲν δὲ κωλύει καὶ διὰ τὴν κύκλῳ φορὰν κωλύεσθαι συνίστασθαι νέφη ἐν τῷ ἀνωτέρω τόπῳ· ῥεῖν γὰρ ἀναγκαῖον ἅπαντα τὸν κύκλῳ ἀέρα, ὅσος μὴ ἐντὸς τῆς περιφερείας λαμβάνεται τῆς ἀπαρτιζούσης ὥστε τὴν γῆν σφαιροειδῆ εἶναι πᾶσαν· φαίνεται γὰρ καὶ νῦν ἡ τῶν ἀνέμων γένεσις ἐν τοῖς λιμνάζουσι τόποις τῆς γῆς, καὶ οὐχ [341a] ὑπερβάλλειν τὰ πνεύματα τῶν ὑψηλῶν ὀρῶν. ῥεῖ δὲ κύκλῳ διὰ τὸ συνεφέλκεσθαι τῇ τοῦ ὅλου περιφορᾷ.

So what is heaviest and coldest, that is, earth and water, separates off at the centre or round the centre: immediately round them are air and what we are accustomed to call fire, though it is not really fire: for fire is an excess of heat and a sort of boiling. But we must understand that of what we call air the part which immediately surrounds the earth is moist and hot because it is vaporous and contains exhalations from the earth, but that the part above this is hot and dry. For vapour is naturally moist and cold and exhalation hot and dry: and vapour is potentially like water, exhalation like fire. We must suppose therefore that the reason why clouds do not form in the upper region is that it contains not air but rather a sort of fire.

At the same time there is no reason why the formation of clouds in the upper region should not also be prevented by the circular motion. For the whole encircling mass of air must necessarily be in motion, except that part of it which is contained within the circumference that makes the earth a perfect sphere. (Thus in fact we find that winds rise in low marshy districts of the earth, and do not [341a] blow above the highest mountains.) It moves in a circle because it is carried round by the motions of the heavens.

The Prime Mover

§ 12—The Unicity and eternity of the Prime Mover
Physics, VIII, 6, 259a6–259a20
English Source: *Physics*, trans. R. P. Hardie and R. K. Gaye in Barnes, *Complete Works of Aristotle* (1984).

[259a6] Εἴπερ οὖν ἀΐδιος ἡ κίνησις, ἀΐδιον καὶ τὸ κινοῦν ἔσται πρῶτον, εἰ ἕν· εἰ δὲ πλείω, πλείω τὰ ἀΐδια. ἓν δὲ μᾶλλον ἢ πολλά, καὶ πεπερασμένα ἢ ἄπειρα, δεῖ νομίζειν. τῶν αὐτῶν γὰρ συμβαινόντων αἰεὶ τὰ πεπερασμένα μᾶλλον ληπτέον· ἐν γὰρ τοῖς φύσει δεῖ τὸ πεπερασμένον καὶ τὸ βέλτιον, ἂν ἐνδέχηται, ὑπάρχειν μᾶλλον. ἱκανὸν δὲ καὶ ἕν, ὃ πρῶτον τῶν ἀκινήτων ἀΐδιον ὂν ἔσται ἀρχὴ τοῖς ἄλλοις κινήσεως.

φανερὸν δὲ καὶ ἐκ τοῦδε ὅτι ἀνάγκη εἶναί τι ἓν καὶ ἀΐδιον τὸ πρῶτον κινοῦν. δέδεικται γὰρ ὅτι ἀνάγκη ἀεὶ κίνησιν εἶναι. εἰ δὲ ἀεί, ἀνάγκη συνεχῆ εἶναι· καὶ γὰρ τὸ ἀεὶ συνεχές, τὸ δ' ἐφεξῆς οὐ συνεχές. ἀλλὰ μὴν εἴ γε συνεχής, μία. μία δ' ἡ ὑφ' ἑνός τε τοῦ κινοῦντος καὶ ἑνὸς τοῦ κινουμένου· εἰ γὰρ ἄλλο καὶ ἄλλο κινήσει, οὐ συνεχὴς ἡ ὅλη κίνησις, ἀλλ' ἐφεξῆς.

[259a6] Motion, then, being eternal, the first mover, if there is but one, will be eternal also; if there are more than one, there will be a plurality of such eternal movers. We ought, however, to suppose that there is one rather than many, and a finite rather than an infinite number. When the consequences of either assumption are the same, we should always assume that things are finite rather than infinite in number, since in things constituted by nature that which is finite and that which is better ought, if possible, to be present rather than the reverse; and here it is sufficient to assume only one mover, the first of unmoved things, which being eternal will be the principle of motion to everything else.

The following argument also makes it evident that the first mover must be something that is one and eternal. We have shown that there must always be motion. That being so, motion must be continuous, because what is always is continuous, whereas what is in succession is not continuous. But further, if motion is continuous, it is one; and it is one only if the mover and the moved are each of them one, since in the event of a thing's being moved now by one thing and now by another the whole motion will not be continuous but successive.

§ 13—The need for the existence of a Prime Mover
Physics, VIII, 6, 259b20–260a18
English Source: *Physics*, trans. R. P. Hardie and R. K. Gaye in Barnes, *Complete Works of Aristotle* (1984).

[259b20] Ἐξ ὧν ἔστιν πιστεῦσαι ὅτι εἴ τί ἐστι τῶν ἀκινήτων μὲν κινούντων δὲ καὶ αὐτὰ κατὰ συμβεβηκός, ἀδύνατον συνεχῆ κίνησιν κινεῖν. ὥστ' εἴπερ ἀνάγκη συνεχῶς εἶναι κίνησιν, εἶναί τι δεῖ τὸ πρῶτον κινοῦν ἀκίνητον καὶ κατὰ συμβεβηκός, εἰ μέλλει, καθάπερ [25] εἴπομεν, ἔσεσθαι ἐν τοῖς οὖσιν ἄπαυστός τις καὶ ἀθάνατος κίνησις, καὶ μενεῖν τὸ ὂν αὐτὸ ἐν αὑτῷ καὶ ἐν τῷ αὐτῷ· τῆς γὰρ ἀρχῆς μενούσης ἀνάγκη καὶ τὸ πᾶν μένειν συνεχὲς ὂν πρὸς τὴν ἀρχήν. οὐκ ἔστιν δὲ τὸ αὐτὸ τὸ κινεῖσθαι κατὰ συμβεβηκὸς ὑφ' αὑτοῦ καὶ ὑφ' ἑτέρου· τὸ μὲν γὰρ ὑφ' [30] ἑτέρου ὑπάρχει καὶ τῶν ἐν τῷ οὐρανῷ ἐνίαις ἀρχαῖς, ὅσα πλείους φέρεται φοράς, θάτερον δὲ τοῖς φθαρτοῖς μόνον.

Ἀλλὰ μὴν εἴ γε ἔστιν τι ἀεὶ τοιοῦτον, κινοῦν μέν τι ἀκίνητον δὲ αὐτὸ καὶ ἀΐδιον, ἀνάγκη καὶ τὸ πρῶτον ὑπὸ τούτου κινούμενον [260a] ἀΐδιον εἶναι. ἔστιν δὲ τοῦτο δῆλον μὲν καὶ ἐκ τοῦ μὴ ἂν ἄλλως εἶναι γένεσιν καὶ φθορὰν καὶ μεταβολὴν τοῖς ἄλλοις, εἰ μή τι κινήσει κινούμενον· τὸ μὲν γὰρ ἀκίνητον [τὴν αὐτὴν] ἀεὶ τὸν

[259b20] Hence we may be sure that if a thing belongs to the class of unmoved things which move themselves accidentally, it is impossible that it should cause continuous motion. So the necessity that there should be motion continuously requires that there should be a first mover that is unmoved even accidentally, if, as [25] we have said, there is to be in the world of things an unceasing and undy ing motion, and the world is to remain self-contained and within the same limits; for if the principle is permanent, the universe must also be permanent, since it is continuous with the principle. (We must distinguish, however, between accidental motion of a thing by itself and such motion by something else, the former being confined to perishable things, whereas [30] the latter belongs also to certain principles of heavenly bodies, of all those, that is to say, that experience more than one locomotion.)

And further, if there is always something of this nature, a mover that is itself unmoved and eternal, then that which is first moved [260a] by it must also be eternal. Indeed this is clear also from the consideration that there would otherwise be no becoming and perishing and no change of any kind in other things, if there were nothing in motion to move them; for the motion imparted by the unmoved will always be imparted in the

αὐτὸν κινήσει τρόπον καὶ μίαν κίνησιν, ἅτε οὐδὲν αὐτὸ μεταβάλλον πρὸς τὸ κινούμενον. τὸ δὲ κινού-μενον ὑπὸ τοῦ κινουμένου μέν, ὑπὸ τοῦ ἀκινήτου δὲ κινουμένου ἤδη, διὰ τὸ ἄλλως καὶ ἄλλως ἔχειν πρὸς τὰ πράγματα, οὐ τῆς αὐτῆς ἔσται κινήσεως αἴτιον, ἀλλὰ διὰ τὸ ἐν ἐναντίοις εἶναι τόποις ἢ εἴδεσιν ἐναντίως παρέξεται κινούμενον ἕκαστον τῶν ἄλλων, καὶ ὁτὲ μὲν ἠρεμοῦν ὁτὲ δὲ κινούμενον.

same way and be one and the same, since the unmoved does not itself change in relation to that which is moved by it. But that which is moved by something that, though it is in motion, is moved directly by the unmoved stands in varying relations to the things that it moves, so that the motion that it causes will not be always the same: by reason of the fact that it occupies contrary positions or assumes contrary forms it will produce contrary motions in each several thing that it moves and will cause it to be at one time at rest and at another time in motion.

§ 14—Atlas, Zeus, and the perpetuity of heavenly motion
Movement of Animals, 2, 699a12–700a6
English Source: *Movement of Animals*, trans. A. S. L. Farquharson in Barnes, *Complete Works of Aristotle* (1984).

[699a12] Ἀπορήσειε δ' ἄν τις, ἆρ' εἴ τι κινεῖ τὸν ὅλον οὐρανόν, εἶναι θέλει ἀκίνητον, καὶ τοῦτο μηθὲν εἶναι τοῦ οὐρανοῦ μόριον μηδ' ἐν τῷ οὐρανῷ. εἴτε γὰρ αὐτὸ κινούμενον κινεῖ αὐτόν, ἀνάγκη τινὸς ἀκινήτου θιγγάνον κινεῖν, καὶ τοῦτο μηδὲν εἶναι μόριον τοῦ κινοῦντος· εἴτ' εὐθὺς ἀκίνητόν ἐστι τὸ κινοῦν, ὁμοίως οὐδὲν ἔσται τοῦ κινουμένου μόριον. καὶ τοῦτό γ' ὀρθῶς λέγουσιν οἱ λέγοντες, ὅτι κύκλῳ φερομένης τῆς σφαίρας οὐδ' ὁτιοῦν μένει μόριον· ἢ γὰρ ἂν ὅλην ἀναγκαῖον ἦν μένειν, ἢ διασπᾶσθαι τὸ συνεχὲς αὐτῆς. ἀλλ' ὅτι τοὺς πόλους οἴονταί τινα δύναμιν ἔχειν, οὐθὲν ἔχοντας

[699a12] Here we may ask the question whether if something moves the whole heavens this mover must be immovable, and moreover be no part of the heavens, nor in the heavens. For either it is moved itself and moves the heavens, in which case it must touch something immovable in order to cause movement, and then this is no part of that which cause movement; or if the mover is from the first immovable it will equally be no part of that which is moved. In this point at least they argue correctly who say that as the sphere is carried round in a circle no single part remains still; for then either the whole would necessarily stand still or its continuity be torn asunder; but they argue less well in supposing that the poles have a certain power,

μέγεθος ἀλλ’ ὄντας ἔσχατα καὶ στιγμάς, οὐ καλῶς. πρὸς γὰρ τῷ μηδεμίαν οὐσίαν εἶναι τῶν τοιούτων μηδενός, καὶ κινεῖσθαι τὴν μίαν κίνησιν ὑπὸ δυοῖν ἀδύνατον· τοὺς δὲ πόλους δύο ποιοῦσιν.

Ὅτι μὲν οὖν ἔχει τι καὶ πρὸς τὴν ὅλην φύσιν οὕτως ὥσπερ ἡ γῆ πρὸς τὰ ζῷα καὶ τὰ κινούμενα δι’ αὐτῶν, ἐκ τῶν τοιούτων ἄν τις διαπορήσειεν. οἱ δὲ μυθικῶς τὸν Ἄτλαντα ποιοῦντες ἐπὶ τῆς γῆς ἔχοντα τοὺς πόδας δόξαιεν ἂν ἀπὸ διανοίας εἰρηκέναι τὸν μῦθον, ὡς τοῦτον ὥσπερ διάμετρον ὄντα καὶ στρέφοντα τὸν οὐρανὸν περὶ τοὺς πόλους· τοῦτο δ’ ἂν συμβαίνῃ καὶ κατὰ λόγον διὰ τὸ τὴν γῆν μένειν, ἀλλὰ τοῖς ταῦτα λέγουσιν ἀναγκαῖον φάναι μηδὲν εἶναι μόριον αὐτὴν τοῦ παντός, πρὸς δὲ τούτοις δεῖ τὴν ἰσχὺν ἰσάζειν τοῦ κινοῦντος καὶ τὴν τοῦ μένοντος. ἔστι γάρ τι πλῆθος ἰσχύος καὶ δυνάμεως καθ’ ἣν μένει τὸ μένον, ὥσπερ καὶ καθ’ ἣν κινεῖ τὸ κινοῦν· καὶ ἔστι τις ἀναλογία ἐξ ἀνάγκης, ὥσπερ τῶν ἐναντίων κινήσεων, οὕτω καὶ τῶν ἠρεμιῶν. καὶ αἱ μὲν ἴσαι ἀπαθεῖς ὑπ’ ἀλλήλων, κρατοῦνται [699b] δὲ κατὰ τὴν ὑπεροχήν. διόπερ εἴτ’ Ἄτλας εἴτε τι τοιοῦτόν ἐστιν ἕτερον τὸ κινοῦν τῶν ἐντός, οὐδὲν μᾶλλον ἀντερείδειν δεῖ τῆς μονῆς ἣν ἡ γῆ τυγχάνει μένουσα· ἢ κινηθήσεται ἡ γῆ ἀπὸ τοῦ μέσου καὶ ἐκ τοῦ αὐτῆς τόπου. ὡς γὰρ τὸ ὠθοῦν ὠθεῖ,

though they have no magnitude, but are merely termini or points. For besides the fact that no such things have any substantial existence it is impossible for a single movement to be initiated by what is twofold; and yet they make the poles two. From a review of these difficulties we may conclude that there is something so related to the whole of nature, as the earth is to animals and things moved by them.

And the mythologists with their fable of Atlas setting his feet upon the earth appear to have based the fable upon intelligent grounds. They make Atlas a kind of diameter twirling the heavens about the poles. Now as the earth remains still this would be reasonable enough, but their theory involves them in the position that the earth is no part of the universe. And further the force of that which initiates movement must be made equal to the force of that which remains at rest. For there is a definite quantity of force or power by dint of which that which remains at rest does so, just as there is of force by dint of which that which initiates movement does so; and as there is a necessary proportion between contrary motions, so there is between states of rest. Now equal forces are unaffected by one another, but are overcome [699b] by a superiority of force. And so Atlas, or whatever similar power initiates movement from within, must exert no more force than will exactly balance the stability of the earth—otherwise the earth will be moved out of her place in the centre of things. For as the pusher pushes

οὕτω τὸ ὠθούμενον ὠθεῖται, καὶ ὁμοίως κατ' ἰσχύν. κινεῖ δὲ τὸ ἠρεμοῦν πρῶτον, ὥστε μᾶλλον καὶ πλείων ἡ ἰσχὺς ἢ ὁμοία καὶ ἴση τῆς ἠρεμίας. ὡσαύτως δὲ καὶ ἡ τοῦ κινουμένου μέν, μὴ κινοῦντος δέ. τοσαύτην οὖν δεήσει τὴν δύναμιν εἶναι τῆς γῆς ἐν τῷ ἠρεμεῖν ὅσην ὅ τε πᾶς οὐρανὸς ἔχει καὶ τὸ κινοῦν αὐτόν. εἰ δὲ τοῦτο ἀδύνατον, ἀδύνατον καὶ τὸ κινεῖσθαι τὸν οὐρανὸν ὑπό τινος τοιούτου τῶν ἐντός. Ἔστι δέ τις ἀπορία περὶ τὰς κινήσεις τῶν τοῦ οὐρανοῦ μορίων, ἣν ὡς οὖσαν οἰκείαν τοῖς εἰρημένοις ἐπισκέψαιτ' ἄν τις. ἐὰν γάρ τις ὑπερβάλλῃ τῇ δυνάμει τῆς κινήσεως τὴν τῆς γῆς ἠρεμίαν, δῆλον ὅτι κινήσει αὐτὴν ἀπὸ τοῦ μέσου. καὶ ἡ ἰσχὺς δ' ἀφ' ἧς αὕτη ἡ δύναμις, ὅτι οὐκ ἄπειρος, φανερόν· οὐδὲ γὰρ ἡ γῆ ἄπειρος, ὥστ' οὐδὲ τὸ βάρος αὐτῆς.

Ἐπεὶ δὲ τὸ ἀδύνατον λέγεται πλεοναχῶς (οὐ γὰρ ὡσαύτως τήν τε φωνὴν ἀδύνατόν φαμεν εἶναι ὁραθῆναι καὶ τοὺς ἐπὶ τῆς σελήνης ὑφ' ἡμῶν· τὸ μὲν γὰρ ἐξ ἀνάγκης, τὸ δὲ πεφυκὸς ὁρᾶσθαι οὐκ ὀφθήσεται), τὸν δ' οὐρανὸν ἄφθαρτον εἶναι καὶ ἀδιάλυτον οἰόμεθα μὲν ἐξ ἀνά γκης [εἶναι], συμβαίνει δὲ κατὰ τοῦτον τὸν λόγον οὐκ ἐξ ἀνάγκης· (πέφυκε γὰρ καὶ ἐνδέχεται εἶναι κίνησιν μείζω καὶ ἀφ' ἧς ἠρεμεῖ ἡ γῆ καὶ ἀφ' ἧς κινοῦνται τὸ πῦρ καὶ τὸ ἄνω σῶμα)· εἰ μὲν οὖν

so is the pushed pushed, and with equal force. But that which initiates movement is to begin with at rest, so that its force is greater, rather than equal and like to the stability. And similarly also than the stability of what is moved but does not initiate movement. Therefore the power of the earth in its immobility will have to be as great as that of the whole heavens, and of that which moves the heavens. But if that is impossible, it follows that the heavens cannot be moved by anything of this kind inside them.

There is a difficulty about the motions of the parts of the heavens which, as akin to what has gone before, may be considered next. For if one could overcome by power of motion the immobility of the earth he would clearly move it away from the centre. And it is plain that the force from which this power would originate will not be infinite; for the earth is not infinite and therefore its weight is not.

Now things are called impossible in several ways; for when we say it is impossible to see a sound, and when we say it is impossible to see the men in the moon, we use the word in different ways: the former is of necessity, the latter, though their nature is to be seen, will not actually be seen by us. Now we suppose that the heavens are of necessity impossible to destroy and to dissolve, whereas the result of the present argument would be to do away with this necessity. For it is natural and possible for a motion to exist greater than that by dint of which the earth is at rest, or than that by dint of which fire and the upper body

εἰσὶν ὑπερέχουσαι κινήσεις, διαλυθήσεται ταῦτα ὑπ' ἀλλήλων· εἰ δὲ μὴ εἰσὶ μέν, ἐνδέχεται δ' εἶναι (ἄπειρον γὰρ οὐκ ἐνδέχεται διὰ τὸ μηδὲ σῶμα ἐνδέχεσθαι ἄπειρον εἶναι), ἐνδέχοιτ' ἂν διαλυθῆναι τὸν οὐρανόν. τί γὰρ κωλύει τοῦτο συμβῆναι, εἴπερ [30] μὴ ἀδύνατον; οὐκ ἀδύνατον δέ, εἰ μὴ τἀντικείμενον ἀναγκαῖον.

Ἀλλὰ περὶ μὲν τῆς ἀπορίας ταύτης ἕτερος ἔστω λόγος· ἆρα δὲ δεῖ τι ἀκίνητον εἶναι καὶ ἠρεμοῦν ἔξω τοῦ κινουμένου, μηδὲν ὂν ἐκείνου μόριον, ἢ οὔ; καὶ τοῦτο πότερον καὶ ἐπὶ τοῦ παντὸς οὕτως ὑπάρχειν ἀναγκαῖον; ἴσως γὰρ ἂν δόξειεν ἄτοπον [35] εἶναι, εἰ ἡ ἀρχὴ τῆς κινήσεως ἐντός. διὸ δόξειεν ἂν τοῖς οὕτως ὑπολαμβάνουσιν εὖ εἰρῆσθαι Ὁμήρῳ ἀλλ' οὐκ ἂν ἐρύσαιτ' ἐξ οὐρανόθεν πεδίονδε [700a]

Ζῆν' ὕπατον πάντων, οὐδ' εἰ μάλα πολλὰ κάμοιτε·
πάντες δ' ἐξάπτεσθε θεοὶ πᾶσαί τε θέαιναι.

Τὸ γὰρ ὅλως ἀκίνητον ὑπ' οὐδενὸς ἐνδέχεται κινηθῆναι. ὅθεν λύεται καὶ ἡ πάλαι λεχθεῖσα ἀπορία, πότερον ἐνδέχεται [5] ἢ οὐκ ἐνδέχεται διαλυθῆναι τὴν τοῦ οὐρανοῦ σύστασιν, εἰ ἐξ ἀκινήτου ἤρτηται ἀρχῆς.

are moved. If then there are superior motions, these will be dissolved by one another; and if there actually are not, but might possibly be (for they cannot be infinite because not even body can be infinite), there is a possibility of the heavens being dissolved. For what is to prevent this coming to pass, unless [30] it be impossible? And it is not impossible unless the opposite is necessary. This difficulty, however, we will discuss elsewhere.

Must there be something immovable and at rest outside of what is moved, and no part of it, or not? And must this necessarily be so also in the case of the universe? Perhaps it would be thought strange [35] were the origin of movement inside. And to those who so conceive it the words of Homer would appear to have been well spoken [700a]:

'Nay, ye would not pull Zeus, highest of all, from heaven to the plain, no not even if ye toiled right hard; come, all ye gods and goddesses! Set hands to the chain'.

For that which is entirely immovable cannot possibly be moved by anything. And herein lies the solution of the difficulty stated just now, [5] the possibility or impossibility of dissolving the system of the heavens, in that it depends from an origin which is immovable.

The Unmoved Movers of Planetary Spheres

§ 15—The number of the Unmoved Movers
Metaphysics, Λ, 8, 1073a3–1074b14
English Source: *Metaphysics*, trans. W. D. Ross in Barnes, *Complete Works of Aristotle* (1984).

[1073a3] ὅτι μὲν οὖν ἔστιν οὐσία τις ἀΐδιος καὶ ἀκίνητος καὶ κεχωρισμένη τῶν αἰσθητῶν, φανερὸν ἐκ τῶν εἰρημένων: δέδεικται δὲ καὶ ὅτι μέγεθος οὐδὲν ἔχειν ἐνδέχεται ταύτην τὴν οὐσίαν ἀλλ' ἀμερὴς καὶ ἀδιαίρετός ἐστιν κινεῖ γὰρ τὸν ἄπειρον χρόνον, οὐδὲν δ' ἔχει δύναμιν ἄπειρον πεπερασμένον: ἐπεὶ δὲ πᾶν μέγεθος ἢ ἄπειρον ἢ πεπερασμένον, πεπερασμένον μὲν διὰ τοῦτο οὐκ ἂν ἔχοι μέγεθος, ἄπειρον δ' ὅτι ὅλως οὐκ ἔστιν οὐδὲν ἄπειρον μέγεθος): ἀλλὰ μὴν καὶ ὅτι ἀπαθὲς καὶ ἀναλλοίωτον: πᾶσαι γὰρ αἱ ἄλλαι κινήσεις ὕστεραι τῆς κατὰ τόπον. ταῦτα μὲν οὖν δῆλα διότι τοῦτον ἔχει τὸν τρόπον.

Πότερον δὲ μίαν θετέον τὴν τοιαύτην οὐσίαν ἢ πλείους, καὶ πόσας, δεῖ μὴ λανθάνειν, ἀλλὰ μεμνῆσθαι καὶ τὰς τῶν ἄλλων ἀποφάσεις, ὅτι περὶ πλήθους οὐθὲν εἰρήκασιν ὅ τι καὶ σαφὲς εἰπεῖν. ἡ μὲν γὰρ περὶ τὰς ἰδέας ὑπόληψις οὐδεμίαν ἔχει σκέψιν ἰδίαν (ἀριθμοὺς γὰρ λέγουσι τὰς ἰδέας οἱ λέγοντες ἰδέας, περὶ δὲ τῶν ἀριθμῶν ὁτὲ μὲν ὡς περὶ ἀπείρων λέγουσιν ὁτὲ δὲ ὡς μέχρι τῆς δεκάδος ὡρισμένων: δι' ἣν δ' αἰτίαν

[1073a3] It is clear then from what has been said that there is a substance which is eternal and unmovable and separate from sensible things. It has been shown also that this substance cannot have any magnitude, but is without parts and indivisible. For it produces movement through infinite time, but nothing finite has infinite power. And, while every magnitude is either infinite or finite, it cannot, for the above reason, have finite magnitude, and it cannot have infinite magnitude because there is no infinite magnitude at all. But it is also clear that it is impassive and unalterable; for all the other changes are posterior to change of place. It is clear, then, why the first mover has these attributes.

We must not ignore the question whether we have to suppose one such substance or more than one, and if the latter, how many; we must also mention, regarding the opinions expressed by others, that they have said nothing that can even be clearly stated about the number of the substances. For the theory of Ideas has no special discussion of the subject; for those who believe in Ideas say the Ideas are numbers, and they speak of numbers now as unlimited, now as limited by the number 10; but as

τοσοῦτον τὸ πλῆθος τῶν ἀριθμῶν, οὐδὲν λέγεται μετὰ σπουδῆς ἀποδεικτικῆς): ἡμῖν δ' ἐκ τῶν ὑποκειμένων καὶ διωρισμένων λεκτέον.

Ἡ μὲν γὰρ ἀρχὴ καὶ τὸ πρῶτον τῶν ὄντων ἀκίνητον καὶ καθ' αὑτὸ καὶ κατὰ [25] συμβεβηκός, κινοῦν δὲ τὴν πρώτην ἀΐδιον καὶ μίαν κίνησιν: ἐπεὶ δὲ τὸ κινούμενον ἀνάγκη ὑπό τινος κινεῖσθαι, καὶ τὸ πρῶτον κινοῦν ἀκίνητον εἶναι καθ' αὑτό, καὶ τὴν ἀΐδιον κίνησιν ὑπὸ ἀϊδίου κινεῖσθαι καὶ τὴν μίαν ὑφ' ἑνός, ὁρῶμεν δὲ παρὰ τὴν τοῦ παντὸς τὴν ἁπλῆν φοράν, ἣν κινεῖν φαμὲν [30] τὴν πρώτην οὐσίαν καὶ ἀκίνητον, ἄλλας φορὰς οὔσας τὰς τῶν πλανήτων ἀϊδίους (ἀΐδιον γὰρ καὶ ἄστατον τὸ κύκλῳ σῶμα: δέδεικται δ' ἐν τοῖς φυσικοῖς περὶ τούτων), ἀνάγκη καὶ τούτων ἑκάστην τῶν φορῶν ὑπ' ἀκινήτου τε κινεῖσθαι καθ' αὑτὴν καὶ ἀϊδίου οὐσίας. ἥ τε γὰρ τῶν ἄστρων φύσις ἀΐδιος [35] οὐσία τις οὖσα, καὶ τὸ κινοῦν ἀΐδιον καὶ πρότερον τοῦ κινουμένου, καὶ τὸ πρότερον οὐσίας οὐσίαν ἀναγκαῖον εἶναι. φανερὸν τοίνυν ὅτι τοσαύτας τε οὐσίας ἀναγκαῖον εἶναι τήν τε φύσιν ἀϊδίους καὶ ἀκινήτους καθ' αὑτάς, καὶ ἄνευ μεγέθους [1073b] διὰ τὴν εἰρημένην αἰτίαν πρότερον. —ὅτι μὲν οὖν εἰσὶν οὐσίαι, καὶ τούτων τις πρώτη καὶ δευτέρα κατὰ τὴν αὐτὴν

for the reason why there should be just so many numbers, nothing is said with any demonstrative exactness.

We however must discuss the subject, starting from the presuppositions and distinctions we have mentioned. The first principle or primary being is not movable either in itself or [25] accidentally, but produces the primary eternal and single movement. And since that which is moved must be moved by something, and the first mover must be in itself unmovable, and eternal movement must be produced by something eternal and a single movement by a single thing, and since we see that besides the simple spatial movement of the universe, which we say the [30] first and unmovable substance produces, there are other spatial movements—those of the planets—which are eternal (for the body which moves in a circle is eternal and unresting; we have proved these points in the Physics), each of these movements also must be caused by a substance unmovable in itself and eternal. For the nature of the stars is [35] eternal, being a kind of substance, and the mover is eternal and prior to the moved, and that which is prior to a substance must be a substance. Evidently, then, there must be substances which are of the same number as the movements of the stars, and in their nature eternal, and in themselves unmovable, and without magnitude, for the reason before mentioned. [1073b]

That the movers are substances, then, and that one of these is first and another second according to the same order as

τάξιν ταῖς φοραῖς τῶν ἄστρων, φανερόν.

Τὸ δὲ πλῆθος ἤδη τῶν φορῶν ἐκ τῆς οἰκειοτάτης φιλοσοφίᾳ τῶν μαθηματικῶν ἐπιστημῶν δεῖ σκοπεῖν, ἐκ τῆς ἀστρολογίας: αὕτη γὰρ περὶ οὐσίας αἰσθητῆς μὲν ἀϊδίου δὲ ποιεῖται τὴν θεωρίαν, αἱ δ' ἄλλαι περὶ οὐδεμιᾶς οὐσίας, οἷον ἥ τε περὶ τοὺς ἀριθμοὺς καὶ τὴν γεωμετρίαν. ὅτι μὲν οὖν πλείους τῶν φερομένων αἱ φοραί, φανερὸν τοῖς καὶ μετρίως ἡμμένοις (πλείους γὰρ ἕκαστον φέρεται μιᾶς τῶν πλανωμένων ἄστρων): πόσαι δ' αὗται τυγχάνουσιν οὖσαι, νῦν μὲν ἡμεῖς ἃ λέγουσι τῶν μαθηματικῶν τινὲς ἐννοίας χάριν λέγομεν, ὅπως ᾖ τι τῇ διανοίᾳ πλῆθος ὡρισμένον ὑπολαβεῖν: τὸ δὲ λοιπὸν τὰ μὲν ζητοῦντας αὐτοὺς δεῖ τὰ δὲ πυνθανομένους παρὰ τῶν ζητούντων, ἄν τι φαίνηται παρὰ τὰ νῦν εἰρημένα τοῖς ταῦτα πραγματευομένοις, φιλεῖν μὲν ἀμφοτέρους, πείθεσθαι δὲ τοῖς ἀκριβεστέροις.

Εὔδοξος μὲν οὖν ἡλίου καὶ σελήνης ἑκατέρου τὴν φορὰν ἐν τρισὶν ἐτίθετ' εἶναι σφαίραις, ὧν τὴν μὲν πρώτην τὴν τῶν ἀπλανῶν ἄστρων εἶναι, τὴν δὲ δευτέραν κατὰ τὸν διὰ μέσων τῶν ζῳδίων, τὴν δὲ τρίτην κατὰ τὸν λελοξωμένον ἐν τῷ πλάτει τῶν ζῳδίων (ἐν μείζονι δὲ πλάτει λελοξῶσθαι καθ' ὃν ἡ σελήνη φέρεται ἢ καθ' ὃν ὁ ἥλιος), τῶν δὲ πλανωμένων ἄστρων ἐν τέτταρσιν ἑκάστου σφαίραις, καὶ τούτων δὲ τὴν μὲν πρώτην

the movements of the stars, is evident.

But in the number of movements we reach a problem which must be treated from the standpoint of that one of the mathematical sciences which is most akin to philosophy—viz. of astronomy; for this science speculates about substance which is perceptible but eternal, but the other mathematical sciences, i.e. arithmetic and geometry, treat of no substance. That the movements are more numerous than the bodies that are moved, is evident to those who have given even moderate attention to the matter; for each of the planets has more than one movement. But as to the actual number of these movements, we now—to give some notion of the subject—quote what some of the mathematicians say, that our thought may have some definite number to grasp; but, for the rest, we must partly investigate for ourselves, partly learn from other investigators, and if those who study this subject form an opinion contrary to what we have now stated, we must esteem both parties indeed, but follow the more accurate.

Eudoxus supposed that the motion of the sun or of the moon involves, in either case, three spheres, of which the first is the sphere of the fixed stars, and the second moves in the circle which runs along the middle of the zodiac, and the third in the circle which is inclined across the breadth of the zodiac; but the circle in which the moon moves is inclined at a greater angle than that in which the sun moves. And the motion of the planets involves, in each case, four spheres, and of these also the first and

καὶ δευτέραν τὴν αὐτὴν εἶναι [25] ἐκείναις (τήν τε γὰρ τῶν ἀπλανῶν τὴν ἁπάσας φέρουσαν εἶναι, καὶ τὴν ὑπὸ ταύτῃ τεταγμένην καὶ κατὰ τὸν διὰ μέσων τῶν ζῳδίων τὴν φορὰν ἔχουσαν κοινὴν ἁπασῶν εἶναι), τῆς δὲ τρίτης ἁπάντων τοὺς πόλους ἐν τῷ διὰ μέσων τῶν ζῳδίων εἶναι, τῆς δὲ τετάρτης τὴν φορὰν κατὰ τὸν λελοξωμένον [30] πρὸς τὸν μέσον ταύτης: εἶναι δὲ τῆς τρίτης σφαίρας τοὺς πόλους τῶν μὲν ἄλλων ἰδίους, τοὺς δὲ τῆς Ἀφροδίτης καὶ τοῦ Ἑρμοῦ τοὺς αὐτούς.

Κάλλιππος δὲ τὴν μὲν θέσιν τῶν σφαιρῶν τὴν αὐτὴν ἐτίθετο Εὐδόξῳ [τοῦτ' ἔστι τῶν ἀποστημάτων τὴν τάξιν], τὸ δὲ πλῆθος τῷ μὲν τοῦ Διὸς καὶ [35] τῷ τοῦ Κρόνου τὸ αὐτὸ ἐκείνῳ ἀπεδίδου, τῷ δ' ἡλίῳ καὶ τῇ σελήνῃ δύο ᾤετο ἔτι προσθετέας εἶναι σφαίρας, τὰ φαινόμενα [37] εἰ μέλλει τις ἀποδώσειν, τοῖς δὲ λοιποῖς τῶν πλανήτων ἑκάστῳ μίαν.

Ἀναγκαῖον δέ, εἰ μέλλουσι συντεθεῖσαι [1074a] πᾶσαι τὰ φαινόμενα ἀποδώσειν, καθ' ἕκαστον τῶν πλανωμένων ἑτέρας σφαίρας μιᾷ ἐλάττονας εἶναι τὰς ἀνελιττούσας καὶ εἰς τὸ αὐτὸ ἀποκαθιστάσας τῇ θέσει τὴν πρώτην σφαῖραν ἀεὶ τοῦ ὑποκάτω τεταγμένου ἄστρου: οὕτω γὰρ μόνως [5] ἐνδέχεται τὴν τῶν πλανήτων φορὰν ἅπαντα ποιεῖσθαι. ἐπεὶ οὖν ἐν αἷς μὲν αὐτὰ φέρεται σφαίραις αἱ μὲν ὀκτὼ αἱ δὲ πέντε καὶ εἴκοσίν εἰσιν, τούτων δὲ μόνας οὐ δεῖ ἀνελιχθῆναι ἐν αἷς τὸ κατωτάτω τεταγμένον φέρεται, αἱ μὲν τὰς τῶν πρώτων δύο ἀνελίττουσαι ἓξ

second are the same [25] as the first two mentioned above(for the sphere of the fixed stars is that which moves all the other spheres, and that which is placed beneath this and has its movement in the circle which bisects the zodiac is common to all), but the poles of the third sphere of each planet are in the circle which bisects the zodiac, and the motion of the fourth sphere is in the circle which is inclined at an angle [30] to the equator of the third sphere; and the poles of the third spheres are different for the other planets, but those of Venus and Mercury are the same.

Callippus made the position of the spheres the same as Eudoxus did, but while he assigned the same number as Eudoxus did to Jupiter and to [35] Saturn, he thought two more spheres should be added to the sun and two to the moon, if we were to explain the phenomena, and one more to each of the other planets.

But it is necessary, if all the spheres combined are to explain the phenomena, [1074a] that for each of the planets there should be other spheres (one fewer than those hitherto assigned) which counteract those already mentioned and bring back to the same position the first sphere of the star which in each case is situated below the star in question; for only thus can [5] all the forces at work produce the motion of the planets. Since, then, the spheres by which the planets themselves are moved are eight and twenty-five, and of these only those by which the lowest-situated planet is moved need not be counteracted, the spheres which counteract those of the first two planets will be six in number, and the spheres

ἔσονται, αἱ δὲ τὰς τῶν ὕστερον τεττάρων ἑκκαίδεκα: ὁ δὴ ἁπασῶν ἀριθμὸς τῶν τε φερουσῶν καὶ τῶν ἀνελιττουσῶν ταύτας πεντήκοντά τε καὶ πέντε. εἰ δὲ τῇ σελήνῃ τε καὶ τῷ ἡλίῳ μὴ προστιθείη τις ἃς εἴπομεν κινήσεις, αἱ πᾶσαι σφαῖραι ἔσονται ἑπτά τε καὶ τεσσαράκοντα.

Τὸ μὲν οὖν πλῆθος τῶν σφαιρῶν ἔστω τοσοῦτον, ὥστε καὶ τὰς οὐσίας καὶ τὰς ἀρχὰς τὰς ἀκινήτους [καὶ τὰς αἰσθητὰς] τοσαύτας εὔλογον ὑπολαβεῖν (τὸ γὰρ ἀναγκαῖον ἀφείσθω τοῖς ἰσχυροτέροις λέγειν): εἰ δὲ μηδεμίαν οἷόν τ᾽ εἶναι φορὰν μὴ συντείνουσαν πρὸς ἄστρου φοράν, ἔτι δὲ πᾶσαν φύσιν καὶ πᾶσαν οὐσίαν ἀπαθῆ καὶ καθ᾽ αὑτὴν τοῦ ἀρίστου τετυχηκυῖαν τέλος εἶναι δεῖ νομίζειν, οὐδεμία ἂν εἴη παρὰ ταύτας ἑτέρα φύσις, ἀλλὰ τοῦτον ἀνάγκη τὸν ἀριθμὸν εἶναι τῶν οὐσιῶν. εἴτε γὰρ εἰσὶν ἕτεραι, κινοῖεν ἂν ὡς τέλος οὖσαι φορᾶς: ἀλλὰ εἶναί γε ἄλλας φορὰς ἀδύνατον παρὰ τὰς εἰρημένας. τοῦτο δὲ εὔλογον ἐκ τῶν φερομένων ὑπολαβεῖν. εἰ γὰρ πᾶν τὸ φέρον τοῦ φερομένου χάριν πέφυκε καὶ φορὰ πᾶσα φερομένου τινός ἐστιν, οὐδεμία φορὰ αὑτῆς ἂν ἕνεκα εἴη οὐδ᾽ ἄλλης φορᾶς, ἀλλὰ τῶν ἄστρων ἕνεκα. εἰ γὰρ ἔσται φορὰ φορᾶς ἕνεκα, καὶ ἐκείνην ἑτέρου δεήσει χάριν εἶναι: ὥστ᾽ ἐπειδὴ οὐχ οἷόν τε εἰς ἄπειρον,

which counteract those of the next four planets will be sixteen, and the number of all the spheres—those which move the planets and those which counteract these—will be fifty-five. And if one were not to add to the moon and to the sun the movements we mentioned, all the spheres will be forty-nine in number.

Let this then be taken as the number of the spheres, so that the unmovable substances and principles may reasonably be taken as just so many; the assertion of necessity must be left to more powerful thinkers. If there can be no spatial movement which does not conduce to the moving of a star, and if further every being and every substance which is immune from change and in virtue of itself has attained to the best must be considered an end, there can be no other being apart from these we have named, but this must be the number of the substances. For if there are others, they will cause change as being an end of movement; but there cannot be other movements besides those mentioned. And it is reasonable to infer this from a consideration of the bodies that are moved; for if everything that moves is for the sake of that which is moved, and every movement belongs to something that is moved, no movement can be for the sake of itself or of another movement, but all movements must be for the sake of the stars. For if a movement is to be for the sake of a movement, this latter also will have to be for the sake of something else; so that since there cannot be an infinite regress, the end of every movement will be

τέλος ἔσται πάσης φορᾶς τῶν φερομένων τι θείων σωμάτων κατὰ τὸν οὐρανόν.

Ὅτι δὲ εἷς οὐρανός, φανερόν. εἰ γὰρ πλείους οὐρανοὶ ὥσπερ ἄνθρωποι, ἔσται εἴδει μία ἡ περὶ ἕκαστον ἀρχή, ἀριθμῷ δέ γε πολλαί. ἀλλ' ὅσα ἀριθμῷ πολλά, ὕλην ἔχει (εἷς γὰρ λόγος καὶ ὁ αὐτὸς πολλῶν, [35] οἷον ἀνθρώπου, Σωκράτης δὲ εἷς): τὸ δὲ τί ἦν εἶναι οὐκ ἔχει ὕλην τὸ πρῶτον: ἐντελέχεια γάρ. ἓν ἄρα καὶ λόγῳ καὶ ἀριθμῷ τὸ πρῶτον κινοῦν ἀκίνητον ὄν: καὶ τὸ κινούμενον ἄρα ἀεὶ καὶ συνεχῶς: εἷς ἄρα οὐρανὸς μόνος.

Παραδέδοται [1074b] δὲ παρὰ τῶν ἀρχαίων καὶ παμπαλαίων ἐν μύθου σχήματι καταλελειμμένα τοῖς ὕστερον ὅτι θεοί τέ εἰσιν οὗτοι καὶ περιέχει τὸ θεῖον τὴν ὅλην φύσιν. τὰ δὲ λοιπὰ μυθικῶς ἤδη προσῆκται πρὸς τὴν πειθὼ τῶν πολλῶν καὶ [5] πρὸς τὴν εἰς τοὺς νόμους καὶ τὸ συμφέρον χρῆσιν: ἀνθρωποειδεῖς τε γὰρ τούτους καὶ τῶν ἄλλων ζῴων ὁμοίους τισὶ λέγουσι, καὶ τούτοις ἕτερα ἀκόλουθα καὶ παραπλήσια τοῖς εἰρημένοις, ὧν εἴ τις χωρίσας αὐτὸ λάβοι μόνον τὸ πρῶτον, ὅτι θεοὺς ᾤοντο τὰς πρώτας οὐσίας εἶναι, θείως ἂν εἰρῆσθαι [10] νομίσειεν, καὶ κατὰ τὸ εἰκὸς πολλάκις εὑρημένης εἰς τὸ δυνατὸν ἑκάστης καὶ τέχνης καὶ φιλοσοφίας καὶ πάλιν φθειρομένων καὶ ταύτας τὰς δόξας ἐκείνων οἷον λείψανα περισεσῶσθαι μέχρι τοῦ νῦν. ἡ μὲν οὖν πάτριος δόξα καὶ ἡ παρὰ τῶν πρώτων ἐπὶ τοσοῦτον ἡμῖν φανερὰ μόνον.

one of the divine bodies which move through the heaven.

Evidently there is but one heaven. For if there are many heavens as there are many men, the moving principles, of which each heaven will have one, will be one in form but in number many. But all things that are many in number have matter. (For one and the same formula applies to many things, [35] e.g. the formula of man; but Socrates is one.) But the primary essence has not matter; for it is fulfillment. So the unmovable first mover is one both in formula and in number; therefore also that which is moved always and continuously is one alone; therefore there is one heaven alone. Our forefathers [1074b] in the most remote ages have handed down to us their posterity a tradition, in the form of a myth, that these substances are gods and that the divine encloses the whole of nature. The rest of the tradition has been added later in mythical form with a view to the persuasion of the multitude [5] and to its legal and utilitarian expediency; they say these gods are in the form of men or like some of the other animals, and they say other things consequent on and similar to these which we have mentioned. But if we were to separate the first point from these additions and take it alone—that they thought the first substances to be gods—we must regard this as an inspired [10] utterance, and reflect that, while probably each art and science has often been developed as far as possible and has again perished, these opinions have been preserved like relics until the present. Only thus far, then, is the opinion of our ancestors and our earliest predecessors clear to us.

§ 16—The need for the existence of other motion(s) apart from the continuous motion of the sphere of the fixed stars to explain generation and corruption
On the Heavens, II, 3, 286a3–286b9
English Source: *On the Heavens*, trans. W. K. C. Guthrie (1939).

[286a3] Ἐπεὶ δ᾽ οὐκ ἔστιν ἐναντία κίνησις ἡ κύκλῳ τῇ κύκλῳ, σκεπτέον διὰ τί πλείους εἰσὶ φοραί, καίπερ πόρρωθεν πειρωμένοις ποιεῖσθαι τὴν ζήτησιν, πόρρω δ᾽ οὐχ οὕτω τῷ τόπῳ, πολὺ δὲ μᾶλλον τῷ τῶν συμβεβηκότων αὐτοῖς περὶ πάμπαν ὀλίγων ἔχειν αἴσθησιν. Ὅμως δὲ λέγωμεν.

Ἡ δ᾽ αἰτία περὶ αὐτῶν ἐνθένδε ληπτέα. Ἕκαστόν ἐστιν, ὧν ἐστιν ἔργον, ἕνεκα τοῦ ἔργου. Θεοῦ δ᾽ ἐνέργεια ἀθανασία· τοῦτο δ᾽ ἐστὶ ζωὴ ἀΐδιος. ὥστ᾽ ἀνάγκη τῷ θεῷ κίνησιν ἀΐδιον ὑπάρχειν. Ἐπεὶ δ᾽ ὁ οὐρανὸς τοιοῦτος (σῶμα γάρ τι θεῖον), διὰ τοῦτο ἔχει τὸ ἐγκύκλιον σῶμα, ὃ φύσει κινεῖται κύκλῳ ἀεί.

Διὰ τί οὖν οὐχ ὅλον τὸ σῶμα τοῦ οὐρανοῦ τοιοῦτον; ὅτι ἀνάγκη μένειν τι τοῦ σώματος τοῦ φερομένου κύκλῳ, τὸ ἐπὶ τοῦ μέσου, τούτου δ᾽ οὐθὲν οἷόν τε μένειν μόριον, οὔθ᾽ ὅλως οὔτ᾽ ἐπὶ τοῦ μέσου. Καὶ γὰρ ἂν ἡ κατὰ φύσιν κίνησις ἦν αὐτοῦ ἐπὶ τὸ μέσον· φύσει δὲ κύκλῳ κινεῖται· οὐ γὰρ ἂν ἦν ἀΐδιος ἡ κίνησις· οὐθὲν γὰρ παρὰ φύσιν ἀΐδιον. Ὕστερον δὲ τὸ παρὰ φύσιν τοῦ κατὰ φύσιν, καὶ ἔκστασίς τίς ἐστιν ἐν τῇ γενέσει τὸ παρὰ φύσιν τοῦ κατὰ φύσιν. Ἀνάγκη

[286a3] Now since there exists no circular motion which is the opposite of another, the question must be asked why there are several different revolutions, although we are far removed from the objects of our attempted inquiry, not in the obvious sense of distance in space, but rather because very few of their attributes are perceptible to our senses. Yet we must say what we can. If we are to grasp their cause, we must start from this, that everything which has a function exists for the sake of that function. The activity of a god is immortality, that is, eternal life. Necessarily, therefore, the divine must be in eternal motion. And since the heaven is of this nature (i.e. is a divine body), that is why it has its circular body, which by nature moves for ever in a circle.

Why, then, is not the whole body of the heaven like this? Because when a body revolves in a circle some part of it must remain still, namely that which is at the centre, but of the body which we have described no part can remain still, whether it be at the centre or wherever it be. If it could, then its natural motion would be towards the centre, whereas in fact its natural motion is circular. Otherwise the motion would not be eternal, for nothing contrary to nature is eternal. The unnatural is subsequent to the natural, being an aberration from the natural in the field of becoming. It

τοίνυν γῆν εἶναι· τοῦτο γὰρ ἠρεμεῖ ἐπὶ τοῦ μέσου. Νῦν μὲν οὖν ὑποκείσθω τοῦτο, ὕστερον δὲ δειχθήσεται περὶ αὐτοῦ. Ἀλλὰ μὴν εἰ γῆν, ἀνάγκη καὶ πῦρ εἶναι· τῶν γὰρ ἐναντίων εἰ θάτερον φύσει, ἀνάγκη καὶ θάτερον εἶναι φύσει, ἐάν περ ᾖ ἐναντίον, καὶ εἶναί τινα αὐτοῦ φύσιν· [25] ἡ γὰρ αὐτὴ ὕλη τῶν ἐναντίων, καὶ τῆς στερήσεως πρότερον ἡ κατάφασις (λέγω δ' οἷον τὸ θερμὸν τοῦ ψυχροῦ), ἡ δ' ἠρεμία καὶ τὸ βαρὺ λέγεται κατὰ στέρησιν κουφότητος καὶ κινήσεως. Ἀλλὰ μὴν εἴπερ ἔστι πῦρ καὶ γῆ, ἀνάγκη καὶ τὰ μεταξὺ αὐτῶν εἶναι σώματα· ἐναντίωσιν γὰρ ἔχει ἕκαστον [30] τῶν στοιχείων πρὸς ἕκαστον. Ὑποκείσθω δὲ καὶ τοῦτο νῦν, ὕστερον δὲ πειρατέον δεῖξαι.

Τούτων δ' ὑπαρχόντων φανερὸν ὅτι ἀνάγκη γένεσιν εἶναι διὰ τὸ μηδὲν οἷόν τ' αὐτῶν εἶναι ἀΐδιον· πάσχει γὰρ καὶ ποιεῖ τἀναντία ὑπ' ἀλλήλων, καὶ φθαρτικὰ ἀλλήλων ἐστίν. Ἔτι δ' οὐκ εὔλογον εἶναί τι κινητὸν ἀΐδιον, [35] οὗ μὴ ἐνδέχεται εἶναι κατὰ φύσιν τὴν κίνησιν ἀΐδιον· [286b] τούτων δ' ἔστι κίνησις. Ὅτι μὲν τοίνυν ἀναγκαῖον εἶναι γένεσιν, ἐκ τούτων δῆλον.

Εἰ δὲ γένεσιν, ἀναγκαῖον καὶ ἄλλην εἶναι φοράν, ἢ μίαν ἢ πλείους· κατὰ γὰρ τὴν τοῦ ὅλου ὡσαύτως ἀναγκαῖον ἔχειν τὰ στοιχεῖα τῶν σωμάτων πρὸς ἄλληλα. [5] Λεχθήσεται δὲ καὶ περὶ τούτου ἐν τοῖς ἑπομένοις σαφέστερον.

Νῦν δὲ τοσοῦτόν ἐστι δῆλον, διὰ τίνα αἰτίαν πλείω τὰ ἐγκύκλιά ἐστι σώματα,

follows that there must be earth, for it is that which remains at rest in the middle. (Let us accept this last statement for the present: it will be dealt with later.) But if there must be earth, there must also be fire; for if one of a pair of contraries exists by nature, so also must the other, if it is truly a contrary, and must have a nature of its own. [25] Contraries have indeed the same matter, and the positive element is prior to the negative, as e.g. hot is prior to cold. And rest and weight are described as the negation of motion and lightness. Again, if earth and fire exist, so also must the intermediate bodies, seeing that each of the [30] elements is in opposition to the others. (Let this too be assumed for the present: later we must try to demonstrate its truth.)

From the existence of the four elements it clearly follows that there must be coming-to-be, for the reason that none of them is eternal, since contraries act upon each other reciprocally, and are destructive of each other. Moreover it is not reasonable that there should be an eternal movable object [35] whose motion cannot be naturally eternal; but these bodies have a motion.

[286b] From these considerations, then, the necessity of coming-to-be is clear.

But if there is coming-to-be, there must be another revolution [*sc.* besides that of the fixed stars], or more than one; for the operation of the revolution of the whole could only result in leaving the relations between the four elements unchanged. [5] This too will be discussed more clearly at a later stage, but for the present it is at least evident why there are more revolving bodies than one, namely,

ὅτι ἀνάγκη γένεσιν εἶναι, γένεσιν δ', εἴπερ καὶ πῦρ, τοῦτο δὲ καὶ τἆλλα, εἴπερ καὶ γῆν· ταύτην δ' ὅτι ἀνάγκη μένειν τι ἀεί, εἴπερ καὶ κινεῖσθαί τι ἀεί.

because there must be coming-to-be: coming-to-be was inevitable if there was fire, fire and the other elements if there was earth; and the existence of the earth followed from the necessity of having something fixed for ever if there was to be something for ever in motion.

§ 17—The eternal causes of the divine things that we perceive
Metaphysics, E, 1, 1026a6–22
English Source: *Metaphysics*, trans. W. D. Ross in Barnes, *Complete Works of Aristotle* (1984).

[1026a6] Ὅτι μὲν οὖν ἡ φυσικὴ θεωρητική ἐστι, φανερὸν ἐκ τούτων· ἀλλ' ἔστι καὶ ἡ μαθηματικὴ θεωρητική· ἀλλ' εἰ ἀκινήτων καὶ χωριστῶν ἐστί, νῦν ἄδηλον, ὅτι μέντοι ἔνια μαθήματα ᾗ ἀκίνητα καὶ ᾗ χωριστὰ θεωρεῖ, δῆλον. εἰ δέ τί ἐστιν ἀΐδιον καὶ ἀκίνητον καὶ χωριστόν, φανερὸν ὅτι θεωρητικῆς τὸ γνῶναι, οὐ μέντοι φυσικῆς γε (περὶ κινητῶν γάρ τινων ἡ φυσική) οὐδὲ μαθηματικῆς, ἀλλὰ προτέρας ἀμφοῖν. ἡ μὲν γὰρ φυσικὴ περὶ χωριστὰ μὲν ἀλλ' οὐκ ἀκίνητα, τῆς δὲ μαθηματικῆς ἔνια περὶ ἀκίνητα μὲν οὐ χωριστὰ δὲ ἴσως ἀλλ' ὡς ἐν ὕλῃ· ἡ δὲ πρώτη καὶ περὶ χωριστὰ καὶ ἀκίνητα. ἀνάγκη δὲ πάντα μὲν τὰ αἴτια ἀΐδια εἶναι, μάλιστα δὲ ταῦτα· ταῦτα γὰρ αἴτια τοῖς φανεροῖς τῶν θείων.

[1026a6] That natural science, then, is theoretical, is plain from these considerations. Mathematics also is theoretical; but whether its objects are immovable and separable from matter, is not at present clear; it is clear, however, that it considers some mathematical objects *qua* immovable and *qua* separable from matter. But if there is something which is eternal and immovable and separable, clearly the knowledge of it belongs to a theoretical science,—not, however, to natural science (for natural science deals with certain movable things) nor to mathematics, but to a science prior to both. For natural science deals with things which are inseparable from matter but not immovable, and some parts of mathematics deal with things which are immovable, but probably not separable, but embodied in matter; while the first science deals with things which are both separable and immovable. Now all causes must be eternal, but especially these; for they are the causes of so much of the divine as appears to us.

Ὥστε τρεῖς ἂν εἶεν φιλοσοφίαι θεωρητικαί, μαθηματική, φυσική, θεολογική (οὐ γὰρ [20] ἄδηλον ὅτι εἴ που τὸ θεῖον ὑπάρχει, ἐν τῇ τοιαύτῃ φύσει ὑπάρχει), καὶ τὴν τιμιωτάτην δεῖ περὶ τὸ τιμιώτατον γένος εἶναι. αἱ μὲν οὖν θεωρητικαὶ τῶν ἄλλων ἐπιστημῶν αἱρετώταται, αὕτη δὲ τῶν θεωρητικῶν.

There must, then, be three theoretical philosophies, mathematics, natural science, and theology, since it is [20] obvious that if the divine is present anywhere, it is present in things of this sort. And the highest science must deal with the highest genus, so that the theoretical sciences are superior to the other sciences, and this to the other theoretical sciences.

§ 18—The subordination of the Unmoved Movers of Planetary Spheres to the Prime Mover

On Generation and Corruption, II, 10, 337a15–25

English Source: *On Generation and Corruption*, trans. H. H. Joachim in Barnes, *Complete Works of Aristotle* (1984).

[337a15] Διότι μὲν οὖν ἔστι γένεσις καὶ φθορὰ καὶ διὰ τίν' αἰτίαν, καὶ τί τὸ γενητὸν καὶ φθαρτόν, φανερὸν ἐκ τῶν εἰρημένων.

Ἐπεὶ δ' ἀνάγκη εἶναί τι τὸ κινοῦν εἰ κίνησις ἔσται, ὥσπερ εἴρηται πρότερον ἐν ἑτέροις, καὶ εἰ ἀεί, ὅτι ἀεί τι δεῖ εἶναι, καὶ εἰ συνεχής, ἓν τὸ αὐτὸ καὶ ἀκίνητον καὶ [20] ἀγένητον καὶ ἀναλλοίωτον, καὶ εἰ πλείους εἶεν αἱ κύκλῳ κινήσεις, πλείους μέν, πάσας δέ πως εἶναι ταύτας ἀνάγκη ὑπὸ μίαν ἀρχήν·

συνεχοῦς δ' ὄντος τοῦ χρόνου ἀνάγκη τὴν κίνησιν συνεχῆ εἶναι, εἴπερ ἀδύνατον χρόνον χωρὶς κινήσεως εἶναι· συνεχοῦς ἄρα τινὸς ἀριθμὸς ὁ χρόνος, τῆς κύκλῳ ἄρα, [25] καθάπερ ἐν τοῖς ἐν ἀρχῇ λόγοις διωρίσθη.

[337a15] It is clear from what been said that coming-to-be and passing-away actually occur, what causes them, and what subject undergoes them.

But if there is to be movement (as we have explained elsewhere, in an earlier work) there must be something which initiates it; if there is to be movement always, there must always be something which initiates it; if the movement is to be continuous, what initiates it must be single, unmoved, [20] ungenerated, and incapable of alteration; and if the circular movements are more than one, they must all of them, in spite of their plurality, be in some way subordinated to a single principle.

Further since time is continuous, movement must be continuous, inasmuch as there can be no time without movement. Time, therefore, is a number of some continuous movement—a number, therefore, of the circular movement, [25] as was established in the discussions at the beginning.

§ 19—The reasons why a single body has many motions in the case of planets and many bodies have a single motion in the case of the sphere of the fixed stars
On the Heavens, II, 12, 291b28–293a14
English Source: *On the Heavens*, trans. W. K. C. Guthrie (1939).

[291b28] Ἔστι δὲ πολλῶν ὄντων τοιούτων οὐχ ἥκιστα θαυμαστόν, διὰ τίνα ποτ᾽ αἰτίαν οὐκ [30] ἀεὶ τὰ πλεῖον ἀπέχοντα τῆς πρώτης φορᾶς κινεῖται πλείους κινήσεις, ἀλλὰ τὰ μεταξὺ πλείστας. Εὔλογον γὰρ ἂν δόξειεν εἶναι τοῦ πρώτου σώματος μίαν κινουμένου φορὰν τὸ πλησιαίτατον ἐλαχίστας κινεῖσθαι κινήσεις, οἷον δύο, τὸ δ᾽ ἐχόμενον τρεῖς ἤ τινα ἄλλην τοιαύτην τάξιν.

Νῦν δὲ συμβαίνει [35] τοὐναντίον· ἐλάττους γὰρ ἥλιος καὶ σελήνη κινοῦνται [292a] κινήσεις ἢ τῶν πλανωμένων ἄστρων ἔνια· καίτοι πορρώτερον τοῦ μέσου καὶ πλησιαίτερον τοῦ πρώτου σώματός εἰσιν αὐτῶν. Δῆλον δὲ τοῦτο περὶ ἐνίων καὶ τῇ ὄψει γέγονεν· τὴν γὰρ σελήνην ἑωράκαμεν διχότομον μὲν οὖσαν, ὑπελθοῦσαν [5] δὲ τῶν ἀστέρων τὸν τοῦ Ἄρεος, καὶ ἀποκρυφέντα μὲν κατὰ τὸ μέλαν αὐτῆς, ἐξελθόντα δὲ κατὰ τὸ φανὸν καὶ λαμπρόν. Ὁμοίως δὲ καὶ περὶ τοὺς ἄλλους ἀστέρας λέγουσιν οἱ πάλαι τετηρηκότες ἐκ πλείστων ἐτῶν Αἰγύπτιοι καὶ Βαβυλώνιοι, παρ᾽ ὧν πολλὰς πίστεις ἔχομεν περὶ ἑκάστου τῶν ἄστρων. [10] Τοῦτό τε δὴ δικαίως ἀπορήσειεν ἄν τις, καὶ διὰ τίνα ποτ᾽ αἰτίαν ἐν μὲν τῇ πρώτῃ φορᾷ τοσοῦτόν ἐστιν

[291b28] These obscurities are many, and one of the most incomprehensible is this: how can we explain the fact that [30] it is not the bodies farthest removed from the primary movement that have the most complex motions, but those which lie in between? Considering that the primary body has only one motion, it would seem natural for the nearest one to it to have a very small number, say two, and the next one three, or some similar proportionate arrangement.

But the opposite is true, [35] for the sun and moon perform simpler motions than some of the planets [292a], although the planets are farther from the centre and nearer the primary body, as has in certain cases actually been seen; for instance, the moon has been observed, when half-full, to approach [5] the planet Mars, which has then been blotted out behind the dark half of the moon, and come out again on the bright side. Similar observations about the other planets are recorded by the Egyptians and the Babylonians, who have watched the stars from the remotest past, and to whom we owe many incontrovertible facts about each of them.

[10] That is one question which it is proper to raise. Another is this: what can be the reason why the primary motion should include such a multitude of

ἄστρων πλῆθος ὥστε τῶν ἀναριθμήτων εἶναι δοκεῖν τὴν πᾶσαν τάξιν, τῶν δ' ἄλλων ἓν χωρὶς ἕκαστον, δύο δ' ἢ πλείω οὐ φαίνεται ἐν τῇ αὐτῇ ἐνδεδεμένα φορᾷ.

Περὶ δὴ τούτων ζητεῖν μὲν καλῶς ἔχει καὶ τὴν ἐπὶ πλεῖον σύνεσιν, καίπερ μικρὰς ἔχοντας ἀφορμὰς καὶ τοσαύτην ἀπόστασιν ἀπέχοντας τῶν περὶ αὐτὰ συμβαινόντων· ὅμως δ' ἐκ τῶν τοιούτων θεωροῦσιν οὐδὲν ἄλογον ἂν δόξειεν εἶναι τὸ νῦν ἀπορούμενον. Ἀλλ' ἡμεῖς ὡς περὶ σωμάτων αὐτῶν μόνον, καὶ μονάδων τάξιν μὲν ἐχόντων, ἀψύχων δὲ πάμπαν, διανοούμεθα· δεῖ δ' ὡς μετεχόντων ὑπολαμβάνειν πράξεως καὶ ζωῆς· οὕτω γὰρ οὐθὲν δόξει παράλογον εἶναι τὸ συμβαῖνον. Ἔοικε γὰρ τῷ μὲν ἄριστα ἔχοντι ὑπάρχειν τὸ εὖ ἄνευ πράξεως, τῷ δ' ἐγγύτατα διὰ ὀλίγης καὶ μιᾶς, τοῖς δὲ πορρωτέρω διὰ πλειόνων, ὥσπερ ἐπὶ σώματος τὸ μὲν οὐδὲ γυμναζόμενον εὖ ἔχει, τὸ δὲ μικρὰ περιπατῆσαν, τῷ δὲ καὶ δρόμου δεῖ καὶ πάλης καὶ κονίσεως, πάλιν δ' ἑτέρῳ οὐδ' ὁποσαοῦν πονοῦντι τοῦτό γ' ἂν ἔτι ὑπάρξαι τἀγαθόν, ἀλλ' ἕτερόν τι.

Ἔστι δὲ τὸ κατορθοῦν χαλεπὸν ἢ τὸ πολλὰ ἢ τὸ πολλάκις, οἷον μυρίους ἀστραγάλους Χίους βαλεῖν ἀμήχανον, ἀλλ' ἕνα ἢ δύο ῥᾷον. Καὶ πάλιν

stars that their whole array seems to be beyond counting, whereas each of the other motions involves one only, and we never see two or more caught in the same revolution?

These are questions on which it is worth while seeking boldly to extend our understanding. It is true that we have very little to start from, and that we are situated at a great distance from the phenomena that we are trying to investigate. Nevertheless if we base our inquiry on what we know, the present difficulty will not appear as anything inexplicable. The fact is that we are inclined to think of the stars as mere bodies or units, occurring in a certain order but completely lifeless; whereas we ought to think of them as partaking of life and initiative. Once we do this, the events will no longer seem surprising. It is reasonable for that which is in the best state to possess the good without taking action, for that which is nearest to the best to obtain it by means of little, or a single, action, and for those things which are farther from it to need more; in the same way as one body may be healthy without any exercise at all, another by means of a little walking, a third may need running and wrestling and violent exertion, and again a fourth despite tremendous efforts cannot preserve this particular good, but only something else.

To succeed in many things, or many times, is difficult; for instance, to repeat the same throw ten thousand times with the dice would be impossible, whereas to make it once or twice

ὅταν τοδὶ μὲν δέῃ τοῦδ᾽ ἕνεκα ποιῆσαι, τοῦτο δ᾽ ἄλλου καὶ τοῦτο ἑτέρου, ἐν μὲν ἑνὶ ἢ δυσὶ ῥᾴδιον ἐπιτυχεῖν, ὅσῳ δ᾽ ἂν διὰ πλειόνων, χαλεπώτερον.

[292b] Διὸ δεῖ νομίζειν καὶ τὴν τῶν ἄστρων πρᾶξιν εἶναι τοιαύτην οἵα περ ἡ τῶν ζῴων καὶ φυτῶν. Καὶ γὰρ ἐνταῦθα αἱ τοῦ ἀνθρώπου πλεῖσται πράξεις· πολλῶν γὰρ τῶν εὖ δύναται τυχεῖν, ὥστε πολλὰ πράττειν, καὶ ἄλλων ἕνεκα. (Τῷ [5] δ᾽ ὡς ἄριστα ἔχοντι οὐθὲν δεῖ πράξεως· ἔστι γὰρ αὐτὸ τὸ οὗ ἕνεκα, ἡ δὲ πρᾶξις ἀεί ἐστιν ἐν δυσίν, ὅταν καὶ οὗ ἕνεκα ᾖ καὶ τὸ τούτου ἕνεκα). Τῶν δ᾽ ἄλλων ζῴων ἐλάττους, τῶν δὲ φυτῶν μικρά τις καὶ μία ἴσως· ἢ γὰρ ἕν τί ἐστιν οὗ τύχοι ἄν, ὥσπερ καὶ ἄνθρωπος, ἢ καὶ τὰ πολλὰ πάντα πρὸ ὁδοῦ [10] ἐστι πρὸς τὸ ἄριστον. Τὸ μὲν οὖν ἔχει καὶ μετέχει τοῦ ἀρίστου, τὸ δ᾽ ἀφικνεῖται [ἐγγὺς] δι᾽ ὀλίγων, τὸ δὲ διὰ πολλῶν, τὸ δ᾽ οὐδ᾽ ἐγχειρεῖ, ἀλλ᾽ ἱκανὸν εἰς τὸ ἐγγὺς τοῦ ἐσχάτου ἐλθεῖν· οἷον εἰ ὑγίεια τέλος, τὸ μὲν δὴ ἀεὶ ὑγιαίνει, τὸ δ᾽ ἰσχνανθέν, τὸ δὲ δραμὸν καὶ ἰσχνανθέν, τὸ δὲ καὶ ἄλλο

is comparatively easy. We must consider too that whenever A must be done as a means to B, B as a means to C, and C as a means to some further end, then if the intermediate steps are one or two, it is easier to attain the end, but the more they are the more difficult it becomes.

[292b] With these considerations in mind, we must suppose the action of the planets to be analogous to that of animals and plants. For here on earth it is the actions of mankind that are the most varied, and the reason is that man has a variety of goods within his reach, wherefore his actions are many, and directed to ends outside themselves. That which is in the best possible state, [5] on the other hand, has no need of action. It is its own end, whereas action is always concerned with two factors, occurring when there is on the one hand an end proposed, and on the other the means towards that end. Yet the animals lower than man have less variety of action than he, and plants might be said to have one limited mode of action only; for either there is only one end for them to attain (as in truth there is for man also), or if there are many, yet they all conduce [10] directly to the best. To sum up, there is one thing which possesses, or shares in, the best, a second which reaches it immediately by few stages, a third which reaches it through many stages, and yet another which does not even attempt to reach it, but is content merely to approach near to the highest. For example, if health is the end, then one creature is always healthy, another by reducing, a third by running in order to

τι [15] πρᾶξαν τοῦ δραμεῖν ἕνεκα, ὥστε πλείους αἱ κινήσεις· ἕτερον δ' ἀδυνατεῖ πρὸς τὸ ὑγιᾶναι ἐλθεῖν, ἀλλὰ πρὸς τὸ δραμεῖν μόνον ἢ ἰσχνανθῆναι, καὶ τούτων θάτερον τέλος αὐτοῖς. Μάλιστα μὲν γὰρ ἐκείνου τυχεῖν ἄριστον πᾶσι τοῦ τέλους· εἰ δὲ μή, ἀεὶ ἄμεινόν ἐστιν ὅσῳ ἂν ἐγγύτερον ᾖ τοῦ ἀρίστου.

Καὶ [20] διὰ τοῦτο ἡ μὲν γῆ ὅλως οὐ κινεῖται, τὰ δ' ἐγγὺς ὀλίγας κινήσεις· οὐ γὰρ ἀφικνεῖται πρὸς τὸ ἔσχατον, ἀλλὰ μέχρι ὅτου δύναται τυχεῖν τῆς θειοτάτης ἀρχῆς. Ὁ δὲ πρῶτος οὐρανὸς εὐθὺς τυγχάνει διὰ μιᾶς κινήσεως. Τὰ δ' ἐν μέσῳ τοῦ πρώτου καὶ τῶν ἐσχάτων ἀφικνεῖται μέν, διὰ πλειόνων δ' ἀφικνεῖται κινήσεων. [25]

Περὶ δὲ τῆς ἀπορίας ὅτι κατὰ μὲν τὴν πρώτην μίαν οὖσαν φορὰν πολὺ πλῆθος συνέστηκεν ἄστρων, τῶν δ' ἄλλων χωρὶς ἕκαστον εἴληφεν ἰδίας κινήσεις, δι' ἓν μὲν ἄν τις πρῶτον εὐλόγως οἰηθείη τοῦθ' ὑπάρχειν· νοῆσαι γὰρ δεῖ τῆς ζωῆς καὶ τῆς ἀρχῆς ἑκάστης πολλὴν ὑπεροχὴν [30] εἶναι τῆς πρώτης πρὸς τὰς ἄλλας, εἴη δ' ἂν ἥδε συμβαίνουσα κατὰ λόγον· ἡ μὲν γὰρ πρώτη μία οὖσα πολλὰ κινεῖ τῶν σωμάτων τῶν θείων, αἱ δὲ πολλαὶ οὖσαι ἓν μόνον [293a] ἑκάστη· τῶν γὰρ πλανωμένων ἓν ὁτιοῦν πλείους φέρεται φοράς. Ταύτῃ τε οὖν ἀνισάζει ἡ φύσις καὶ ποιεῖ τινὰ τάξιν, τῇ μὲν μιᾷ φορᾷ πολλὰ

reduce, a fourth by doing something else to prepare itself for running, and so going through a larger number of motions: another creature [15] cannot attain to health, but only to running or reducing. To such creatures one of these latter is the end. To attain the ultimate end would be in the truest sense best for all; but if that is impossible, a thing gets better and better the nearer it is to the best.

[20] This then is the reason why the earth does not move at all, and the bodies near it have only few motions. They do not arrive at the highest, but reach only as far as it is within their power to obtain a share in the divine principle. But the first heaven reaches it immediately by one movement, and the stars that are between the first heaven and the bodies farthest from it reach it indeed, but reach it through a number of movements. [25]

Concerning the difficulty that in the primary movement, in spite of its uniqueness, a whole host of stars is involved, whereas each of the other stars has separate motions of its own, there is one thing which may be thought of first of all as supplying a satisfactory cause for this. In considering each of these living principles, we must bear in mind that the primary one has an immense superiority [30] over the rest, and this falls in with our argument. The primary principle, we say, though one, moves many of the divine bodies, but the others, which are many, move only [293a] one each, for any one of the planets moves with several motions. This then is Nature's way of equalizing things and introducing

ἀποδοῦσα σώματα, τῷ δ' ἑνὶ σώματι πολλὰς φοράς.

Καὶ ἔτι διὰ τόδε ἓν ἔχουσι σῶμα αἱ ἄλλαι φοραί, ὅτι πολλὰ σώματα κινοῦσιν αἱ πρὸ τῆς τελευταίας καὶ τῆς ἓν ἄστρον ἐχούσης· ἐν πολλαῖς γὰρ σφαίραις ἡ τελευταία σφαῖρα ἐνδεδεμένη φέρεται, ἑκάστη δὲ σφαῖρα σῶμά τι τυγχάνει ὄν. Ἐκείνης ἂν οὖν κοινὸν εἴη τὸ ἔργον· αὐτῇ μὲν γὰρ ἑκάστῃ ἡ ἴδιος φύσει φορά, αὕτη δὲ οἷον πρόσκειται, παντὸς δὲ πεπερασμένου σώματος πρὸς πεπερασμένον ἡ δύναμίς ἐστιν.

Ἀλλὰ περὶ μὲν τῶν τὴν ἐγκύκλιον φερομένων κίνησιν ἄστρων εἴρηται ποῖ' ἄττα κατά τε τὴν οὐσίαν ἐστὶ καὶ κατὰ τὸ σχῆμα, περί τε τῆς φορᾶς καὶ τῆς τάξεως αὐτῶν.

order, by assigning many bodies to the one motion, and to the one body many motions.

Here is a second reason why the other motions carry one body: the motions before the last one, which carries the one star, move many bodies, for the last sphere moves round embedded in a number of spheres, and each sphere is corporeal. The work of the last one, therefore, will be shared by the others. Each one has its own proper and natural motion, and this one is, as it were, added. But every limited body has limited powers.

Here we finish the subject of the revolving stars, their substance and their shape, their motion and their order.

Theophrastus and Plotinus's objections to the doctrine on the plurality of the Unmoved Movers

§ 20—Doubts (of Theophrastus) about the multi-substance theology of Aristotle
Theophrastus, *On First Principles*, I, 1–II, 8
English Source: Theophrastus *On First Principles* (known as his *Metaphysics*), ed. and trans. D. Gutas (2010). (Bracketed inserts in the English are Gutas's).

[I, 1] Πῶς ἀφορίσαι δεῖ καὶ ποίοις τὴν ὑπὲρ τῶν πρώτων θεωρίαν; ἡ γὰρ δὴ τῆς φύσεως πολυ χουστέρα, καὶ ὥς γε δή τινές φασιν, ἀτακτοτέρα, μεταβολὰς ἔχουσα παντοίας· ἡ δὲ τῶν πρώτων ὡρισμένη καὶ ἀεὶ κατὰ ταὐτά· διὸ δὴ καὶ ἐν

[I, 1] How and with what sort of [things] should one mark the boundaries of the study of the first [things]? For surely the [study] of nature is more multifarious and, at least as some actually say, more lacking in order, involving as it does all sorts of changes; but the [study] of the first [things] is bounded and always the same, for which reason, indeed, they even place it among the intelligibles

νοητοῖς, οὐκ αἰσθητοῖς, αὐτὴν τιθέασιν ὡς ἀκινήτοις καὶ ἀμεταβλήτοις, καὶ τὸ ὅλον δὲ σεμνοτέραν καὶ μείζω νομίζουσιν αὐτήν.

but not the sensibles, on the ground that the [intelligibles] are unmovable and unchangeable, and on the whole consider it more venerable and more important.

[2] Ἀρχὴ δέ, πότερα συναφή τις καὶ οἷον κοινωνία πρὸς ἄλληλα τοῖς τε νοητοῖς καὶ τοῖς τῆς φύσεως, ἢ οὐδεμία ἀλλ' ὥσπερ ἑκάτερα κεχωρισμένα συνεργοῦντα δέ πως εἰς τὴν πᾶσαν οὐσίαν. εὐλογώτερον δ' οὖν εἶναί τινα συναφὴν καὶ μὴ ἐπεισοδιῶδες τὸ πᾶν, ἀλλ' οἷον τὰ μὲν πρότερα τὰ δ' ὕστερα, καὶ ἀρχὰς τὰ δ' ὑπὸ τὰς ἀρχάς, ὥσπερ καὶ τὰ ἀίδια τῶν φθαρτῶν.

[2] The starting point is, whether [there is] some connection and something like a mutual association between intelligibles and the [things] of nature or [there is] none, but the two are, as it were, separated, though somehow both contributing to [bring about] all of existence. At any rate, it is more reasonable that there is some connection and that the universe is not episodic, but rather that the former are, as it were, prior and the latter posterior—and first principles, too, and the latter subordinate to the first principles—just as eternal [things] too are to the perishable.

[3] εἰ δ' οὖν οὕτω, τίς ἡ φύσις αὐτῶν καὶ ἐν ποίοις; εἰ μὲν γὰρ ἐν τοῖς μαθηματικοῖς μόνον τὰ νοητά, καθά πέρ | τινές φασιν, οὔτ' ἄγαν εὔσημος ἡ συναφὴ τοῖς αἰσθητοῖς, οὔθ' ὅλως ἀξιόχρεα φαίνεται παντός· οἷον γὰρ μεμηχανημένα δοκεῖ δι' ἡμῶν εἶναι σχήματά τε καὶ μορφὰς καὶ λόγους περιτιθέντων, αὐτὰ δὲ δι' αὐτῶν οὐδεμίαν ἔχειν φύσιν· εἰ δὲ μή, οὐχ οὕτως γε συνάπτειν τοῖς τῆς φύσεως ὥστ' ἐμποιῆσαι καθά περ ζωὴν καὶ κίνησιν αὐτοῖς· οὐδὲ γὰρ αὐτὸς ὁ ἀριθμός, ὅν περ δὴ πρῶτον καὶ κυριώτατόν τινες τιθέασιν.

[3] If so, then, what is their nature and among what sort of [things are they]? For if, on the one hand, [it is] among the mathematicals only that the intelligibles are, as some say, neither is [their] connection with the sensibles very conspicuous nor do they appear to be at all serviceable with regard to the universe. For they seem, as it were, to have been devised by us as figures, shapes, and proportions that we ascribe [to things], while they in themselves have no nature at all; or, if not, they are not able to have a connection with the [things] of nature that would produce in them something like life and motion—no, not even number itself, the very one which some people rank as first and most dominant.

[4] εἰ δ᾽ ἑτέρα τις οὐσία προτέρα καὶ κρείττων ἐστίν, ταύτην πειρατέον λέγειν, πότερον μία τις κατ᾽ ἀριθμὸν ἢ κατ᾽ εἶδος ἢ κατὰ γένος. εὐλογώτερον δ᾽ οὖν ἀρχῆς φύσιν ἐχούσαν ἐν ὀλίγοις εἶναι καὶ περιττοῖς, εἰ μὴ ἄρα καὶ πρώτοις καὶ ἐν τῷ πρώτῳ. τίς δ᾽ οὖν αὕτη καὶ τίνες, εἰ πλείους, πειρατέον ἐμφαίνειν ἁμῶς γέ πως εἴτε κατ᾽ ἀναλογίαν εἴτε κατ᾽ ἄλλην ὁμοίωσιν. ἀνάγκη δ᾽ ἴσως δυνάμει τινὶ καὶ ὑπεροχῇ τῶν ἄλλων λαμβάνειν, ὥσπερ ἂν εἰ τὸν θεόν· θεία γὰρ ἡ πάντων ἀρχή, δι᾽ ἧς ἅπαντα καὶ ἔστιν καὶ διαμένει. τάχα μὲν οὖν ῥᾴδιον τὸ οὕτως ἀποδοῦναι, χαλεπὸν δὲ σαφεστέρως ἢ πειστικωτέρως.

[4] But if, on the other hand, some other substance is prior and more powerful, one ought to try to say whether it is one such in number, or in species, or in genus. At any rate, it is more reasonable that, having the nature of a first principle, they are among few and extraordinary [things]—if not, indeed, even among the first [things] as well as in the first.

[4 cont. (Gutas 5)] What, at any rate, this is, or what they are, if they are more, one ought to try to reveal somehow or other, whether by analogy or by some other [procedure through] similarity. Perhaps it is necessary to apprehend [it or them] by means of some power and superiority over others, as if [we were apprehending] god, for divine is the first principle of all, through which all [things] both are and abide. And yes, maybe it is easy to provide an explanation in this manner—but more explicitly or more convincingly, difficult.

[5] Τοιαύτης δ᾽ οὔσης τῆς ἀρχῆς, ἐπεί περ συνάπτει τοῖς αἰσθητοῖς, ἡ δὲ φύσις ὡς ἁπλῶς εἰπεῖν ἐν κινήσει καὶ τοῦτ᾽ αὐτῆς τὸ ἴδιον, δῆλον ὡς αἰτίαν θετέον ταύτην τῆς κινήσεως· ἐπεὶ δ᾽ ἀκίνητος καθ᾽ αὑτήν, φανερὸν ὡς οὐκ ἂν εἴη τῷ κινεῖσθαι τοῖς τῆς φύσεως αἰτία, ἀλλὰ λοιπὸν ἄλλῃ τινὶ δυνάμει κρείττονι καὶ προτέρᾳ· τοιαύτη δ᾽ ἡ τοῦ ὀρεκτοῦ φύσις, ἀφ᾽ ἧς ἡ κυκλικὴ ἡ συνεχὴς καὶ ἄπαυστος. ὥστε κατ᾽ ἐκεῖνο λύοιτο ἂν τὸ μὴ εἶναι κινήσεως ἀρχὴν ἢ εἰ κινούμενον κινήσει.

[5 (Gutas 5.1)] Such being the first principle, then, since it is connected with sensibles, and nature is, to put it simply, in motion (which is the property unique to it), it is evident that it should be posited as the cause of movement. But since it is in itself motionless, it is obvious that it could not be a cause to the [things] of nature by being in motion but, as the remaining [alternative], by some other, superior and prior, power; and such is the nature of the desirable, from which [there springs] the circular, continuous and unceasing [motion]. And so on this [basis] also the difficulty] could be resolved that there can be no beginning of motion unless something in motion initiate it.

[6] Μέχρι μὲν δὴ τούτων οἷον ἄρτιος ὁ λόγος, ἀρχήν τε ποιῶν μίαν πάντων καὶ τὴν ἐνέργειαν καὶ τὴν οὐσίαν ἀποδιδούς, ἔτι δὲ μὴ διαιρετὸν μηδὲ ποσόν τι λέγων ἀλλ' ἁπλῶς ἐξαίρων εἰς κρείττω τινὰ μερίδα καὶ θειοτέραν· οὕτω γὰρ μᾶλλον ἀποδοτέον ἢ τὸ διαιρετὸν καὶ μεριστὸν ἀφαιρετέον. ἅμᾳ γὰρ ἐν ὑψηλοτέρῳ τε καὶ ἀληθινωτέρῳ λόγῳ τοῖς λέγουσιν ἡ ἀπόφασις.

[6] Now up to these [arguments], the account has all its parts in place, so to speak: it both posits a single first principle for all [things] and provides [its] actualised state and essence, and it further says that it is neither something divisible nor quantifiable, but exalts it in an absolute sense to some better and more divine rank (for it is better that one should provide such an account rather than that one should remove [from it] divisibility and partitionability, because those who make the negative statement [above make it] in an argument that is at the same time loftier and closer to truth);

[II, 7] Τὸ δὲ μετὰ ταῦτ' ἤδη λόγου δεῖται πλείονος περὶ τῆς ἐφέσεως, ποία καὶ τίνων, ἐπειδὴ πλείω τὰ κυκλικὰ καὶ αἱ φοραὶ τρόπον τινὰ ὑπεναντίαι, καὶ τὸ ἀνήνυτον καὶ οὗ χάριν ἀφανές. εἴτε γὰρ ἓν τὸ κινοῦν, ἄτοπον τὸ μὴ πάντα τὴν αὐτήν· εἴτε καθ' ἕκαστον ἕτερον αἵ τ' ἀρχαὶ πλείους, ὥστε τὸ σύμφωνον αὐτῶν εἰς ὄρεξιν ἰόντων τὴν ἀρίστην οὐθαμῶς φανερόν. [8] τὸ δὲ κατὰ τὸ πλῆθος τῶν σφαιρῶν τῆς αἰτίας μείζονα ζητεῖ λόγον· οὐ γὰρ <ἀρκεῖ> ὅ γε τῶν ἀστρολόγων. ἄπορον δὲ καὶ πῶς ποτε φυσικὴν ὄρεξιν ἐχόντων οὐ τὴν ἠρεμίαν διώκουσιν ἀλλὰ τὴν κίνησιν.

[II, 7] but what [comes] immediately after these [arguments] needs further discussion about the impulsion—of what kind and towards which [things] it is—because the rotating [bodies] are more than one and the[ir] motions are somehow opposed, while [their] interminableness and "[that] for the sake of which" are not apparent. For if the mover is one, it makes no sense that all not with the same [motion]; and if it is different for each and the first principles are more than one, [the consequence is such] that the concord of [the rotating bodies], as they proceed in fulfillment of the best desire, is by no means obvious. [8 (Gutas 7)] As for the [matter] of the great number of spheres, it calls for a fuller account of the reason [for it]; for the one [given] by the astronomers, at any rate, is not pertinent.

§ 21—The shortfall (according to Theophrastus) in the discourse of astrologers about the first movers
Theophrastus, *On First Principles*, IV, 27–28
English Source: Theophrastus *On First Principles* (known as his *Metaphysics*), ed. and trans. D. Gutas (2010). (Bracketed inserts in the English are Gutas's).

[IV] Ὅσοι <δὲ> τὸν οὐρανὸν ἀΐδιον
[27] ὑπολαμβάνουσιν, ἔτι δὲ τὰ κατὰ τὰς φορὰς καὶ τὰ μεγέθη καὶ τὰ σχήματα καὶ τὰς ἀποστάσεις καὶ ὅσα ἄλλα ἀστρολογία δείκνυσιν, τούτοις κατάλοιπον τά τε πρῶτα κινοῦντα καὶ τὸ τίνος ἕνεκα λέγειν καὶ τίς ἡ φύσις ἑκάστου καὶ ἡ πρὸς ἄλληλα θέσις καὶ ἡ τοῦ σύμπαντος οὐσία καὶ ὑποβαίνοντι δὴ πρὸς τὰ ἄλλα καθ' ἕκαστον τῶν εἰδῶν ἢ μερῶν ἄχρι ζῴων καὶ φυτῶν. εἰ οὖν ἀστρολογία συνεργεῖ μέν, οὐκ ἐν τοῖς πρώτοις δὲ τῆς φύσεως, ἕτερα τὰ κυριώτατ' ἂν εἴη καὶ πρότερα· καὶ γὰρ δὴ καὶ ὁ τρόπος, ὡς οἴονταί τινες, οὐ φυσικὸς ἢ οὐ πᾶς. καίτοι τό γε κινεῖσθαι καὶ ἁπλῶς τῆς φύσεως οἰκεῖον καὶ μάλιστα τοῦ οὐρανοῦ. διὸ καὶ εἰ ἐνέργεια τῆς οὐσίας ἑκάστου καὶ τὸ καθ' ἕκαστον ὅταν ἐνεργῇ καὶ κινεῖται, καθά περ ἐν τοῖς ζῴοις καὶ φυτοῖς (εἰ δὲ μή, ὁμώνυμα), δῆλον ὅτι κἂν ὁ οὐρανὸς

[IV, 27 (Gutas 20c2)] … those who assume the heavens to be eternal and, further, what has to do with the movements and the sizes and the figures and the distances and whatever else astronomy shows—for these [people] it remains to state both the first movers and "that for the sake of which," as well as what the nature of each is, the position of the one relative to the other, and the substance of the universe; and then, for someone progressing downwards to the rest [of the things, there remains to state the same] for each species or part individually down to animals and plants.

[(Gutas 21)] If, then, astronomy helps, but not with regard to the first [things], the principal [things] will be other than and prior to nature; and certainly also the method, as some think, is not that of nature, or not entirely. In fact, being in motion, at least, is proper both to nature in general and to the heavens in particular. Hence, (if) the actualised state of each [thing] is of its essence, and each individual [thing] is also in motion when in an actualised state, as in the case of animals and plants (otherwise [they would be animals and plants] in name [only]), then it is obvious that the heavens,

ἐν τῇ περιφορᾷ κατὰ τὴν οὐσίαν εἴη, χωριζόμενος δὲ καὶ ἠρεμῶν ὁμώνυμος· οἷον γὰρ ζωή τις ἡ περιφορὰ τοῦ παντός.

[28] Ἀρ' οὖν εἴ γε μηδ' ἐν τοῖς ζῴοις τὴν ζωὴν ἢ ὡδὶ ζητητέον, οὐδ' ἐν τῷ οὐρανῷ καὶ τοῖς οὐρανίοις τὴν φορὰν ἢ τρόπον τινὰ ἀφωρισμένον; συνάπτει δέ πως ἡ νῦν ἀπορία καὶ πρὸς τὴν ὑπὸ τοῦ ἀκινήτου κίνησιν.

too, in [their] rotation, would be in accordance with their essence, but when divorced [from movement] and at rest, [they would be the heavens] in name [only]—for the rotation of the universe is, as it were, a kind of life.

[28] So I wonder, if even in the case of animals life is not to be investigated except in this way, isn't then also movement, in the case of the heavens and the heavenly [bodies, not to be investigated] except in some manner whose boundaries have been marked? The present aporia is in a way also connected with the movement [caused] by the unmoved.

§ 22—Criticism (of Plotinus) to the impossibility of individuation (due to the absence of matter) of Aristotle's Unmoved Movers
Plotinus, *Enneads*, V, 1, 8–9
English Source: Plotinus, *Enneads*, V. 1–9, trans. A. H. Armstrong (1984)

[V, 1, 8] Ἥπτετο μὲν οὖν καὶ Παρμενίδης πρότερον τῆς τοιαύτης δόξης καθόσον εἰς ταὐτὸ συνῆγεν ὂν καὶ νοῦν, καὶ τὸ ὂν οὐκ ἐν τοῖς αἰσθητοῖς ἐτίθετο «<τὸ γὰρ αὐτὸ νοεῖν ἐστί τε καὶ εἶναι>» λέγων. καὶ <ἀκίνητον> δὲ λέγει τοῦτο—καίτοι προστιθεὶς τὸ νοεῖν—σωματικὴν πᾶσαν κίνησιν ἐξαίρων ἀπ' αὐτοῦ, ἵνα μένῃ ὡσαύτως, καὶ <ὄγκῳ σφαίρας> ἀπεικάζων, ὅτι πάντα ἔχει περιειλημμένα καὶ ὅτι τὸ νοεῖν οὐκ ἔξω, ἀλλ' ἐν ἑαυτῷ. ἓν δὲ λέγων ἐν τοῖς ἑαυτοῦ συγγράμμασιν αἰτίαν εἶχεν ὡς τοῦ ἑνὸς τούτου πολλὰ εὑρισκομένου. ὁ δὲ παρὰ Πλάτωνι Παρμενίδης ἀκριβέστερον λέγων διαιρεῖ ἀπ'

[V, 1, 8] And Parmenides also, before Plato, touched on a view like this, in that he identified Being and Intellect and that it was not among things perceived by the senses that he placed Being, when he said "Thinking and Being are the same." And he says that this Being is unmoved—though he does attach thinking to it—taking all bodily movement from it that it may remain always in the same state, and likening it to "the mass of a sphere," because it holds all things in its circumference and because its thinking is not external, but in itself. But when he said it was one, in his own works, he was open to criticism because this one of his was discovered to be many. But Parmenides in Plato speaks more accurately, and distinguishes from

ἀλλήλων τὸ πρῶτον ἕν, ὃ κυριώτερον ἕν, καὶ δεύτερον <ἓν πολλά> λέγων, καὶ τρίτον <ἓν καὶ πολλά>. καὶ σύμφωνος οὕτως καὶ αὐτός ἐστι ταῖς φύσεσι ταῖς τρισίν.

[V, 1, 9] Ἀναξαγόρας δὲ νοῦν καθαρὸν καὶ ἀμιγῆ λέγων ἁπλοῦν καὶ αὐτὸς τίθεται τὸ πρῶτον καὶ χωριστὸν τὸ ἕν, τὸ δ' ἀκριβὲς δι' ἀρχαιότητα παρῆκε. καὶ Ἡράκλειτος δὲ τὸ ἓν οἶδεν ἀίδιον καὶ νοητόν· τὰ γὰρ σώματα γίγνεται ἀεὶ καὶ ῥέοντα. τῷ δὲ Ἐμπεδοκλεῖ τὸ <νεῖκος> μὲν διαιρεῖ, ἡ δὲ <φιλία> τὸ ἕν—ἀσώματον δὲ καὶ αὐτὸς τοῦτο—τὰ δὲ στοιχεῖα ὡς ὕλη.

Ἀριστοτέλης δὲ ὕστερον <χωριστὸν> μὲν τὸ πρῶτον καὶ <νοητόν, νοεῖν> δὲ αὐτὸ <ἑαυτὸ> λέγων πάλιν αὖ οὐ τὸ πρῶτον ποιεῖ· πολλὰ δὲ καὶ ἄλλα νοητὰ ποιῶν καὶ τοσαῦτα, ὁπόσαι ἐν οὐρανῷ σφαῖραι, ἵν' ἕκαστον ἑκάστην κινῇ, ἄλλον τρόπον λέγει τὰ ἐν τοῖς νοητοῖς ἢ Πλάτων, τὸ εὔλογον οὐκ ἔχον ἀνάγκην τιθέμενος. ἐπιστήσειε δ' ἄν τις, εἰ καὶ εὐλόγως· εὐλογώτερον γὰρ πάσας πρὸς μίαν σύνταξιν συντελούσας πρὸς ἓν καὶ τὸ πρῶτον βλέπειν. ζητήσειε δ' ἄν τις τὰ πολλὰ νοητὰ εἰ ἐξ ἑνός ἐστιν αὐτῷ τοῦ πρώτου, ἢ πολλαὶ αἱ ἐν τοῖς νοητοῖς ἀρχαί· καὶ εἰ μὲν ἐξ ἑνός, ἀνάλογον δηλονότι ἕξει ὡς ἐν τοῖς αἰσθητοῖς αἱ

each other the first One, which is more properly called One, and the second which he calls "One-Many" and the third, "One and Many". In this way he too agrees with the doctrine of the three natures.

[V, 1, 9] And Anaxagoras also, when he says that Intellect is pure and unmixed, posits that the first principle is simple and the that One is separate, but he neglects to give an accurate account because of his antiquity. Heraclitus also knows that the One is eternal and intelligible: for bodies are always coming into being and flowing away. And for Empedocles Strife divides, but Love is the One—he too makes it incorporeal—and the elements serve as matter.

Later, Aristotle makes the first principle separate and intelligible, but when he says that it knows itself, he goes back again and does not make it the first principle; and by making many other intelligible realities, as many as the heavenly spheres, that each particular intelligible may move one particular sphere, he describes the intelligible world in a different way from Plato, making a probable assumption which has no philosophical necessity. But one might doubt whether it is even probable: for it would be more probable that all the spheres, contributing their several movements to a single system, should look to one principle, the first. And one might enquire whether Aristotle thinks that the many intelligibles derive from one, the first, or whether there are many primary principles in the intelligible world; and if they derive from one, the situation will clearly be analogous to that of

σφαῖραι ἄλλης ἄλλην περιεχούσης, μιᾶς δὲ τῆς ἔξω κρατούσης· ὥστε περιέχοι ἂν κἀκεῖ τὸ πρῶτον καὶ κόσμος [20] νοητὸς ἔσται· καὶ ὥσπερ ἐνταῦθα αἱ σφαῖραι οὐ κεναί, ἀλλὰ μεστὴ ἄστρων ἡ πρώτη, αἱ δὲ ἔχουσιν ἄστρα, οὕτω κἀκεῖ τὰ κινοῦντα πολλὰ ἐν αὐτοῖς ἕξει καὶ τὰ ἀληθέστερα ἐκεῖ. εἰ δὲ ἕκαστον ἀρχή, κατὰ συντυχίαν αἱ ἀρχαὶ ἔσονται· καὶ διὰ τί συνέσονται καὶ πρὸς ἓν ἔργον τὴν τοῦ παντὸς [25] οὐρανοῦ συμφωνίαν ὁμονοήσει; πῶς δὲ ἴσα πρὸς τὰ νοητὰ καὶ κινοῦντα τὰ ἐν οὐρανῷ αἰσθητά; πῶς δὲ καὶ πολλὰ οὕτως ἀσώματα ὄντα ὕλης οὐ χωριζούσης;

the heavenly spheres in the sense-world, where each contains the other and one, the outermost, dominates; so that there too the first would contain the others and there will be an intelligible [20] universe; and, just as here in the sense-world the spheres are not empty, but the first is full of heavenly bodies and the others have heavenly bodies in them, so there also the moving principles will have many realities in them, and the realities there will be truer. But if each is primary principle, the primary principles will be a random assembly; and why will they be a community and in agreement on one work, [25] the harmony of the whole universe? And how can the perceptible beings in heaven be equal in number to the intelligible movers? And how can the intelligible even be many, when they are incorporeal, as they are, and matter does not divide them?

The Astronomical System of Aristotle

§ 23—The influence of the superlunary on the sublunary
Meteorology, I, 2, 339a19–339a32
English Source: *Meteorology*, trans. E.W. Webster in Barnes, *Complete Works of Aristotle* (1984).

[339a19] Ὁ δὴ περὶ τὴν γῆν [20] ὅλος κόσμος ἐκ τούτων συνέστηκε τῶν σωμάτων· περὶ οὗ τὰ συμβαίνοντα πάθη φαμὲν εἶναι ληπτέον. ἔστιν δ' ἐξ ἀνάγκης συνεχὴς οὗτος ταῖς ἄνω φοραῖς, ὥστε πᾶσαν αὐτοῦ τὴν δύναμιν κυβερνᾶσθαι ἐκεῖθεν· ὅθεν γὰρ ἡ τῆς κινήσεως ἀρχὴ πᾶσιν, ἐκείνην αἰτίαν νομιστέον πρώτην. πρὸς δὲ τούτοις [25] ἡ μὲν ἀίδιος καὶ τέλος

[339a19] The whole world surrounding the earth, [20] then, the affections of which are our subject, is made up of these bodies. This world necessarily has a certain continuity with the upper motions; consequently all its power is derived from them. (For the originating principle of all motion must be deemed the first cause. Besides, [25] that element is eternal

οὐκ ἔχουσα τῷ τόπῳ τῆς κινήσεως, ἀλλ' ἀεὶ ἐν τέλει· ταῦτα δὲ τὰ σώματα πάντα πεπερασμένους διέστηκε τόπους ἀλλήλων. ὥστε τῶν συμβαινόντων περὶ αὐτὸν πῦρ μὲν καὶ γῆν καὶ τὰ συγγενῆ τούτοις ὡς ἐν ὕλης εἴδει τῶν γιγνομένων αἴτια χρὴ νομίζειν (τὸ γὰρ ὑποκείμενον καὶ πάσχον τοῦτον προσαγορεύομεν τὸν τρόπον), τὸ δ' οὕτως αἴτιον ὅθεν ἡ τῆς κινήσεως ἀρχή, τὴν τῶν ἀεὶ κινουμένων αἰτιατέον δύναμιν.

and its motion has no limit in space, but is always complete; whereas all these other bodies have separate regions which limit one another.) So we must treat fire and earth and the elements like them as the material causes of the events in this world (meaning by material what is subject and is affected), but must assign causality in the sense of the originating principle of motion to the power of the eternally moving bodies.

§ 24—The simultaneity and contiguity in the mover-moved series
Physics, VII, 2, 243a32–243a40; 243a11–243a18*
English Source: *Physics*, trans. R. P. Hardie and R. K. Gaye in Barnes, *Complete Works of Aristotle* (1984).

[243a32] Τὸ δὲ πρῶτον κινοῦν, μὴ ὡς τὸ οὗ ἕνεκεν, ἀλλ' ὅθεν ἡ ἀρχὴ τῆς κινήσεως, ἅμα τῷ κινουμένῳ ἐστί (λέγω δὲ τὸ ἅμα, ὅτι οὐδέν ἐστιν αὐτῶν μεταξύ)· τοῦτο γὰρ κοινὸν ἐπὶ παντὸς κινουμένου καὶ κινοῦντός ἐστιν. ἐπεὶ δὲ τρεῖς αἱ κινήσεις, ἥ τε κατὰ τόπον καὶ ἡ κατὰ τὸ ποιὸν καὶ ἡ κατὰ τὸ ποσόν, ἀνάγκη καὶ τὰ κινοῦντα τρία εἶναι, τό τε φέρον καὶ τὸ ἀλλοιοῦν καὶ τὸ αὖξον ἢ φθῖνον. πρῶτον οὖν εἴπωμεν περὶ τῆς φορᾶς· πρώτη γὰρ αὕτη τῶν κινήσεων.

[243a32] That which is the first mover of a thing—in the sense that it supplies not that for the sake of which but the source of the motion—is always together with that which is moved by it (by 'together' I mean that there is nothing between them). This is universally true wherever one thing is moved by another. And since there are three kinds of motion, local, qualitative, and quantitative, there must also be three kinds of mover, that which causes locomotion, that which causes alteration, and that which causes increase or decrease. Let us begin with locomotion, for this is the primary motion.

* The main critical editions of ΤΗΣ ΦΥΣΙΚΗΣ ΑΚΡΟΑΣΙΣ reflect the apparent inconsistency in the Bekkerian numbering we reproduce here. The reason for this inconsistency is that the Immanuel Bekker's edition (*Aristoteles graece ex recensione Immanuelis Bekkeri*, edidit Academia Regia Borussica, Berolini, apud Georgium Reimerum, 1831, vol. I) was improved, in the order of the texts by W. D. Ross (*Aristotelis Physica*, Oxford, The Clarendon Press, 1950), which collates specific manuscripts of two different traditions of the *Physics* text: Simplicius and the so-called τὸ ἕτερον.

[243a11] Ἅπαν δὴ τὸ φερόμενον ἢ ὑφ' ἑαυτοῦ κινεῖται ἢ ὑπ' ἄλλου. ὅσα μὲν οὖν αὐτὰ ὑφ' αὑτῶν κινεῖται, φανερὸν ἐν τούτοις ὅτι ἅμα τὸ κινούμενον καὶ τὸ κινοῦν ἐστιν· ἐνυπάρχει γὰρ αὐτοῖς τὸ πρῶτον κινοῦν, ὥστ' [15] οὐδέν ἐστιν ἀναμεταξύ· ὅσα δ' ὑπ' ἄλλου κινεῖται, τετραχῶς ἀνάγκη γίγνεσθαι· τέτταρα γὰρ εἴδη τῆς ὑπ' ἄλλου φορᾶς, ἕλξις, ὦσις, ὄχησις, δίνησις. ἅπασαι γὰρ αἱ κατὰ τόπον κινήσεις ἀνάγονται εἰς ταύτας·

[243a11] Everything that is in locomotion is moved either by itself or by something else. In the case of things that are moved by themselves it is evident that the moved and the mover are together; for they contain within themselves their first mover, so that there is nothing in between. The motion of things that are moved by something else must proceed in one of four ways; for there are four kinds of locomotion caused by something other than that which is in motion, viz. pulling, pushing, carrying, and twirling. All forms of locomotion are reducible to these.

§ 25—The force produced by the attraction of the Prime Mover on the circumference of the sphere (rather than on its axis)
Physics, VIII, 10, 267a21–267b9
English Source: *Physics*, trans. R. P. Hardie and R. K. Gaye in Barnes, *Complete Works of Aristotle* (1984).

[267a21] Ἐπεὶ δ' ἐν τοῖς οὖσιν ἀνάγκη κίνησιν εἶναι συνεχῆ, αὕτη δὲ μία ἐστίν, ἀνάγκη δὲ τὴν μίαν μεγέθους τέ τινος εἶναι (οὐ γὰρ κινεῖται τὸ ἀμέγεθες) καὶ ἑνὸς καὶ ὑφ' ἑνός (οὐ γὰρ ἔσται συνεχής, ἀλλ' ἐχομένη ἑτέρα ἑτέρας καὶ διῃρημένη), τὸ δὴ [25] κινοῦν εἰ ἕν, ἢ κινούμενον κινεῖ ἢ ἀκίνητον ὄν. εἰ μὲν δὴ κινούμενον, συνακολουθεῖν δεήσει καὶ μεταβάλλειν αὐτό, ἅμα δὲ [267b] κινεῖσθαι ὑπό τινος, ὥστε στήσεται καὶ ἥξει εἰς τὸ κινεῖσθαι ὑπὸ ἀκινήτου. τοῦτο γὰρ

[267a21] Since there must be continuous motion in the world of things, and this is a single motion, and a single motion must be a motion of a magnitude (for that which is without magnitude cannot be in motion), and of a single magnitude moved by a single mover (for otherwise there will not be continuous motion but a consecutive series of separate motions), then if the [25] mover is a single thing, it is either in motion or unmoved: if, then, it is in motion, it will have to keep pace with that which it moves and itself be in process of change, [267b] and it will also have to be moved by something: so we have a series that must come to an end, and a point will be reached at which motion is imparted by something that is

οὐκ ἀνάγκη συμμεταβάλλειν, ἀλλ' ἀεί τε δυνήσεται κινεῖν (ἄπονον γὰρ τὸ οὕτω κινεῖν) καὶ ὁμαλὴς αὕτη ἡ κίνησις ἢ μόνη ἢ μάλιστα· οὐ γὰρ ἔχει μεταβολὴν τὸ κινοῦν οὐδεμίαν. δεῖ δὲ οὐδὲ τὸ κινούμενον πρὸς ἐκεῖνο ἔχειν μεταβολήν, ἵνα ὁμοία ᾖ ἡ κίνησις. ἀνάγκη δὴ ἢ ἐν μέσῳ ἢ ἐν κύκλῳ εἶναι· αὗται γὰρ αἱ ἀρχαί. ἀλλὰ τάχιστα κινεῖται τὰ ἐγγύτατα τοῦ κινοῦντος. τοιαύτη δ' ἡ τοῦ κύκλου κίνησις· ἐκεῖ ἄρα τὸ κινοῦν.

unmoved. Thus we have a mover that has no need to change along with that which it moves but will be able to cause motion always (for the causing of motion under these conditions involves no effort); and this motion alone is regular, or at least it is so in a higher degree than any other, since the mover is never subject to any change. So, too, in order that the motion may continue to be of the same character, the moved must not be subject to change in relation to it. So it must occupy either the centre or the circumference, since these are the principles. But the things nearest the mover are those whose motion is quickest, and in this case it is the motion of the circumference that is the quickest: therefore the mover occupies the circumference.

§ 26—The simultaneous pulling and pushing (on the surface of a sphere) in the rotation produced by the Unmoved Movers
Physics, VII, 2, 244a2–6
English Source: *Physics*, trans. R. P. Hardie and R. K. Gaye in Barnes, *Complete Works of Aristotle* (1984).

[244a2] Ἡ δὲ δίνησις σύγκειται ἐξ ἕλξεώς τε καὶ ὤσεως· ἀνάγκη γὰρ τὸ δινοῦν τὸ μὲν ἕλκειν τὸ δ' ὠθεῖν· τὸ μὲν γὰρ ἀφ' αὑτοῦ τὸ δὲ πρὸς αὑτὸ ἄγει. ὥστ' εἰ τὸ ὠθοῦν καὶ τὸ ἕλκον ἅμα τῷ ὠθουμένῳ καὶ τῷ ἑλκομένῳ, φανερὸν ὅτι τοῦ κατὰ τόπον κινουμένου καὶ κινοῦντος οὐδέν ἐστι μεταξύ.

[244a2] And twirling is a compound of pulling and pushing; for that which is twirling a thing must be pulling one part of the thing and pushing another part, since it impels one part away from itself and another part towards itself. If, therefore, it can be shown that that which is pushing and that which is pulling are together with that which is being pushed and that which is being pulled, it will be evident that in all locomotion there is nothing between moved and mover.

§ 27—The lineal velocity increase (from centre to periphery) in the circular motion of the spheres that carry the stars

On the Heavens, II, 8, 289b1–290a7

English Source: *On the Heavens*, trans. W. K. C. Guthrie (1939).

[289b1] Ἐπεὶ δὲ φαίνεται καὶ τὰ ἄστρα μεθιστάμενα καὶ ὅλος ὁ οὐρανός, ἀναγκαῖον ἤτοι ἠρεμούντων ἀμφοτέρων γίγνεσθαι τὴν μεταβολήν, ἢ κινουμένων, ἢ τοῦ μὲν ἠρεμοῦντος τοῦ δὲ κινουμένου.

Ἀμφότερα μὲν τοίνυν ἠρεμεῖν ἀδύνατον ἠρεμούσης [5] γε τῆς γῆς· οὐ γὰρ ἂν ἐγίγνετο τὰ φαινόμενα. Τὴν δὲ γῆν ὑποκείσθω ἠρεμεῖν. Λείπεται δὴ ἀμφότερα κινεῖσθαι, ἢ τὸ μὲν κινεῖσθαι τὸ δ' ἠρεμεῖν.

Εἰ μὲν οὖν ἀμφότερα κινήσεται, ἄλογον τὸ ταὐτὰ τάχη τῶν ἄστρων εἶναι καὶ τῶν κύκλων· ἕκαστον γὰρ δὴ ὁμοταχὲς ἔσται τῷ κύκλῳ καθ' ὃν φέρεται. [10] Φαίνεται γὰρ ἅμα τοῖς κύκλοις καθιστάμενα πάλιν εἰς τὸ αὐτό. Συμβαίνει οὖν ἅμα τό τε ἄστρον διεληλυθέναι τὸν κύκλον καὶ τὸν κύκλον ἐνηνέχθαι τὴν αὑτοῦ φοράν, διεληλυθότα τὴν περιφέρειαν. Οὐκ ἔστι δ' εὔλογον τὸ τὸν αὐτὸν λόγον ἔχειν τὰ τάχη τῶν ἄστρων καὶ τὰ μεγέθη τῶν [15] κύκλων. Τοὺς μὲν γὰρ κύκλους οὐθὲν ἄτοπον ἀλλ' ἀναγκαῖον ἀνάλογον ἔχειν τὰ τάχη τοῖς μεγέθεσι, τῶν δ' ἄστρων ἕκαστον τῶν ἐν τούτοις οὐθαμῶς εὔλογον· εἴτε γὰρ ἐξ ἀνάγκης τὸ τὸν μείζω κύκλον φερόμενον θᾶττον ἔσται, δῆλον ὅτι

[289b1] Change is apparent in the position both of the stars and of the whole heaven, and this change must be reconciled with one of three possibilities. Either both are at rest, or both are in motion, or one is at rest and the other in motion.

For both to be at rest is impossible, [5] if the earth is at rest, for that would not produce the phenomena; and the immobility of the earth shall be our hypothesis. There remain the alternatives that both move or that one moves and the other is at rest.

If both move, we have the improbable result that the speeds of the stars and the circles are the same, for each star would then have the same speed as the circle in which it moves, [10] seeing that they may be observed to return to the same spot simultaneously with the circles. This means that at the same moment the star has traversed the circle and the circle has completed its own revolution, having traversed its own circumference. But it is not reasonable to suppose that the speeds of the stars are related to one another as the size of [15] their circles. That the circles should have their speeds proportional to their magnitudes is no absurdity, indeed it is a necessity, but that each of the stars in them should show the same proportion is not reasonable. If it is by necessity that the one which moves in the path of the larger circle is the swifter,

κἂν μετατεθῇ τὰ ἄστρα εἰς τοὺς ἀλλήλων κύκλους, τὸ μὲν ἔσται θᾶττον τὸ δὲ βραδύτερον (οὕτω δ' οὐκ ἂν ἔχοιεν οἰκείαν κίνησιν, ἀλλὰ φέροιντ' ἂν ὑπὸ τῶν κύκλων), εἴτε ἀπὸ ταὐτομάτου συνέπεσεν, οὐδ' οὕτως εὔλογον ὥστ' ἐν ἅπασιν ἅμα τόν τε κύκλον εἶναι μείζω καὶ τὴν φορὰν θάττω τοῦ ἐν αὐτῷ ἄστρου· τὸ μὲν γὰρ ἓν ἢ δύο τοῦτον τὸν τρόπον ἔχειν οὐθὲν ἄτοπον, τὸ δὲ πάνθ' ὁμοίως πλάσματι ἔοικεν. Ἅμα δὲ καὶ οὐκ ἔστιν ἐν τοῖς φύσει τὸ ὡς ἔτυχεν, οὐδὲ τὸ πανταχοῦ καὶ πᾶσιν ὑπάρχον τὸ ἀπὸ τύχης.

Ἀλλὰ μὴν πάλιν εἰ οἱ μὲν κύκλοι μένουσιν, αὐτὰ δὲ τὰ ἄστρα κινεῖται, τὰ αὐτὰ καὶ ὁμοίως ἔσται ἄλογα· συμβήσεται γὰρ θᾶττον κινεῖσθαι τὰ ἔξω, καὶ τὰ τάχη εἶναι κατὰ τὰ μεγέθη τῶν κύκλων. Ἐπεὶ τοίνυν οὔτ' ἀμφότερα κινεῖσθαι εὔλογον οὔτε τὸ ἕτερον μόνον, λείπεται τοὺς μὲν κύκλους κινεῖσθαι, τὰ δὲ ἄστρα ἠρεμεῖν καὶ ἐνδεδεμένα τοῖς κύκλοις φέρεσθαι· μόνως γὰρ οὕτως οὐθὲν ἄλογον συμβαίνει· τό τε γὰρ θᾶττον εἶναι τοῦ μείζονος κύκλου τὸ τάχος εὔλογον περὶ τὸ αὐτὸ κέντρον ἐνδεδεμένων [290a] (ὥσπερ γὰρ ἐν τοῖς ἄλλοις τὸ μεῖζον σῶμα θᾶττον φέρεται τὴν οἰκείαν φοράν, οὕτως καὶ ἐν τοῖς

then it is clear that even if the stars were transposed into each others' circles, still the one in the larger circle would be swifter, and the other slower; but in that case they would possess no motion of their own, but be carried by the circles. If on the other hand it has happened by chance, yet it is equally unlikely that chance should act so that in every case the larger circle is accompanied by a swifter movement of the star in it. That one or two should show this correspondence is conceivable, but that it should be universal seems fantastic. In any case, chance is excluded from natural events, and whatever applies everywhere and to all cases is not to be ascribed to chance.

But again, if the circles are at rest and the stars move by themselves, the same absurdity arises and in the same way: for the effect will be that the stars which he far out will move faster, and the speeds will correspond to the size of the circles.

Since then neither the motion of both nor the motion of the star alone can be defended, we are left with the conclusion that the circles move and that the stars stay still and are carried along because fixed in the circles. This is the only hypothesis that does not lead to an absurdity. That the larger circle should have the higher speed is reasonable, seeing that the stars are dotted around one and the same centre. [290a] Among the other elements, the larger a body the more swiftly it performed its proper motion, and the same is true of the bodies whose motion is circular. If arcs are

ἐγκυκλίοις· μεῖζον γὰρ τῶν ἀφαιρουμένων ὑπὸ τῶν ἐκ τοῦ κέντρου τὸ τοῦ μείζονος κύκλου τμῆμα, ὥστ' εὐλόγως ἐν τῷ ἴσῳ χρόνῳ ὁ μείζων περιοισθήσεται κύκλος), τό τε μὴ διασπᾶσθαι τὸν οὐρανὸν διά τε τοῦτο συμβήσεται καὶ ὅτι δέδεικται συνεχὲς ὂν τὸ ὅλον.

cut off by lines radiating from the centre, that of the larger circle will be larger, and it is natural therefore that the larger circle should take the same time as the others to revolve. This too is one reason why the heaven does not spring apart, and another is that the whole has been demonstrated to be continuous.

§ 28—The influence of the motion of the first heaven on the zodiacal revolution of the planets

On the Heavens, II, 10, 291a32–291b10

English Source: *On the Heavens*, trans. W. K. C. Guthrie (1939).

[291a32] Συμβαίνει δὲ κατὰ λόγον γίγνεσθαι τὰς ἑκάστου κινήσεις τοῖς ἀποστήμασι τῷ τὰς μὲν εἶναι θάττους τὰς δὲ βραδυτέρας· ἐπεὶ γὰρ ὑπόκειται τὴν μὲν ἐσχάτην τοῦ οὐρανοῦ περιφορὰν ἁπλῆν τ' εἶναι [291b] καὶ ταχίστην, τὰς δὲ τῶν ἄλλων βραδυτέρας τε καὶ πλείους (ἕκαστον γὰρ ἀντιφέρεται τῷ οὐρανῷ κατὰ τὸν αὑτοῦ κύκλον), εὔλογον ἤδη τὸ μὲν ἐγγυτάτω τῆς ἁπλῆς καὶ πρώτης περιφορᾶς ἐν πλείστῳ χρόνῳ διιέναι τὸν αὑτοῦ κύκλον, τὸ δὲ πορρωτάτω ἐν ἐλαχίστῳ, τῶν δ' ἄλλων τὸ ἐγγύτερον ἀεὶ ἐν πλείονι, τὸ δὲ πορρώτερον ἐν ἐλάττονι. Τὸ μὲν γὰρ ἐγγυτάτω μάλιστα κρατεῖται, τὸ δὲ πορρωτάτω πάντων ἥκιστα διὰ τὴν ἀπόστασιν· τὰ δὲ μεταξὺ κατὰ λόγον ἤδη τῆς ἀποστάσεως, ὥσπερ καὶ δεικνύουσιν οἱ μαθηματικοί.

[291a32] One characteristic is that their movements are faster or slower according to their distances. That is, once it has been admitted that the outermost revolution of the heaven is simple [291a] and is the swiftest of all, whereas that of the inner spheres is slower and composite (for each in performing its own revolution is going against the motion of the heaven), then it becomes natural for the star nearest to the simple and primary revolution to complete its own circle in the longest time, and the one farthest away in the shortest, and so with the others—the nearer in a longer time, the farther in a shorter. This is because the nearest one is most strongly counteracted by the primary motion, and the farthest least, owing to its distance. The others are influenced in proportion to their distances, and how this works out is demonstrated by the mathematicians.

§ 29—The influence of ecliptic obliquity on generation and corruption processes
On Generation and Corruption, II, 10, 336a23–336b14
English Source: *On Generation and Corruption*, trans. H. H. Joachim in Barnes, *Complete Works of Aristotle* (1984).

[336a23] Ἐπεὶ δ' ὑπόκειται καὶ δέδεικται συνεχὴς οὖσα τοῖς πράγμασι καὶ [25] γένεσις καὶ φθορά, φαμὲν δ' αἰτίαν εἶναι τὴν φορὰν τοῦ γίνεσθαι, φανερὸν ὅτι μιᾶς μὲν οὔσης τῆς φορᾶς οὐκ ἐνδέχεται γίνεσθαι ἄμφω διὰ τὸ ἐναντία εἶναι· τὸ γὰρ αὐτὸ καὶ ὡσαύτως ἔχον ἀεὶ τὸ αὐτὸ πέφυκε ποιεῖν. Ὥστε ἤτοι γένεσις ἀεὶ ἔσται ἢ φθορά.

Δεῖ <δὴ> πλείους εἶναι τὰς [30] κινήσεις καὶ ἐναντίας, ἢ τῇ φορᾷ ἢ τῇ ἀνωμαλίᾳ· τῶν γὰρ ἐναντίων τἀναντία αἴτια· διὸ καὶ οὐχ ἡ πρώτη φορὰ αἰτία ἐστὶ γενέσεως καὶ φθορᾶς, ἀλλ' ἡ κατὰ τὸν λοξὸν κύκλον· ἐν ταύτῃ γὰρ καὶ τὸ συνεχές ἐστι καὶ τὸ κινεῖσθαι δύο κινήσεις· ἀνάγκη γάρ, εἴ γε ἀεὶ ἔσται συνεχὴς γένεσις καὶ [336b] φθορά, ἀεὶ μέν τι κινεῖσθαι, ἵνα μὴ ἐπιλείπωσιν αὗται αἱ μεταβολαί, δύο δ', ὅπως μὴ θάτερον συμβαίνῃ μόνον.

[336a23] We have assumed, and have proved, that coming-to-be and passing-away happen to things continuously; and we assert that motion causes coming-to-be. [25] That being so, it is evident that, if the motion be single, both processes cannot occur since they are contrary to one another; for nature by the same cause, provided it remain in the same condition, always produces the same effect, so that either coming-to-be or passing-away will always result. The movements must be [30] more than one, and they must be one another either by the sense of their motion or by its irregularity; for contrary effects demand contraries as their causes.

This explains why it is not the primary motion that causes coming-to-be and passing-away, but the motion along the inclined circle; for this motion not only possesses the necessary continuity, but includes a duality of movements as well. For if coming-to-be and passing-away are always to be continuous, [336b] there must be some body always being moved (in order that these changes may not fail) and moved with a duality of movements (in order that both changes, not one only, may result).

Τῆς μὲν οὖν συνεχείας ἡ τοῦ ὅλου φορὰ αἰτία, τοῦ δὲ προσιέναι καὶ ἀπιέναι ἡ ἔγκλισις· συμβαίνει γὰρ ὁτὲ μὲν πόρρω γίνεσθαι ὁτὲ δ' ἐγγύς. Ἀνίσου δὲ τοῦ διαστήματος ὄντος ἀνώμαλος ἔσται ἡ κίνησις· ὥστ' εἰ τῷ προσιέναι καὶ ἐγγὺς εἶναι γεννᾷ, τῷ ἀπιέναι ταὐτὸν τοῦτο καὶ πόρρω γίνεσθαι φθείρει, καὶ εἰ τῷ πολλάκις προσελθεῖν γεννᾷ, καὶ τῷ πολλάκις ἀπελθεῖν φθείρει· τῶν γὰρ ἐναντίων τἀναντία αἴτια. Καὶ ἐν ἴσῳ χρόνῳ καὶ ἡ φθορὰ καὶ ἡ γένεσις ἡ κατὰ φύσιν.

Διὸ καὶ οἱ χρόνοι καὶ οἱ βίοι ἑκάστων ἀριθμὸν ἔχουσι καὶ τούτῳ διορίζονται· πάντων γάρ ἐστι τάξις, καὶ πᾶς βίος καὶ χρόνος μετρεῖται περιόδῳ, πλὴν οὐ τῇ αὐτῇ πάντες, ἀλλ' οἱ μὲν ἐλάττονι οἱ δὲ πλείονι· τοῖς μὲν γὰρ ἐνιαυτός, τοῖς δὲ μείζων, τοῖς δὲ ἐλάττων ἡ περίοδός ἐστι τὸ μέτρον.

Φαίνεται δὲ καὶ κατὰ τὴν αἴσθησιν ὁμολογούμενα τοῖς παρ' ἡμῶν λόγοις· ὁρῶμεν γὰρ ὅτι προσιόντος μὲν τοῦ ἡλίου γένεσίς ἐστιν, ἀπιόντος δὲ φθίσις, καὶ ἐν ἴσῳ χρόνῳ ἑκάτερον· ἴσος γὰρ ὁ χρόνος τῆς φθορᾶς καὶ τῆς γενέσεως τῆς κατὰ φύσιν.

Now the continuity of this movement is caused by the motion of the whole; but the approaching and retreating of the moving body are caused by the inclination. For the consequence of the inclination is that the body becomes alternately remote and near; and since its distance is thus unequal, its movement will be irregular. Therefore, if it generates by approaching and by its proximity, it—this very same body—destroys by retreating and becoming remote; and if it generates by many successive approaches, it also destroys by many successive retirements. For contrary effects demand contraries as their causes; and the natural processes of passing-away and coming-to-be occupy equal periods of time.

Hence, too, the times—i.e. the lives—of the several kinds of things have a number by which they are distinguished; for there is an order for all things, and every time (i.e. every life) is measured by a period. Not all of them, however, are measured by the same period, but some by a smaller and others by a greater one; for to some of them the period, which is their measure, is a year, while to some it is longer and to others shorter.

And there are facts of observation in manifest agreement with our theories. Thus we see that coming-to-be occurs as the sun approaches and decay as it retreats; and we see that the two processes occupy equal times.

§ 30—On how the cyclicality of natural processes mirrors the circular motions of heaven

On Generation and Corruption, II, 10, 336b26–337a7

English Source: *On Generation and Corruption*, trans. H. H. Joachim in Barnes, *Complete Works of Aristotle* (1984).

[336b26] Τοῦτο δ' εὐλόγως συμβέβηκεν· ἐπεὶ γὰρ ἐν ἅπασιν ἀεὶ τοῦ βελτίονος ὀρέγεσθαί φαμεν τὴν φύσιν, βέλτιον δὲ τὸ εἶναι ἢ τὸ μὴ εἶναι (τὸ δ' εἶναι ποσαχῶς λέγομεν ἐν ἄλλοις εἴρηται), [30] τοῦτο δ' ἀδύνατον ἐν ἅπασιν ὑπάρχειν διὰ τὸ πόρρω τῆς ἀρχῆς ἀφίστασθαι, τῷ λειπομένῳ τρόπῳ συνεπλήρωσε τὸ ὅλον ὁ θεός, ἐνδελεχῆ ποιήσας τὴν γένεσιν· οὕτω γὰρ ἂν μάλιστα συνείροιτο τὸ εἶναι διὰ τὸ ἐγγύτατα εἶναι τῆς οὐσίας τὸ γίνεσθαι ἀεὶ καὶ τὴν γένεσιν.

Τούτου δ' αἴτιον, ὥσπερ [337a] εἴρηται πολλάκις, ἡ κύκλῳ φορά· μόνη γὰρ συνεχής. Διὸ καὶ τἆλλα ὅσα μεταβάλλει εἰς ἄλληλα κατὰ τὰ πάθη καὶ τὰς δυνάμεις, οἷον τὰ ἁπλᾶ σώματα, μιμεῖται τὴν κύκλῳ φοράν· ὅταν γὰρ ἐξ ὕδατος ἀὴρ γένηται καὶ ἐξ ἀέρος [5] πῦρ καὶ πάλιν ἐκ πυρὸς ὕδωρ, κύκλῳ φαμὲν περιεληλυθέναι τὴν γένεσιν διὰ τὸ πάλιν ἀνακάμπτειν. Ὥστε καὶ ἡ εὐθεῖα φορὰ μιμουμένη τὴν κύκλῳ συνεχής ἐστιν.

[336b26] For in all things, as we affirm, nature always strikes after the better. Now being (we have explained elsewhere the variety of meanings we recognize in this term) is better than not-being; [30] but not all things can possess being, since they are too far removed from the principle. God therefore adopted the remaining alternative, and fulfilled the perfection of the universe by making coming-to-be uninterrupted; for the greatest possible coherence would thus be secured to existence, because that coming-to-be should itself come-to-be perpetually is the closest approximation to eternal being.

The cause of this as [337a] we have often said, is circular motion; for that is the only motion which is continuous. That, too, is why all the other things—the things, I mean, which are reciprocally transformed in virtue of their qualities and their powers, e.g. the simple bodies—imitate circular motion. For when Water is transformed into Air, Air [5] into Fire, and the Fire back into Water, we say the coming-to-be has completed the circle, because it reverts again to the beginning. Hence it is by imitating circular motion that rectilinear motion too is continuous.

§ 31—The continuity and eternity of circular motion

Physics, VIII, 8, 264b9–19

English Source: *Physics*, trans. R. P. Hardie and R. K. Gaye in Barnes, *Complete Works of Aristotle* (1984).

[264b9] ἡ δ' ἐπὶ τῆς περιφεροῦς ἔσται μία καὶ συνεχής· οὐθὲν [10] γὰρ ἀδύνατον συμβαίνει· τὸ γὰρ ἐκ τοῦ Α κινούμενον ἅμα κινήσεται εἰς τὸ Α κατὰ τὴν αὐτὴν πρόθεσιν (εἰς ὃ γὰρ ἥξει, καὶ κινεῖται εἰς τοῦτο), ἀλλ' οὐχ ἅμα κινήσεται τὰς ἐναντίας οὐδὲ τὰς ἀντικειμένας· οὐ γὰρ ἅπασα ἡ εἰς τοῦτο τῇ ἐκ τούτου ἐναντία οὐδ' ἀντικειμένη, ἀλλ' ἐναντία μὲν ἡ κατ' εὐθεῖαν [15] (ταύτῃ γάρ ἐστιν ἐναντία κατὰ τόπον, οἷον τὰ κατὰ διάμετρον· ἀπέχει γὰρ πλεῖστον), ἀντικειμένη δὲ ἡ κατὰ τὸ αὐτὸ μῆκος. ὥστ' οὐδὲν κωλύει συνεχῶς κινεῖσθαι καὶ μηδένα χρόνον διαλείπειν· ἡ μὲν γὰρ κύκλῳ κίνησίς ἐστιν ἀφ' αὑτοῦ εἰς αὑτό, ἡ δὲ κατ' εὐθεῖαν ἀφ' αὑτοῦ εἰς ἄλλο·

[264b9] On the other hand, motion on a circular line will be one and continuous; for here we are met by no impossible consequence: [10] that which is in motion from A will in virtue of the same direction of energy be simultaneously in motion to A (since it is in motion to the point at which it will finally arrive), and yet will not be undergoing two contrary or opposite motions; for a motion to a point and a motion from that point are not always contraries or opposites: they are contraries only if they are on the same straight line [15] (for this has points contrary in place, e.g. the points on a diameter—for they are furthest from one another), and they are opposites only if they are along the same line. Therefore there is nothing to prevent the motion being continuous and free from all intermission; for rotatory motion is motion of a thing from its place to its place, whereas rectilinear motion is motion from its place to another place.

APPENDIX 2

Eudoxus's System: Additional Resources (Year 2023)

Mobile Reconstruction of Eudoxus's Model by Henry Mendell
https://web.calstatela.edu/faculty/hmendel/Ancient%20Mathematics/Eudoxus/Astronomy/EudoxusHomocentricSpheres.htm

Mobile Reconstruction of Eudoxus's Model by the Museum Galileo of Florence
http://brunelleschi.imss.fi.it/museum/esim.asp?c=500052

© GERARDO BOTTERI AND ROBERTO CASAZZA, 2023 | DOI:10.1163/9789004525535_013

APPENDIX 3

The *Grupo de Estudio del Cielo*

This study group, created in 2008 at the Escuela de Filosofía of the Facultad de Humanidades y Artes (FHyA) of the Universidad Nacional de Rosario (Argentina), is primarily devoted to naked-eyed observation and understanding of the celestial sphere (its coordinates and main constellations and asterisms) and to the study of ancient, medieval and Renaissance philosophical-astronomical texts. The study group's activities seek to promote a return to astronomy within our current understanding of life, based on an admiration of the heavens and its phenomena, and the study of human efforts made throughout history to understand them.

A patient inquiry into apparent celestial motions causes central philosophical issues to surface, such as the origin of the universe and life, the place of humankind in the cosmos, the mathematical-geometrical structure of human understanding, etc. An approach to these issues, together with readings of relevant classical philosophical and astronomical texts, allows study-group members to learn and grow on an ongoing basis.

The *Grupo de Estudio del Cielo* studies, among other topics, the motions of the Earth, the apparent motions of the heaven, the celestial coordinates (the celestial poles, equator, and axis, the ecliptic, the zodiac, right ascension, declination, the prime meridian, the horizon, etc.), and the use of star charts and celestial sphere simulation software. Some of the astronomical-philosophical texts read within the study group are the works of Plato, Aristotle, Cicero, Manilius, Grosseteste, Nicholas of Cusa, Copernicus, Kepler, and Giordano Bruno.

© GERARDO BOTTERI AND ROBERTO CASAZZA, 2023 | DOI:10.1163/9789004525535_014

FIGURE 51 Regular solids made of wood by Gerardo Botteri for the Grupo de Estudio del Cielo (Facultad de Humanidades y Artes, Universidad Nacional de Rosario). The five solids present the same edge length, resulting in significantly different volumes. These five pieces were part of the exhibition *Lecturas del cielo* at the Biblioteca Nacional (November 2009 to April 2010) organised by the Asociación Argentina Amigos de la Astronomía (UBA), the Instituto de Astronomía y Física del Espacio, the Observatorio de La Plata (UNLP), and the Biblioteca Nacional. For a full description of the works exhibited see *Lecturas del cielo—Libros de astronomía en la Biblioteca Nacional* (Buenos Aires: Biblioteca Nacional, 2011).

FIGURE 52 Regular solids made from clay by Roberto Casazza

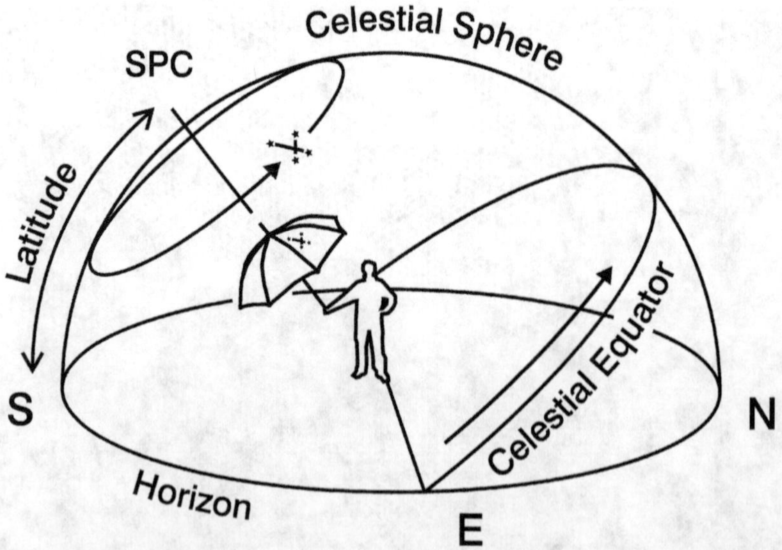

FIGURE 53 The *Southern Astronomical Umbrella* (Botteri–Casazza, 2010) shows the celestial map of the Southern Hemisphere and replicates celestial motion a.

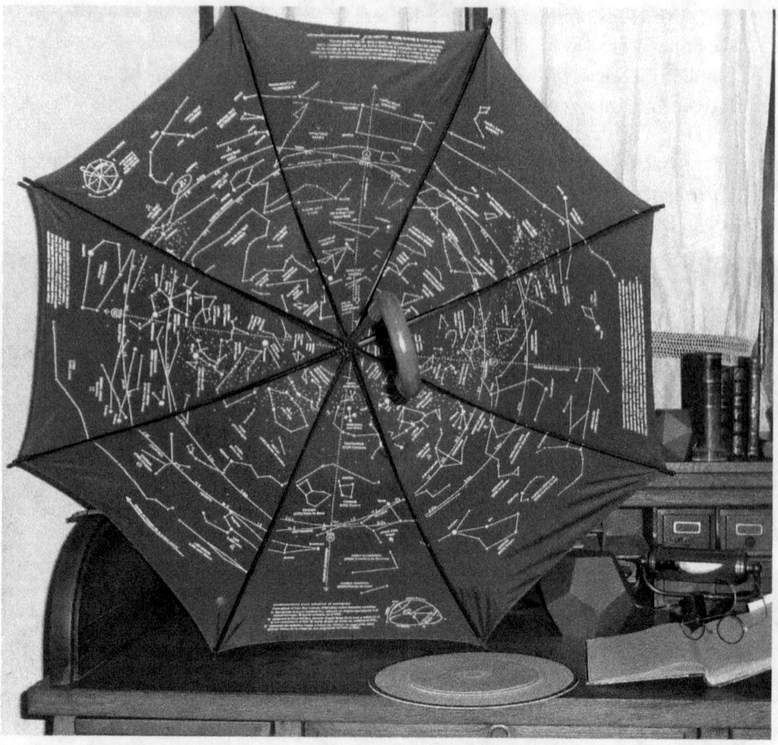

FIGURE 54 *Southern Astronomical Umbrella*, designed by Gerardo Botteri & Roberto Casazza.

Sources

Curd, Patricia, ed. *A Presocratics Reader: Selected Fragments and Testimonia*. 2nd ed. Translations by Patricia Curd and Richard D. McKirahan. Indianapolis: Hackett Publishing Company, 2011.

Diels, Hermann and Walther Kranz, eds. *Die Fragmente der Vorsokratiker*. 3 vols. 6th ed., revised and expanded. Berlin: Weidmann, 1951–1952 (first published 1903). (DK).

Fairbanks, Arthur, ed. and trans. *The First Philosophers of Greece*. London: Kegan Paul, Trench, Trübner and Co., 1898.

Los filósofos presocráticos. 3 vols. Madrid: Gredos, 1997–3002; vol. 1 (2000), edited by Conrado Eggers Lan and Victoria E. Juliá; vol. 2 (2003), edited by Néstor Luis Cordero, Francisco J. Olivieri, Ernesto La Croce and Conrado Eggers Lan; vol. 3 (1997), edited by Armando Poratti, Conrado Eggers Lan, María Isabel Santa Cruz de Prunes and Néstor Luis Cordero. (LFP).

García Bacca, Juan David, ed. and trans. *Los presocráticos*. Vol. 1: *Jenófanes, Parménides, Empédocles*. Pánuco: El colegio de México, 1934.

Graham, Daniel W., ed. and trans. *The Texts of Early Greek Philosophy: The Complete Fragments and Selected Testimonies of the Major Presocratics*. 2 vols. Cambridge: Cambridge University Press, 2010.

Die Griechischen Christlichen Schriftsteller der ersten Jahrhunderte. Leipzig: Hinrichs, 1897–1969. (GCS).

Kirk, Geoffrey Stephen, John E. Raven, and Malcolm Schofield. *The Presocratic Philosophers: A Critical History with a Selection of Texts*. 2nd ed. Cambridge: Cambridge University Press, 1983. (KRS).

Liddell, Henry George and Robert Scott. *A Greek-English Lexicon*. Revised and augmented throughout by Sir Henry Stuart Jones, with the assistance of Roderick McKenzie. Oxford: Clarendon Press, 1940. Available online through the Perseus Digital Library.

Suidae Lexicon. Edited by Ada Adler. Lexicography Graeci 1. Leipzig: Teubner, 1928–1938.

Taylor, Christopher C. W. *The Atomists: Leucippus and Democritus. Fragments: A Text and Translation with a Commentary*. Toronto: University of Toronto Press, 1999.

Aristotle. *The Complete Works of Aristotle, Revised Oxford Translation*. 2 vols. Edited by Jonathan Barnes. (Bollingen series LXX–2). Princeton: Princeton University Press, 1984; Past Masters Electronic Edition, InteLex Corp, 1992.

Aristotle. *Ética eudemia*. Translated into Spanish by Antonio Gómez Robledo. Bibliotheca Scriptorvm graecorvm et romanorvm mexicana. Mexico: UNAM, 1994.

Aristotle. *Eudemian Ethics*. Translated into English by Joseph Solomon. In vol. 2 of *The Complete Works of Aristotle, Revised Oxford Translation*, edited by Jonathan Barnes.

Aristotle. *Fragmentos*. Translated into Spanish by Álvaro Vallejo Campos. Madrid: Gredos, 2005.

Aristotle. [*Fragments*] *Aristotelis qui ferebantur librorum fragmenta*. Edited by Valentin Rose. Leipzig: Teubner, 1886.

Aristotle. *Metafísica*. Trilingual edition (Greek, Latin, Spanish). Translated into Spanish by Valentín García Yebra. Madrid: Gredos, 1981.

Aristotle. *Metafísica*. Translated into Spanish by Tomás Calvo Martínez. Madrid: Gredos, 1994.

Aristotle. *Metaphysics*. Translated into English by William David Ross. In vol. 2 of *The Complete Works of Aristotle, Revised Oxford Translation*, edited by Jonathan Barnes.

Aristotle. [*Metaphysics*] *Aristotelis Metaphysica*. 2 vols. Edited by Hermann Bonitz. Bonn: A Marcus, 1848–1849.

Aristotle. [*Metaphysics*] *Aristotle's Metaphysics. A Revised Text with Introduction and Commentary*. 2 vols. Edited by William David Ross. Oxford: Clarendon Press, 1924; reprinted 1997.

Aristotle. *La Metaphysique*. 2 vols. Translated into French by Jules Tricot. Paris: Vrin, 1974.

Aristotle. *Meteorologica*. Translated into English by Henry Desmond Pritchard Lee. Loeb Classical Library 397. London: W. Heinemann; Cambridge, MA: Harvard University Press, 1952.

Aristotle. *Meteorology*. Translated into English by Erwin Wentworth Webster. In vol. 1 of *The Complete Works of Aristotle, Revised Oxford Translation*, edited by Jonathan Barnes.

Aristotle. *Movement of Animals*. Translated into English by Arthur Spenser Loat Farquharson. In vol. 1 of *The Complete Works of Aristotle, Revised Oxford Translation*, edited by Jonathan Barnes.

Aristotle. [*Nicomachean Ethics*] *Ética nicomáquea*. Translated into Spanish by Julio Pallí Bonet. Barcelona: RBA (Gredos); Buenos Aires: Del Nuevo Extremo, 2008.

Aristotle. *Nicomachean Ethics*. Translated into English by William David Ross, revised by James O. Urmson. In vol. 2 of *The Complete Works of Aristotle, Revised Oxford Translation*, edited by Jonathan Barnes.

Aristotle. *On Generation and Corruption*. Translated into English by Harold H. Joachim. In vol. 1 of *The Complete Works of Aristotle, Revised Oxford Translation*, edited by Jonathan Barnes.

Aristotle. [*On Generation and Corruption*] *Acerca de la generación y la corrupción: Tratados breves de historia natural*. Translated into Spanish by Ernesto La Croce and Alberto Bernabé Pajares. Madrid: Gredos, 1998.

Aristotle. *On the Heavens*. Translated into English by William K. C. Guthrie. Loeb Classical Library 338. London: W. Heinemann; Cambridge, MA: Harvard University Press, 1939.

Aristotle. [*On the Heavens*] *Acerca del cielo; Meteorológicos*. Translated into Spanish by Miguel Candel. Madrid: Gredos, 1996.

Aristotle. *On Sophistical Refutations; On Coming-to-Be and Passing Away* (translated into English by Edward Seymour Forster); *On the Cosmos* (translated by David J. Furley). Loeb Classical Library 400. London: W. Heinemann; Cambridge, MA: Harvard University Press, 1955.

Aristotle. [*On the Soul*] *Acerca del alma*. Translated into Spanish by Marcelo Boeri. Buenos Aires: Colihue, 2010.

Aristotle. *On the Soul*. Translated into English by John Alexander Smith. In vol. 1 of *The Complete Works of Aristotle, Revised Oxford Translation*, edited by Jonathan Barnes.

Aristotle. [*Parts of Animals; Progression of Animals; Movement of Animals*] *Las partes de los animales* (*Partes de los animales; Marcha de los animales; Movimiento de los animales*). Translated into Spanish by Elvira Jiménez Sánchez-Escariche and Almudena Alonso Miguel. Madrid: Gredos, 2000.

Aristotle. [*Physics*] *Física*. Translated into Spanish by Guillermo R. de Echandía. Madrid: Gredos, 1995.

Aristotle. [*Physics*] *Física. Libros I y II*. Translated into Spanish by Marcelo D. Boeri. Buenos Aires: Biblos, 1993.

Aristotle. [*Physics*] *Física. Libros III y IV*. Translated into Spanish by Alejandro Vigo. Buenos Aires: Biblos, 1995.

Aristotle. [*Physics*] *Física. Libros VII–VIII*. Translated into Spanish by Marcelo D. Boeri. Buenos Aires: Biblos, 2003.

Aristotle. *Parts of Animals*. Translated into English by William Ogle. In vol. 1 of *The Complete Works of Aristotle, Revised Oxford Translation*, edited by Jonathan Barnes.

Aristotle. *Physica*. Edited by William David Ross. Oxford: Clarendon Press, 1936.

Aristotle. *Physics*. Translated into English by Philip H. Wicksteed and Francis M. Cornford. 2 vols. Loeb Classical Library 228 (*Books I–IV*) and 255 (*Books V–VIII*). London: W. Heinemann; Cambridge, MA: Harvard University Press, 1929 and 1934.

Aristotle. *Physics*. Translated into English by Robert Purves Hardie and Russell Kerr Gaye. In vol. 1 of *The Complete Works of Aristotle, Revised Oxford Translation*, edited by Jonathan Barnes.

Aristotle. [Ibn Rushd]. *Ibn Rushd's Metaphysics: A Translation with Introduction of Ibn Rushd's Commentary on Aristotle's* Metaphysics *Book Lām*. Translated into English by C. Genequand. Islamic Philosophy, Theology and Science, Texts and Studies 1, series editor Hans Daiber. Leiden: E. J. Brill, 1986.

Cicero. *De Natura Deorum*. Translated by Harris Rackham, Loeb Classical Library 268. London: W. Heinemann; Cambridge, MA: Harvard University Press, 1933.

Cicero. *Sobre la naturaleza de los dioses*. In vol. 1 of *Obras filosóficas*. Translated into Spanish by Víctor-José Herrero Llorente and Ángel Escobar. Barcelona: Gredos, 2009.

Descartes, René. *Le Monde, Ou Traité De La Lumière*. With English translation by Michael Sean Mahoney. New York: Abaris Books, 1979.

Descartes, *El mundo o el tratado de la luz*. Translated into Spanish by Ana Rioja. Madrid: Alianza, 1991.

Diogenes Laertius. *Lives of Eminent Philosophers*. 2 vols. Translated into English by Robert Drew Hicks. Loeb Classical Library 184 and 185. London: W. Heinemann; Cambridge, MA: Harvard University Press, 1975 [1925]).

[Empedocles.] Inwood, Brad. *The Poem of Empedocles: A Text and Translation with a Commentary*. Revised ed. Toronto: University of Toronto Press, 2001.

[Eudoxus] Blass, Friedrich, ed. *Eudoxi ars astronomica qualis charta aegyptiaca superest*. Kiel: Schmidt and Klaunig, 1887.

Eusebius. *Eusebii praeparatio evangelica*, 5 vols. Edited and translated into English by Edwin H. Gifford. Oxford: Oxford University Press, 1903.

Grosseteste, Robert. *Astronomía*. Translated into Spanish by Celina A. Lértora Mendoza. Buenos Aires: Ediciones del Rey, 1988.

Hesiod. *The Homeric Hymns and Homerica*. Translated into English by Hugh G. Evelyn-White. Loeb Classical Library 57. London: W. Heinemann; Cambridge, MA: Harvard University Press, 1914.

Hippocrates, Heracleitus. *Nature of Man. Regimen in Health. Humours. Aphorisms. Regimen 1–3. Dreams. Heracleitus: On the Universe*. Translated into English by William H. S. Jones. Loeb Classical Library 150. London: W. Heinemann; Cambridge, MA: Harvard University Press, 1931.

Hippolytus. *Refutation of All Heresies, Translated with an Introduction and Notes*. Edited and translated into English by M. David Litwa. Atlanta: SBL Press, 2016.

Homer. *The Iliad*. 2 vols. Translated into English by Augustus T. Murray. Loeb Classical Library 170 and 171. London: W. Heinemann; Cambridge, MA: Harvard University Press, 1924.

Homer. *The Iliad*. Translated into English by Robert Fitzgerald. Garden City, NY: Anchor Press, 1974.

Homer. *The Odyssey*. 2 vols. Translated into English by Augustus T. Murray. Loeb Classical Library 104 and 105. London: W. Heinemann; Cambridge, MA.: Harvard University Press, 1919.

Iamblichus. *Iamblichus' Life of Pythagoras*. Translated into English by Thomas Taylor. London: J. M. Watkins, 1818.

Iamblichus. *De vita Pythagorica liber*. Edited by Ludwig Deubner. Bibliotheca Scriptorum Graecorum et Romanorum Teubneriana. Leipzig: Teubner, 1937; revised ed. by U. Klein, Stuttgart: Teubner, 1975.

Kepler, Johannes. *New Astronomy* [*Astronomia Nova* (1609)]. Translated into English by William Donahue. Cambridge: Cambridge University Pres, 1992.

Newton, Isaac. *The Principia: Mathematical Principles of Natural Philosophy*. Translated into English by I. Bernard Cohen and Anne Whitman, assisted by Julia Budenz. Berkeley and Los Angeles: University of California Press, 1999.

Newton, Isaac. [*Principia*] *Principios matemáticos de la filosofía natural*. Translated into Spanish by Eloy Rada García. Madrid: Alianza, 2011.

[Parmenides.] Coxon, Allan H. *The Fragments of Parmenides. A Critical Text with Introduction and Translation, the Ancient Testimonia and a Commentary*. Expanded ed. with new English translations by R. McKirahan. Las Vegas: Parmenides Pub., 2009.

Parménides, Zenón, Meliso, Heráclito—Fragmentos. Translated into Spanish by José Antonio Míguez. Madrid: Orbis, 1983.

Plato. *Banquete*. Translated into Spanish by Marcos Martínez Hernández. Barcelona: RBA; Buenos Aires: Del Nuevo Extremo, 2008.

Plato. *Diálogos IV. República*. Translated into Spanish by Conrado Eggers Lan. Madrid: Gredos, 1986.

Plato. *Diálogos VI. Filebo, Timeo, Critias*. Translated into Spanish by María de los Ángeles Durán and Francisco Lisi. Madrid: Gredos, 1992.

Plato. *The Laws of Plato*. Translated into English by Thomas L. Pangle. Chicago: University of Chicago Press, 1980.

Plato. *Leyes*. Translated into Spanish by José Manuel Pabón and Manuel Fernández Galiano. Madrid: Instituto de Estudios Políticos, 1999.

Plato. *La República*. Bilingual edition. Translated into Spanish by Antonio Gómez Robledo. Mexico: UNAM, 1971.

Plato. *La República*. Translated into Spanish by José Manuel Pabón y Manuel Fernández-Galiano. Madrid: Centro de Estudios Constitucionales, 1981.

Plato. *La Repubblica*. 2 vols. Translated into Italian by Francesco Gabrieli. Milan: Rizzoli, 1991.

Plato. *The Republic of Plato*. 2 vols. Edited by James Adam. Cambridge: Cambridge University Press, 1902.

Plato. *The Republic of Plato*. Translated into English by Francis M. Cornford. New York and London: Oxford University Press, 1941.

Plato. *Theaetetus. Sophist*. Translated into English by Harold North Fowler. Loeb Classical Library 123. London: W. Heinemann; Cambridge, MA: Harvard University Press, 1921.

Plato. [*Timaeus*] *Plato's Cosmology: The Timaeus of Plato*. Translated into English by Francis M. Cornford. Indianapolis: Hackett Publishing Company, 1997 [1935].

Plato. [*Timaeus*] *Timeo*. Translated into Spanish by Conrado Eggers Lan and Ivana Costa. Buenos Aires: Colihue, 1999.

Pliny the Elder. *Historia natural. Libros I–II*. Translated into Spanish by Antonio Fontán, Ana María Moure Casas and others. Madrid: Gredos, 1995.

Pliny the Elder. *Natural History. Volume II: Books 3–7*. Translated into English by Harris Rackham. Loeb Classical Library 352. Cambridge, MA: Harvard University Press, 1942.

Plotinus. *Enéadas V–VI*. Translated into Spanish by Jesús Igal. Madrid: Gredos, 1998.

Plotinus. *The Enneads*. Translated into English by Stephen MacKenna, revised by Bertram Samuel Page, and introduced by Paul Henry. London: Faber, 1969.

Plotinus. *Enneads*. Vol. 5: *Enneads*, V. 1–9. Translated into English by Arthur Hilary Armstrong. Loeb Classical Library 444. London: W. Heinemann; Cambridge, MA: Harvard University Press, 1984.

Pseudo-Aristotle. *Aristotelis qui fertur de Melisso, Xenophane, Gorgia libellus*. Edited by Hermann Diels. Berlin: Abhandlungen der Königlichen Akademie der Wissenschaften, 1900.

Simplicius. *In Aristotelis de Caelo commentaria*. Edited by Johan Ludvig Heiberg. Commentaria in Aristotelem graeca 7. Berlin: Reimer, 1894.

Simplicius. *In Aristotelis Physica commentaria*. Edited by Hermann Diels. Commentaria in Aristotelem graeca 9 and 10. Berlin: Reimer, 1882 and 1895.

Simplicius. *Simplicii Commentarius in IV libros Aristotelis de Caelo*. Edited by Simon Karsten. Utrecht: Kemink and Son, 1865.

Theodoret of Cyrrhus. *Thérapeutiques des maladies helleniques* (*Graecarum affectuum curatio*). Edited and translated into French by P. Canivet. Sources Chrétiennes 57. Paris: Cerf, 1958.

Theon of Smyrna. *Theonis Smyrnaei platonici liber de astronomia cum Sereni fragmento*. Edited by Thomas-Henri Martin. Paris: Reipublicae Typographeo, 1849.

Theophrastus. *Algunas cuestiones de metafísica*. Edited and translated into Spanish by Miguel Candel. Barcelona: Anthropos, 1991.

Theophrastus. *Theophrastus On First Principles* (*known as his Metaphysics*). *Greek Text and Medieval Arabic Translation, Edited and Translated with Introduction, Commentaries and Glossaries, as well as the Medieval Latin Translation, and with an Excursus on Graeco-Arabic Editorial Technique*. Edited and translated into English and by Dimitri Gutas. Philosophia Antiqua 119. Leiden and Boston: Brill, 2010.

Thomas Aquinas. *Comentario a la Física de Aristóteles* [*Commentaria in octo libros physicorum Aristotelis*]. Translated into Spanish by Celina A. Lértora Mendoza. Navarra: Eunsa, 2001.

Thomas Aquinas. *Comentario al libro XII* (*lambda*) *de la Metafísica de Aristóteles* [*Sententia libri Metaphysicae*]. Translated into Spanish by Ana Mallea and Marta Daneri-Rebok. Buenos Aires: Agape, 2011.

Thomas Aquinas. *Commentary on the Metaphysics of Aristotle* [*Sententia libri Metaphysicae*]. 2 vols. Translated into English by John Patrick Rowan. Chicago: Regnery, 1961.

Thomas Aquinas. *In duodecim libros Metaphysicorum Aristotelis expositio* [*Sententia libri Metaphysicae*]. Edited by Raimondo M. Spiazzi. Turin-Rome: Marietti, 1964.

Thomas Aquinas. *Sententia libri Metaphysicae.* Edited by Roberto Busa and Enrique Alarcón. Turin: Marietti, 1950. Available online from the Corpus Thomisticum.

Thomas Aquinas. *Summa theologiae*. Madrid: Biblioteca de Autores Cristianos, 1964.

Thomas Aquinas. *The Summa Theologiæ of St. Thomas Aquinas.* Second and revised ed., 1920. Translated into English by the Fathers of the English Dominican Province (Online Edition from New Advent, 2017).

Xenophanes. *Xenophanes of Colophon: Fragments: A Text and Translation with Commentary*. Edited and translated into English by James H. Lesher. Toronto: University of Toronto Press, 1992.

Bibliography

Afnan, Soheil M. *Avicenna: His Life and Works*. London: Allen and Unwin, 1958.

Asín, Fernando M. *Astronomía*. Madrid: Paraninfo, 1997.

Bachelard, Gaston. *The Poetics of Space*. Boston: Beacon Press, 1994.

Baikouzis, Constantino. *Carta celeste y calculador astronómico para el hemisferio sur*. Buenos Aires: Universidad Nacional de La Plata, 2005.

Bailly, Jean Sylvain. *Histoire de l'astronomie ancienne*. Paris: De Bure, 1781.

Berger, Ernst Hugo. *Geschichte der wissenschaftlichen Erdkunde der Griechen*. Leipzig: Veit, 1903.

Berti, Enrico. "Unmoved mover(s) as efficient cause(s) in Metaphysics Λ 6." In *Nuovi studi aristotelici*, vol. 2: *Fisica, Antropologia, Metafisica*, 181–206. Brescia: Morcelliana, 2005. Previously published in *Aristotle's Metaphysics Lambda: Symposium Aristotelicum*, edited by Michael Frede and David Charles (Oxford: Oxford University Press, 2000).

Berti, Enrico. "La causalità del primo Motore immobile secondo Aristotele." In *Nuovi studi aristotelici*, vol. 2: *Fisica, Antropologia, Metafisica*, 453–469. Brescia: Morcelliana, 2005. Previously published in *Gregorianum* 83 (2002): 637–654.

Berti, Enrico. *Ser y tiempo en Aristóteles* [Text transcribed and translated from a course given in Argentina in 2008]. Translated by Patricio Perkins. Buenos Aires, Biblos, 2011.

Boeckh, August. *Gesammelte kleine Schriften*. 7 vols. Leipzig: Teubner, 1858–72; esp. vol. 3, 1866.

Boeckh, August. *Untersuchungen über das kosmische System des Platon*. Berlin: Veit, 1852.

Boeri, Marcelo D., "*'Entre motor y movido debe haber contacto': una dificultad en la teoría aristotélica del movimiento* (Física, *266b27–267a12*)." Revista Latinoamericana de Filosofía, *24, no. 2 (1998): 251–262*.

Bonitz, Hermann. *Observationes criticae in Aristotelis libros metaphysicos*. Berlin: Bethge, 1842.

Borges, Jorge Luis. "Pascal's sphere." In *Other Inquisitions, 1937–1952*, 6–9. Translated by Ruth L. C. Simms. Austin: University of Texas Press, 1964. Originally published as "La esfera de Pascal," in *Otras inquisiciones* (Buenos Aires: Sur, 1952).

Botteri, Gerardo and Roberto Casazza. *Paraguas Astronómico Austral*. Buenos Aires, 2010.

Botteri, "El sistema astronómico de Aristóteles: *Metafísica*, λ, 8, 1073a14–1074a34." *Actas del XXI Simposio Nacional de Estudios Clásicos* (2010). Digital publication from Universidad Nacional del Litoral (Santa Fe).

Bodnár, István. "Aristotle's Natural Philosophy." In *The Stanford Encyclopedia of Philosophy*. Edited by Edward N. Zalta. Online Spring 2012 edition.

Bowen, Alan C. *Simplicius on the Planets and Their Motions*. Boston: Brill, 2013.

Bradshaw, David. "A New Look at the Prime Mover." *Journal of the History of Philosophy* 39, no. 1 (2001): 1–22.

Broadie, Sarah. "Que fait le premier moteur d'Aristote?" *Revue Philosophique de la France et de l'Étranger* 183 (1993): 375–411.

Carone, Gabriela Roxana. *Plato's Cosmology and its Ethical Dimensions*. Cambridge: Cambridge University Press, 2005.

Casazza, Roberto. "Las figuras universal (U) y paradigmática (P) del *De coniecturis* de Nicolás de Cusa, a la luz de la interpretación cosmológica (figura U-P) de Athanasius Kircher." In *Identidad y alteridad en Nicolás de Cusa. Actas del II Congreso Cusano de Latinoamérica*, 201–223. Buenos Aires: Biblos, 2010.

Casazza, Roberto. "El círculo y la esfera." *La Biblioteca* 13. Buenos Aires: Biblioteca Nacional, 2013.

Casazza, Roberto. "Iconology of the Medieval and Renaissance Iconography of voluntas." MA Thesis, The Warburg Institute, University of London, 1995.

Casazza, Roberto and Gangui, Alejandro. "La explicación de los eclipses en la Antigüedad grecolatina." *Revista de Estudios Clásicos* 39 (Mendoza, Universidad Nacional de Cuyo, 2012): 79–103.

Castello Dubra, Julio Antonio. "Creación, cambio y eternidad del mundo en Tomás de Aquino." In *Tempo e Eternidade na Idade Média*, 102–108. Compiled by Jan G. J. Ter Reegen, Luis de Boni and Marcos Costa. Porto Alegre: EST, 2007.

Cherniss, Harold, *Aristotle's Criticism of Presocratic Philosophy*. Baltimore: Johns Hopkins University Press, 1935.

Clagett, Marshall. *Greek Science in Antiquity*. New York: Abelard-Schuman, 1955.

"Condemnation of 219 Propositions." Translated by Ernest L. Fortin and Peter D. O'Neill. In *Medieval Political Philosophy: A Sourcebook*, 335–54, edited by Ralph Lerner and Muhsin Mahdi. New York: Free Press, 1963. Published in Latin in *Siger de Brabant et l'averroïsme latin au XIIIme siècle*, edited by P. Mandonnet (Louvain: Institut Supérieur de Philosophie de l'Université, 1908).

Corbin, Henry. *Avicenna and the Visionary Recital*. Translated by Willard R. Trask. New York: Pantheon Books, 1960. Originally published as *Avicenne et le recit visionnaire* (Teheran: Institut Franco Iranien, 1952).

Cornelius, Geoffrey. *The Starlore Handbook: An Essential Guide to the Night Sky*. San Francisco, CA: Chronicle Books, 1997.

De Filippo, Joseph G. "Aristotle's Identification of the Prime Mover as God." *Classical Quarterly* 44, no. 2, (1994): 393–409.

Dreyer, John Louis Emil. *A History of Astronomy from Thales to Kepler*. Revised by William Harris Stahl, Cambridge: Cambridge University Press, 1953.

Duhem, Pierre. *Le système du monde: Histoire des doctrines cosmologiques*. Vol. 5: *Troisième partie: La Crue de l'Aristotélisme* (suite). Paris: Hermann, 1917.

Durán Guardeño, Antonio. "Ciencia y Renacimientos: Thãbit ibn Qurra y Gerolamo Cardano." In *Matemáticas y matemáticos*, edited by José Ferreirós and Antonio Durán Guardeño. Seville: Universidad de Sevilla, 2003.

Düring, Ingemar. *Aristoteles: Darstellung und Interpretation seines Denkens*. Heidelberg: C. Winter, 1966.

Eliade, Mircea. *A History of Religious Ideas*. Vol. 1: *From the Stone Age to the Eleusinian Mysteries*. Translated by Willard R. Trask. Chicago: University of Chicago Press, 1978.

Faitanin, Paolo. "La individuación de las sustancias separadas según Tomás de Aquino—1." *Aquinate* 1, no. 6 (2011): 20–49.

Flasch, Kurt. *Aufklärung im Mittelalter? Die Verurteilung von 1277*. Mainz: Dieterich, 1989.

Gangui, Alejandro. *El Big Bang—La génesis de nuestra cosmología actual*. Buenos Aires: Eudeba, 2005.

Gangui, Alejandro. *Poética astronómica—El cosmos de Dante Alighieri*. Buenos Aires: Fondo de Cultura Económica, 2008.

Ghika, Matila. *Esthétique des proportions dans la nature et dans les arts*. Paris: Gallimard, 1927. Published in Spanish as *Estética de las proporciones en la naturaleza y en el arte* (Barcelona: Poseidón, 1977).

Gill, Mary Louise. "Aristotle on Self-Motion." In *Self-motion: From Aristotle to Newton*, edited by Mary Louise Gill and James G. Lennox, 15–34. Princeton: Princeton University Press, 1994.

Gilson, Étienne. *God and Philosophy*. New Haven: Yale University Press, 1941.

Granada, Miguel Ángel, "Introducción" to Giordano Bruno, *Del infinito: el universo y los mundos*, Madrid, Alianza Forma, 1993.

Grant, Edward. *Planets, Stars, and Orbs: The Medieval Cosmos, 1200–1687*. Cambridge: Cambridge University Press, 1994.

Guazzoni-Foà, Virginia. "Un ripensamento sulla ΣΦΑΙΡΗ di Parmenide." *Giornale di Metafisica* 21 (1966): 344–354.

Guerrero, Rafael Ramón. *El pensamiento filosófico árabe*. Historia de la filosofía 8. Madrid: Cincel, 1985.

Gurev, Grigorii Abramovich. *Los sistemas del mundo desde la antigüedad hasta Newton*. Translated from the Russian by N. Caplán. Buenos Aires: Problemas, 1947.

Guthrie, William Keith Chambers. *A History of Greek Philosophy*. Vol. 1: *The earlier Presocratics and the Pythagoreans*, Cambridge: Cambridge University Press, 1962.

Guthrie, William Keith Chambers. *A History of Greek Philosophy*. Vol. 2: *The Presocratic Tradition from Parmenides to Democritus*, Cambridge: Cambridge University Press, 1965.

Guthrie, William Keith Chambers. "The Development of Aristotle's Theology—I." *The Classical Quarterly* 27, nos. 3/4 (July–October 1933): 162–171.

Guthrie, William Keith Chambers. "The Development of Aristotle's Theology—II." *The Classical Quarterly* 28, no. 2 (April 1934): 90–98.

Hanson, Norwood Russell. *Constellations and Conjectures*. Edited by Willard C. Humphreys, Jr. Dordrecht/Boston: D. Reidel, 1973. Spanish edition metioned in text: *Constelaciones y conjeturas*, translated by Carlos Solís (Madrid: Alianza, 1978).

Heath, Thomas Little. *Aristarchus of Samos, the Ancient Copernicus: A History of Greek Astronomy to Aristarchus. Together with Aristarchus's Treatise on the Sizes and Distances of the Sun and Moon: A New Greek Text*. Oxford: Clarendon Press, 1913.

Heisenberg, Werner. "Classical Education, Science and the West." In *The Physicist's Conception of Nature*, 51–67. Translated by Arnold J. Pomerans (London: Hutchinson and Co., 1958). Originally published as *Das Naturbild der heutigen Physik* (Hamburg: Rowohlt, 1955).

Jaeger, Werner. *Aristotle: Fundamentals of the History of His Development*. Translated by Richard Robinson. Oxford: Clarendon Press, 1934. Originally published as *Aristoteles: Grundlegung einer Geschichte seiner Entwicklung* (Berlin: Weidmann, 1923).

Jaeger, Werner. *Theology of the Early Greek* Philosophers. Translated from the manuscript by Edward S. Robinson for The Gifford Lectures, 1936. Oxford: Clarendon Press, 1947.

Judson, Lindsay. "Heavenly Motion and the Unmoved Mover." In *Self-motion: From Aristotle to Newton*, edited by Mary Louise Gill and James G. Lennox, 155–171. Princeton: Princeton University Press, 1994.

Kelley, David H. and Milone, Eugene F. *Exploring Ancient Skies: A survey of Ancient and Cultural Astronomy*. New York: Springer, 2011.

Kosman, Aryeh. "Aristotle's Prime Mover." In *Self-motion: From Aristotle to Newton*, edited by Mary Louise Gill and James G. Lennox, 135–153. Princeton: Princeton University Press, 1994.

Kouermenos, Theokritos. *Heavenly Stuff: The Constitution of the Celestial Objects and the Theory of Homocentric Spheres in Aristotle's Cosmology*. Stuttgart: Franz Steiner, 2010.

Koyré, Alexandre. *From the Closed World to the Infinite Universe*. Baltimore: The Johns Hopkins Press, 1957.

Kuhn, Thomas S. *The essential Tension: Selected Studies in Scientific Tradition and Change*. Chicago/London: University of Chicago Press, 1997.

Lang, Helen S. *Aristotle's Physics and Its Medieval Varieties*. Albany: State University of New York, 1992.

Lepsius, Richard. *Die Chronologie der Aegypter*. Berlin: Nicolai, 1849.

Levinas, Marcelo Leonardo. *Las imágenes del universo—Una historia de las ideas del cosmos*. Buenos Aires: Siglo XXI, 2006.

Lloyd, Geoffrey Ernest Richard. *Aristotle: The Growth and Structure of his Thought*. Cambridge: Cambridge University Press, 1968.

Mandonnet, Pierre. *Siger de Brabante et l'averroïsme latin au XIIIme siècle*. 2 vols. Louvain: Institut Supérieur de Philosophie de l'Université, 1908–1911.

Martin, Thomas-Henri. "Mémoire sur les hypothèses astronomiques chez les Grecs et les Romains. Hypothèse astronomique de Platon." *Mémoires de l' Académie des Inscriptions et Belles-Lettres*, 30, no. 1 (1881): 35–39.

Martin, Thomas-Henri. "Mémoire sur les hypothèses astronomiques d'Eudoxe, de Callippe, d'Aristote et de leur école." *Mémoires de l' Académie des Inscriptions et Belles-Lettres*, 30, no. 1 (1881): 153–302.

Merlan, Philip. "Aristotle's Unmoved Movers." *Traditio* 4 (1946): 1–30.

Míguez, Gustavo, Roberto Casazza et alii, eds. *Lecturas del cielo—Libros de astronomía en la Biblioteca Nacional*. Buenos Aires: Biblioteca Nacional, 2012.

Mondolfo, Rodolfo. *El pensamiento antiguo*. 2 vols. Buenos Aires: Losada, 1942.

Mugnier, René. *La théorie du Premier Moteur et l'evolution de la pensée aristotélicienne*. Paris: Vrin, 1930.

Neugebauer, Otto. *The Exact Sciences in Antiquity*. New York: Dover, 1969.

O'Brien, Denis. *Empedocles' Cosmic Cycle: A Reconstruction from the Fragments and Secondary Sources*. Cambridge: Cambridge University Press, 1969.

Owens, Joseph. "The Reality of the Aristotelian Separate Movers." *Review of Metaphysics* 3 (1949/1950): 319–337.

Pastor, Julio Rey and José Babini. *Historia de las matemáticas*. 2 vols. Barcelona: Gedisa, 2000.

Paulus, Jean. "La Théorie du Premier Motor chez Aristote." *Revue de Philosophie*, Nouvelle Serie 4, no. 3 (May–June 1933): 259–294; no. 4–5 (July–October 1933): 395–424.

Proust, Christine and John Steele, eds. *Scholars and Scholarship in Late Babylonian Uruk*. Berlin: Springer, 2019.

Puig Montada, Joseph. "Averroes y el problema de la eternidad del movimiento." *La ciudad de Dios* 212, no. 1 (1999): 231–244.

Quiles, Ismael. *Aristóteles—Vida, escritos, doctrina*. Buenos Aires: Desalma, 1986.

Reale, Giovanni. *The Concept of First Philosophy and the Unity of the Metaphysics of Aristotle*. Translated by John R. Catan. Albany: State University of New York, 1980. This translation is from the 3rd ed. of *Il concetto di filosofia prima e l'unità della Metafisica di Aristotele* (Milan: Società Editrice Vita e Pensiero, 1967).

Reale, *Guía de lectura de la "Metafísica" de Aristóteles*. Barcelona: Herder, 1999.

Rey Pastor, Julio and José Babini. *Historia de las matemáticas*. 2 vols. Barcelona: Gedisa, 2000.

Ross, William David. *Aristotle*. London: Methuen, 1923.

Ross, William David. "The Development of Aristotle's Thought." In *Aristotle and Plato in the Mid-Fourth Century. Papers of the Symposium Aristotelicum held at Oxford in August, 1957*, edited by Ingemar Düring and Gwilym Ellis Lane Owen. Studia Graeca et Latina Gothoburgensia 11. Göteborg: Almquist & Wiksell, 1960.

Ross Hernández, Alberto. "De la biología a la cosmología: ¿es el primer motor un alma?" *Themata, Revista de filosofía* 43 (2010): 373–384.

Ross Hernández, Alberto. "La causalidad del Primer Motor en Metafísica XII." *Diánoia* 52, n. 59 (2007): 3–26.

Ruvituso, Marcos. "Conocimiento y conjetura en los fragmentos de Jenófanes." PhD diss. Universidad de La Plata, Grupo de Investigaciones Nova Lectio Antiquitatis, 1998.

Schiaparelli, Giovanni. *Scritti sulla storia della astronomia antica.* 3 vols. Bologna: Nicola Zanichelli, 1925–1927.

Schopenhauer, Arthur. *Essay on the Freedom of the Will.* Translated by Konstantin Kolenda. New York: Bobbs-Merrill, 1960. Original German title: *Preisschrift über die Freiheit des Willens* (1839).

Sloterdijk, Peter. *Spheres*. 3 vols. Cambridge: MIT Press, 2011, 2014 and 2016.

Sorabji, Richard. *Matter, Space and Motion: Theories in Antiquity and Sequel.* London: Duckworth, 1988.

Steele, John M. "Eclipse Prediction in Mesopotamia." *Archive for the History of Exact Sciences* 54 (2000): 421–454.

Stewart, Ian. *Taming the Infinite: The Story of Mathematics*. London: Quercus, 2007.

Stewart, John Alexander. *The Myths of Plato*. Translated, with introductory and other observations. London: Macmillan and Co., 1905.

Taylor, Alfred Edward. *A Commentary on Plato's Timaeus*. Oxford: Clarendon Press, 1928.

Thompson, D'Arcy Wentworth. "On Plato's 'Theory of the Planets,' Republic X. 616 E." *The Classical Review* 24, no. 5 (August 1910): 137–142. First published in *The Observatory* (1904).

Thorndike, Lynn. *The Sphere of Sacrobosco and its Commentators*. Chicago: University of Chicago Press, 1949.

Vigo, Alejandro. *Aristóteles. Una introducción*. Santiago de Chile: Instituto de Estudios de la Sociedad, 2007.

Weisheipl, James A. "The Celestial Movers in Medieval Physics." *Thomist* 24 (1961): 286–326.

Wolfson, Harry. "The Plurality of Immovable Movers in Aristotle and Averroes." *Harvard Studies in Classical Philology* 63 (1958): 233–253.

Yavetz, Ido. "On the Homocentric Spheres of Eudoxus." *Archive for History of Exact Sciences* 52 (1998): 221–278.

Yavetz, Ido. "A New Role for the Hippopede of Eudoxus." *Archive for History of Exact Sciences* 56 (2001): 69–93.

Subject Index

Author Index

www.ingramcontent.com/pod-product-compliance
Lightning Source LLC
LaVergne TN
LVHW010543300926
849162LV00002B/216